Foreword

The term 'tropical agriculture' conjures up the idea of institutes in industrialised countries, research centres in the Third World, and an army of agricultural specialists and instructors, imposing the fruits of modernization, its methods, tools, seeds and fertilizers on a backward people.

Hugues Dupriez and Philippe De Leener, both with considerable experience of tropical agriculture, have approached the subject from a different, more complex angle. Their starting-point is not the 'ignorant' peasant but is the experience and knowledge of the African farmer. They reject the approach calling for a 'clean sweep' and also that based on the primitive ways of the African farmer, and choose to explore history and ecology to help us to appreciate all the different aspects of African rural settlements. Water, soil and plants form the varied agricultural landscape, so often shaped by prolonged reflection and by trial and error.

The authors might have been tempted to plead strongly in favour of the old farming systems and leave it at that. On the contrary, while stressing vigorously the great importance of the knowledge, practices and techniques built up over the centuries, they are careful not to canonise the status quo.

To encapsulate the authors' approach, we could speak of a cultural blending because the authors keep environmental factors to the fore while stressing the strains imposed by increased productivity and the need to produce food for all. They offer the African farmer the benefit of up-to-date agricultural scientific knowledge by suggesting substitutes or complements for long-established practices.

As I read the book, I thought how certain chapters and illustrations could be translated or adapted to the farming methods of the Serer, the Baoulé, the Bobo and the Bashi peoples, and be used as a working document in schools, for adult extension programmes and for farmers' associations.

The book's outstanding appeal lies in that, on the one hand, it is rooted in the traditional outlook of African peoples, and on the other, it can be of use to the literate farmer. The village school-teacher will also find it a handy teaching aide for child and adult groups alike.

The authors' aim is in complete agreement with the philosophy of many institutions and organizations, including ENDA, for whom development is increasingly synonymous with promoting the diversification of people's response to the wide variety of eco-cultures in Africa.

We are convinced that this approach will be pursued and examined in depth in the future.

Cheikh Hamidou Kane
President of ENDA Tiers-Monde

🔲🔲🔲🔲🔲🔲🔲🔲🔲🔲🔲🔲🔲🔲🔲🔲🔲🔲🔲🔲

Introduction

Translated into English from the first edition of 'Agriculture tropicale en milieu paysan africain' (Agriculture in African Rural Communities), H. Dupriez and P. De Leener, Editions Terres et Vie, 1983, 282 pages.

Agriculture is a difficult subject. There is much to observe, to explain, to understand, and a single book can only be a tiny step in the right direction.

Nature has its own laws which man must respect if he wants to gather nature's fruits and to grow food. Man must tame but never abuse nature.

Over the centuries, African farmers, men and women alike, have learned to control nature so as to win all kinds of useful produce by ways and means often quite different from those of other people.

Growing crops in association is a special feature of African farming and the starting-point for this book, the first in a series which will attempt to study, understand and explain farming practices on small holdings. Although certain traditional values and structures are being eroded, the small farmer still manages to feed many mouths. Yet, he will have to improve his farming methods and output if he is to cope with the growth in population.

We must face the fact that if agriculture is to flourish, produce enough food and provide employment within African rural communities, then it will have to be based on the knowledge and know-how of these same communities. They alone can come to grips with their environment.

The farmer's job is to farm the land. Consequently, his responsibility and that of the technician are quite separate. The function of the technician is not to cause upheaval but to face the day-to-day realities of farm-life and to help the cultivator to try out new, hopefully progressive, ideas.

The reader should approach this book with an open frame of mind, considering the ideas and examining the illustrations, but also asking himself what is relevant to his own land.

The farmers of Africa must bear one thing in mind. Often they are told to look at the farmers in Europe, America or China, and copy them. As well as looking at other continents, they must also look carefully at what is going on around them, on their own farms and on nearby farms. There is much to learn but it is possible that a system of better farming is being practised on their own doorstep. They must not rely on outsiders to improve their farming methods. They must know their own land, the kind of soil and the weather they have to contend with. The technician's word is like a seed. Who can say if it is a good seed until he has sown it in his field ?

Land and Life

Agriculture in African Rural Communities

Hugues Dupriez
Philippe De Leener

Translated by
Bridaine O'Meara

...LISHERS

TERRES et VIE and CTA

Terres et Vie is indebted to the CTA who financed the translation of the book into English and actively promotes its distribution. The **CTA, Technical Centre for Agricultural and Rural Co-operation**, has its headquarters at Ede-Wageningen, Netherlands. It was founded in 1983 by decision of the Second ACP-EEC Convention of Lomé.

The Convention was renewed in 1984 and associates sixty-six ACP and twelve European countries.

The aim of the CTA is to collect and circulate scientific and technical information on agricultural and rural development, facilitate the exchange of information, promote technical vulgarization, and encourage research and training. It contributes to studies and publications, organizes specialist meetings, assists the documentation centres of the ACP States and has a question-answer service at their disposal.

Postal address : CTA, P.O.B. 380, 6700 AJ Wageningen, Netherlands. International telephone number : (31) (8380) 20484. Telex : 30169 CTANL.

Published in association with **TERRES et VIE**, rue Laurent Delvaux 13, 1400 Nivelles, Belgium, from the French title **Agriculture tropicale en milieu paysan africain**, edited by TERRES et VIE in association with HARMATTAN and ENDA.

First English edition published 1988
Reprinted 1992

Published by THE MACMILLAN PRESS LTD
London and Basingstoke
*Associated companies and representatives in Accra,
Auckland, Delhi, Dublin, Gaborone, Hamburg, Harare,
Hong Kong, Kuala Lumpur, Lagos, Manzini, Melbourne,
Mexico City, Nairobi, New York, Singapore, Tokyo.*

ISBN 0 – 333 – 44595 – 3   ISBN 2 – 87105 – 006 – 6

Printed in China

(M)630.096 D

In agriculture, no single topic may be studied out of context. The forty-nine lessons in this book are each on a different subject, but they form an interrelated whole, like a spider's web.

Certain basic ideas shaped this book and dictated its framework, as outlined in the table of contents.

- First, agriculture is **life** with its many facets, as it unfolds in a non-living environment.

- It is also a field of **human activity** in which men and women experiment, learn and cooperate in order to make nature serve their interests. Moreover, like every human activity, agriculture is exposed to the contradictory forces of change and tradition, and also to conflicting economic interests and social and political tensions.

- Good agriculture is like the spider's web whose main threads are fastened to fixed points while the minor threads are woven into a solid, coherent network. No two spider's webs are exactly the same, but all are structured in the same way.

This book does not treat agriculture simply as a technical activity which needs only specialists trained in institutes of higher education. If this approach were followed, the people of Africa would not be fed and the enormous agricultural potential of the continent would not be developed. Feeding the people is the business of all sections of the community, and each member of the community has a job to do. Some plant and sow, others breed cattle, others make tools, others organize transport and supplies, or encourage and experiment. This interrelation between the different segments of society enables progress and modernization to take place without necessarily destroying valuable traditional skills. The purpose of technicians is to encourage such interaction and to give the farmers information, not to dictate rules of behaviour as though they were the supreme, omniscient authority.

We therefore encourage the insertion of the economic goals of agriculture into regional and village structures, and visits and exchanges between the inhabitants of the same region.

Agriculture is an activity which does not readily accept pressures from outside the farming community. So this book emphasizes what the farmers can do simply by relying on the resources of their own or their neighbours' lands : the soil, the crops, the water supplies, the work force. Only when these local resources have been fully exploited and farmers have organized themselves properly, should investment in new resources, buying from the outside, adapting new ideas to local requirements, be considered. The farmer's priorities should be : autonomy, self-reliance, and a critical awareness of the extent to which the farming community relies on its own resources and on outside resources.

This book is about tropical agriculture in general, although stockraising and the keeping of poultry and small animals are not covered. The book does not overemphasize how best the grower might improve his or her farming methods. When he has read the book and studied the illustrations, the reader must draw his own conclusions about the practical steps to be taken. It is more important to understand and think over one's farming methods than to imitate slavishly. That is why we stress the need to examine the illustrations very carefully. They show ways of working the land. Whether these ways are good or bad, appropriate or inappropriate for a particular farmer, only that farmer knows.

millet granaries in Burkina Faso

Contents

page

iii **Foreword**

iv **Introduction**

vi **Contents**

x **To the Reader**

1 *Lessons 1 and 2 : Fields and farming*

1   **1 : Visiting the fields**

5   After these first steps on African soil, there are many important points to remember

5   **2 : Farming : goals, means and limitations**

6   What the farmer wants to do
7   What the farmer can do
7   What the farmer cannot do

11 *Lessons 3 to 5 : The environment*

11   **3 : The living environment**

15   **4 : The physical environment**

18   **5 : The economic and social environment**

25 *Lessons 6 to 8 : Fields change with the seasons*

25   **6 : The story of a field in a dry region**

27   In this particular part of Senegal, agriculture is faced with difficulties and limitations

28   **7 : The story of a field in a wet forest region**

32   What can be learned from the story of this field ?

35   **8 : The rotation of crops and fallow lands**

35   Why is this rotation necessary ?
36   Why does the soil become exhausted ?

37 *Lesson 9 : Legume crops*

40 *Lessons 10 to 15 : Some features of rural settlements*

40   **10 : Mapping rural settlements**

41   The art of observing and representing our own rural settlements

44   **11 : Land relief**

45   Agricultural land falls into four categories where land relief is concerned

47   **12 : Water resources in rural settlements**

47   The water resources in rural settlements and their uses
49   Living matter and organic residues are an important source of moisture

50   **13 : Land allotment**

54   **14 : A rural settlement near Bobo-Dioulasso (Burkina Faso)**

54   Village and temporary farmstead
55   How the family transforms the landscape
58   How man can exercise control over this settlement

60   **15 : Land management in Serer country (Senegal)**

60   The foundations of the economic and social structures in Serer country

61   Outside Serer Settlements - What happens when the land
        authority has ceased to exist
62   The management of Serer settlements in practice
65   Basic lessons handed down by the Serer elders

66   *Lessons 16 and 17 : The need for trees*

67   **16 : Why trees are useful in the landscape**

67   Trees mitigate the effects of climate
68   Trees preserve soil life and transform it
69   Trees are a source of many useful consumer products

71   **17 : Trees act as windbreaks**

72   How do trees break the force of the wind ?
75   The composition of a good hedge
76   Other functions of hedges
77   Hedges and fields

78   *Lessons 18 to 22 : Soil and water*

80   **18 : Soil composition**

80   Part of the soil is solid
81   Part of the soil is gaseous
81   Part of the soil is liquid
82   Sand and clay are two very different substances

83   **19 : Soil structure**

84   Some soil structures are bad for agriculture

86   **20 : Soil horizons**

87   Every soil horizon has its own level of fertility
88   The interaction of soil and water
90   Soil laterization is disastrous for farming

92   **21 : Water movements**

92   Rain falls on the ground
93   Rainwater falls off sloping ground
96   Water infiltrates the soil
97   Soil evaporation and plant transpiration
97   Water in valley bottoms

98   **22 : Altering water movements for the benefit of farming**

99   The role of plants
99   Rainsplash and soil degradation
102   Using plants to stop runoff
103   Mechanical ways of fighting runoff and erosion
106   The fight against erosion, undertaken collectively for the whole
        rural settlement, is the one most likely to give the best results
107   Some farming practices to limit evaporation
108   Some farming practices for promoting rainwater infiltration

110   *Lessons 23 to 27 : Plant environment*

111   **23 : Plants**

111   Plant parts and their uses
113   Stages in the life of a plant

116   **24 : Plants need light**

120   **25 : The proper use of light**

120   How to give light to cultivated plants
123   Ways of shading

125   **26 : Agricultural calendars ; rain calendars**

125   The agricultural calendar
125   Rainfall and rain measurement

130   Defining climates on the basis of rainfall graphs

134   **27 : When do plants need water ?**

134   Plants and the water they need
138   The life cycles of cultivated plants and rain calendars
141   Labour availability

142   *Lessons 28 to 30 : Plant varieties and seeds for farming*

142   **28 : Plant varieties**

143   Determining the characteristics of the plant varieties in use
145   Selection or improvement of varieties

147   **29 : Planting material**

148   Seeds
152   Vegetative seeds
155   Good seeds and how to obtain them
157   Ways of procuring good seeds on the farm
159   Grain seed must be fresh when sown

160   **30 : Seeds and farm economy**

162   Should seeds be produced on the farm ?
163   Farm production of seeds is not worthwhile in every case

164   *Lessons 31 and 32 : How plants use the soil*

164   **31 : Roots**

164   Roots and root systems
170   Root systems respond to soil structures
172   Root extent is modified by the presence of other plants
172   Another way of representing the root systems of plants
172   Cultural practices

174   **32 : Knowing root zones**

174   How best to sow and plant
175   Good intercropping practices
176   How best to occupy the soil surface
177   How best to occupy the volume of available soil
177   The age factor in crop mixtures
178   Cultivated plants confronted with weeds

180   *Lessons 33 to 36 : Soil life and humus*

180   **33 : Animals, fungi and microorganisms in the soil**

180   Everything living in the soil
185   Humus
186   Soil life is fragile

189   **34 : Plants transform the soil**

190   Plants separate the soil
190   Roots change soil composition and structure
191   Roots improve water circulation in the soil
192   Associating useful species in relation to their role in the soil

193   **35 : Fallow land**

193   Why leave cultivated land fallow ?
194   Duration of fallow intervals
194   Variety in fallow vegetation
196   How to make fallow land as productive as possible

198   **36 : Cultural practices for producing humus**

204   **Lesson 37 : Working the soil**

204   There are many reasons for tilling
204   There are many ways of working the land
205   Some common ways of working the land

213	Implements for mechanical tilling
215	Distinctive signs of good tillage
217	Dangers of tilling
218	Plants also help to till the soil
219	Summing up of cultural practices connected with soils

220   Lessons 38 and 39 : Cutting the farmer's risks, increasing his chances of success

220   38 : Weather and biological risks in farming

220	Circumventing weather risks to rainfed farming
225	Biological risks : plant diseases and pests
229	Seizing opportunities

230   39 : Economic and technical risks

230	Adapting and improving agricultural methods
231	Building up reserves
232	Promoting mutual aid
232	Diversifying economic activities on farms and in villages

237   Lessons 40 to 44 : Feeding the soil with fertilizers

237   40 : Plant nutrients

238	Cycle of organic matter and mineral elements
238	Decomposition of vegetable waste
239	The cycle of organic matter and the removal of mineral elements from farms
240	Some thoughts on farm economy

241   41 : Organic fertilizers

241	The role of organic fertilizers
242	Some organic fertilizers are more useful than others
242	Green manures

245   42 : Mineral fertilizers

249   43 : The proper use of mineral fertilizers

255   44 : Making the most of field space

256	The shape of plants
257	Ways of sowing and planting
260	Arranging plants and exploiting field space

268   Lessons 45 to 49 : Agricultural returns

269   45 : Units of measurement and farm output

272   46 : Production and return

272	Overall farm production
273	Return from production factors
275	How to estimate returns from cultivated fields
276	Receipts and expenditure

279   47 : The value of the food we eat

279	Why is one food different from another ?

283   48 : Measuring return from fertilizers

286   49 : Returns from intercropping

289   Bibliography

291   Botanical Names in English and Latin

292   General Index

To the Reader

Before starting on the first lesson, the reader should become familiar with the language of this book, how it is expressed in the illustrations and in the text.

Photographs have been used to show everything that can be photographed. However, they in themselves are not enough and have been supplemented by drawings and tables to explain ideas that cannot be photographed, for instance, air, wind, rain, the sun's rays and things happening underground. It is important to understand the exact meaning of the line representing the surface of the ground which occurs in many drawings.

The lay-out of the text is such that it flows down the page like water over rocks from left to right, from right to left. The reader does not have to go back up the page except to consult illustrations or to read a boxed text. The figures and tables are numbered consecutively throughout the book. References to illustrations and tables are shown like this : **(figure 10), (table 11)**.

When a technical word is used for the first time, it is printed in **bold letters** and explained .

Some words and sentences are also in heavy print because they are key words and ideas.

A final point : masculine nouns take precedence in the English language. This may cause confusion especially in the context of African agricultural systems for which women are responsible in most cases. However, the book has been conceived for men and women alike. Words like 'cultivator', 'worker', 'farmer' are used in a general way to connote every active agricultural worker, whether man or woman.

This, the first English language edition, is imperfect in many ways. The authors would appreciate any criticisms, remarks or suggestions which would help them to publish an improved version at a later stage. Information on smallholders' farming practices and know-how would be specially welcome.

Lessons 1 and 2

Fields and farming

Lesson 1

Visiting the fields

A visit to the fields will be our first step on the vast continent of Africa. (Look at the map on page 10.) We shall visit our own fields, those of other members of the family, our friends and neighbours. A desire for knowledge will lead us to observe everything that lives around us and fulfils a purpose.

Figure 1 shows a field where maize and groundnut grow in association. Growing crops in association (in mixtures) is common in regions with good rainfall. **Figure 3** illustrates the same field in more detail.

In drier regions, millet or sorghum and cowpea are often grown together because they require less water than maize and groundnut **(figure 2)**.

Maize, millet, sorghum, finger millet *(Eleusine)*, wheat, fonio (hungry rice or acha), teff, etc. are **cereals** providing mealy grain.

Groundnut produces **oil-seeds**, i.e. seeds rich in oil.

Groundnuts and cowpeas, beans, peas, soya beans and bambara groundnuts are **legumes**. They all fertilize the soil and provide seeds full of nutrients. Lesson 9 covers these crops in detail.

All over Africa, **the associated cropping of cereals and legumes is common practice** and is valuable for reasons which will be explained later.

Sweet potato and cassava (manioc) are sometimes grown together **(figure 4)**. Both are **tubers**. The sweet potato grows quickly and produces tubers after four to eight months while cassava takes from fourteen to twenty-four months.

1

2

3

This diagram represents the field shown in **figure 1**, and will help to make the explanations clear. Particular notice should be taken of the line indicating the level of the ground.

maize

groundnut ground

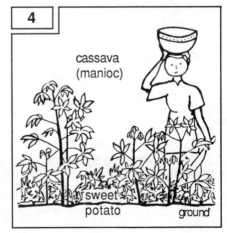

4

cassava (manioc)

sweet potato ground

Months later, in this same field, when the potatoes have been harvested, all kinds of vegetables - tomatoes, eggplants, okra, gourds and chillies - will be grown in the space between the cassava stalks.

2

Tubers are **starchy foods** low in other nutrients. Relish is therefore needed to make a balanced meal from tubers which are not particularly nourishing. Sweet potato, cassava and other vegetables can be added to make a satisfying dish.

In some parts of Africa, potatoes and beans are associated crops grown on mounds (**figure 5**). The potato contributes high starch content to the diet, whereas the bean gives more protein. Moreover, the bean fertilizes the soil and helps the potato grow.

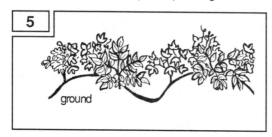

5

ground

6 | peas | potato | maize

In mountainous Central Africa, intercropping of potatoes and maize is practised (**figure 6**). The potato crop provides the tubers, and the maize adds the cereals. Peas are sometimes also associated and their climbers bear highly nutritious seeds.

Figure 7 shows a pineapple crop in well-watered grassland. The farmer was inconvenienced by **weeds**, which overran his crop, especially at the beginning of the growing season. To reduce his labour in the field, he planted sweet potato between the rows of pineapple. Sweet potato covers the ground quickly, protecting the topsoil from sun and rain. It also chokes the weeds which would otherwise spread among the pineapple. By associating sweet potato and pineapple, the farmer is killing two birds with the one stone. He has reduced his labour input and secured an extra crop, the sweet potato.

8 | pineapple

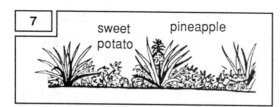

7

sweet potato   pineapple

Other farmers would rather cultivate pineapple on its own (monocropping) and combat weeds by **hoeing** (**figure 8**).

Here is a cotton field in monocrop cultivation except for a few shea butter trees. The field has been weeded to eliminate all other plants (**figure 9**).

9 | cotton | shea butter tree

Bashi farmers living near Lake Kivu in Zaire have been known to associate up to five or six different crops and sometimes several varieties of each species.

In this type of field, maize, sorghum, cassava, sweet potato and many kinds of bean and other vegetables are associated **(figure 10)**. But the composition of the field changes from month to month, and its changing pattern can be observed by regular inspections over a period of weeks.

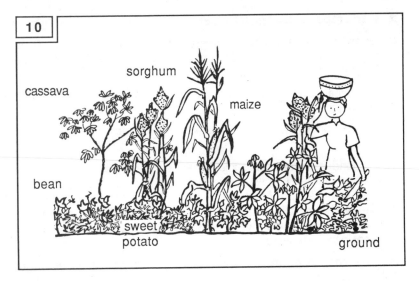

This field of beans in a banana grove was photographed in Rwanda **(figure 11)**. The beans were sown close together but the banana plants were fairly widely spaced to allow the beans enough light.

Fields like the one in **figure 12** are common in wet forest zones. Farmers practising this type of cropping are never short of good food for their families.

After forest clearance, only a few big trees are left standing. Bush fires have got rid of the branches and creepers that obstructed work in the fields. The farmer then sows several species in the field. **Some of these grow at the same time, while others take root and grow over a longer period.** Lesson 7 considers this type of field in detail.

Some of the crops sown are propagated by **cuttings**, for example, cassava, banana and sugar cane, while others are grown from **seeds**, for example, groundnut, maize, pawpaw, cucumber, okra, black nightshade, eggplant and sorrel.

The field in **figure 13** is also typical of a forest zone with abundant rain throughout the year. This leads to **two annual cropping seasons**. Intercropping of a wide variety of species is practised :

- **tubers** : the sweet potato covering the ground, the tall-stalked cassava and the cocoyam (tannia) ;

4

- **seed crops** : cucumber, maize, etc.;
- **crops grown for their leaves, their stalks or their fruit** : sugar cane for home consumption, tobacco, chillies, tomatoes, eggplants, gourds ;
- **fruit trees** : pawpaw growing behind the sugar cane, oil palm, lemon, etc.

Here is a coffee plantation protected by the shady *Leucaena* **(figure 14)**.

The ground is covered with thick green grass called paspalum. Trees have been planted in rows between the coffee shrubs and their leaves provide some shade. Most shade trees belong to the legume family whose species fertilize the soil. They ensure good growth conditions for the coffee trees by protecting them from the hot sun and the winds frequent in these regions.

Banana and cocoyam are often associated in forest zones **(figure 15)**.

The banana is a **fruit crop** whose leaves spread out several metres above the ground. Dead leaves and stalks are plentiful. These fertilize the soil by rotting and also keep in the moisture.

The cocoyam is a tuber with edible leaves. It thrives on shady, moist soil, under banana plants, for instance. These two crops in association provide large quantities of starchy foods.

Millet and *Acacia albida* are often associated **(figure 16)**. The cereal benefits from the proximity of the tree which is also a source of animal fodder. *Acacia albida* is a legume and like other species of its kind, it fertilizes the soil.

Other trees are associated with cereal crops, for instance, the shea butter tree and the African locust bean. Both bear edible fruit and act as windbreaks.

The way associated crops are **layered** may be the result of careful planning as in the case of the coffee shrubs and the shady *Leucaena*, but it may also be accidental as shown in **figures 12 to 16**, and result from the felling of trees or the spontaneous reproduction of seeds.

shade trees

Leucaena

coffee

14

banana

cocoyam

15

Acacia albida

millet

16

Notes

After these first steps on African soil, there are many important points to remember

To understand what farming is really about, we must look from the ground up. If our eyes travel from the ground to the treetops, **the layering of plants** is clearly noticeable. Some plants are creepers, some are erect, others are arborescent.

The life cycles of associated field plants usually vary in length. Some crops live for the duration of a season and they are **seasonal** crops. Other crops have a longer life cycle. These are called **perennial** if they live for many years, or **semi-perennial** if their life cycle, though more than one season, is not so long.

Among the seasonal crops we find in the fields are maize, sorghum, millet, groundnut, beans and cucumber. Among the perennial plants, are pawpaw, palm, coffee, *Leucaena*, shea butter and *Acacia albida* and some semi-perennials are banana, yam and sugar cane.

We have seen **the associated cropping of seasonal, perennial and semi-perennial plants**. If we visit the same field over a period of months, we will see that some plants are still in the ground, others have disappeared completely, while others have just begun to sprout.

The wide range of uses to which these plants are put is striking. In the same field it is common to find associated crops each with its own function. Some are grown for **human consumption**, others provide **animal fodder**. There are different kinds of food crop : flour of various sorts, oil, sugar, etc. Some crops have other consumer uses, for example, tobacco, cotton, medicinal plants, dyes for cloth, and trees grown for timber. However, in some fields, **there are plants whose sole purpose is to be useful to other plants**, the *Leucaena* , for example, which fertilizes the soil and provides shade for the coffee shrubs.

In monocropped fields only one plant is grown. In **figure 8** for instance, the farmer only wants to grow pineapple and has removed all other plants. The same is true for the cotton field in **figure 9**, although there are trees growing here and there to form a canopy over the cotton shrubs.

Finally, we have seen that **legume plants play an important role in the fields observed**. These plants are specially valuable for two reasons : firstly, their seeds are highly nutritious, and secondly, their fertilizing potential far exceeds that of other plants. We shall look at these plants in greater detail in Lesson 9.

A brief comparison

*In the tropics, and especially in Africa south of the Sahara, a great number of agricultural plants are sown by **cuttings**.*

***Seeds** from flower fertilization are also used.*

The difference between farming systems using cuttings and those using seeds is significant. In temperate or cool climates, propagation by seed predominates, but in wet tropical climates, the use of cuttings is widespread. These two methods require different farming practices.

But why be a farmer anyway ? What are the farmer's aims ? What are the reasons behind the growing methods we have been discussing ? Lesson 2 will attempt to answer these basic questions.

Lesson 2

Farming : goals, means, limitations

Each of the fields visited so far has different characteristics which depend on many factors, such as the climate, the type of soil, farming techniques, the crops grown, the needs of the farmer and his family. To understand what farming is all about, these factors must be examined in detail. Some distinctions must be made from the start.

Farming is influenced by :   what the farmer **wants to do** ;

what the farmer **can do** ;

what the farmer **cannot do**.

What the farmer wants to do

Every farmer wants to live off the land and, if possible, to live well. How he pursues this goal will depend on his wealth, the place where he lives and the style of life in that part of the country.

■ He may be trying to satisfy the needs of his family, in which case we speak about **family self-sufficiency** or **family consumption**.

■ He may be interested in the money from the sale of his crops, in which case he is looking for **cash income**.

■ He may want to make a **secure living** for himself, his family and his workmen, or contribute to family or community harmony.

■ He may want to **protect the land and the other resources** at his disposal for farming.

The choice of crops and growing methods will depend on the goals pursued.

In the field in **figure 17**, cocoyam and okra are associated to meet family consumption.

Here is another field cultivated to meet family consumption **(figure 18)**.

17

18

19

After simultaneous crops of maize and groundnut for one cultural season, the field is planted with several other food crops : yam, sweet potato, cassava, taro and legumes. Some trees have been preserved to act as windbreaks and provide branches used locally for making hoe-handles.

All the produce of this field is consumed in the village. Little is available for sale.

In contrast to the fields geared to family consumption, here are two situations where the farmer is only keen on selling produce to make a lot of money. The pineapples grown in **figure 19** are for export to Europe. The crop covers the field (monocropping) except for a few bananas lining the paths.

The same is true for the pawpaw plantation in **figure 20**. The produce is for export. For both pineapple and pawpaw production, the planter is involved in expenditure. He must invest in machinery, fertilizers, and so on. He must cover the cost of these by selling a lot of produce. This kind of farming is based on money.

The orchard **in figure 21** contrasts with the monocropping patterns in **figures 19 and 20**. The farmer has associated many trees to produce a diversified harvest. Banana, palm, coffee, cocoa, avocado, lemon and herbaceous plants are mixed, so at all times of the year there is produce for consumption and

20 selected pawpaw

native pear

banana

bitter leaf

palm

coffee

21

subsistence crops

cash crops

22

for sale. There is security in being able to obtain a cash income throughout the year, especially when unexpected expenses have to be met.

By practising mixed cropping in his orchard, the farmer hopes to ensure **regular supplies** for his family and/or for the market.

Given the opportunity and motivation, the farmer will try to achieve all the goals mentioned and, will manage his farm with these goals in mind.

The farmer's goals sometimes compete with each other. In **figure 22**, a pineapple crop is growing on the low, fertile, well-watered land. The fruit is sold abroad to earn money. The badly eroded land on the slopes is left for subsistence crops. The pineapple planter is only interested in marketing his produce for money. He is not aiming at family sufficiency.

We are led to the conclusion that the system of agriculture followed is determined by the goals pursued.

What the farmer can do

The farmer has specific resources at his disposal called **production factors**, and the main ones are listed in **table 23** (page 8). Production factors must be combined on farms to ensure agricultural production.

Farming involves suiting production factors as carefully as possible to the goals chosen, but at the same time, being wary of methods which simply exploit production factors and ignore the need to preserve and regenerate them for the future.

What the farmer cannot do

The farmer is subject to many constraints. This is what he means when he says he hasn't got the means to do this or that, or he doesn't know how to do this or that. Here are a few examples of these constraints.

- If **rainfall** is inadequate at the beginning of the rainy season, seeds dry up and the harvest is bound to be poor. Water is the restricting factor in this case.

- If the harvest is plentiful and there are not enough **workers** to bring it in, there is a labour shortage or restriction, meaning that the workers on the farm are unable to harvest all the produce.

- Perhaps a machine would speed up the work of the harvesters, but there is no money to buy it. This is a **money** constraint.

8

Land

Farm production depends on both the **extent** and the **quality** of the farmland.

Farmlands can only be cultivated if they are **watered** and properly **fed**.

Soil feeding is ensured by manure from organic waste, by chemical fertilizers sold commercially or by the decomposition of green crops buried in the ground.

Money is needed to buy chemical fertilizers, so money is also a production factor. This point will be discussed later in the book.

Water

Water is essential for life and for the development of agriculture. It must be present **at the right time** in the life cycle of plants and **in sufficient quantity**. There is no life without water.

In most cases, **rain** provides water for agriculture. Sometimes the farmer controls the water supply to his fields. He **irrigates** the land - he taps water at some distance from his fields and uses it to water his crops.

In many parts of Africa, more especially in the Sahel, the lack of **water limits agricultural production** and rainfall is inadequate and erratic.

Plants

Farming is organized round the plants grown. **Everything the farmer does is aimed at satisfying the needs of the plants, to obtain the highest possible yields.**

Agriculture is characterized by the number of plants cultivated and by the diversity and quality of the produce destined for consumption.

The type of agriculture practised de-pends, for instance, on whether the plants cultivated are **trees** or **herbaceous plants**, on whether they live for many years, **perennial** or long-living plants, or are **seasonal** or annual crops.

The farmer must realise how important it is to discover which plant **varieties** suit him best in quantity and quality. He must therefore look for good **seeds**.

Work

Farming the land is demanding work which depends on the seasons and the life cycle of the crops grown. **Farm work is seasonal.** At certain times of the year, the farmer has a great deal of work, for example, during the sowing season or at harvest time. At other times, there is not much work to be done in the fields.

The cultural methods used and the degree of sophistication of the implements used may influence the **availability, quality and ease of work**.

When power is needed, the work can be done by **human**, **animal** or **machine** energy.

However, the worker must also use his **intelligence** to benefit agriculture. The quality of his decision-making determines the value of his farming.

The work can be done by the **family**, or by **paid labour** from outside the family circle.

Farm Implements

The farmer cannot do his work without implements. These have different levels of sophistication. They can be operated manually, by animal traction or by a motor or an engine.

The more sophisticated the imple-ments, the less likely they are to be produced by farm or village inhabi-tants. Simpler implements can be made by these people or by local craftsmen.

Implements bought commercially presuppose reserve funds.

Money

Money is needed to buy all the production factors which cannot be found or made on the farm, for instance, chemical fertilizers, pes-ticides, machinery, traction power, fuel for machines.

Self-sufficient farming, carried out by the farmer for family needs, requires **little money**. On the other hand, some **intensive farming** systems, organized solely for the **sale of produce**, require **large amounts of money** be-cause they use many machines, fertilizers, pesticides and other production factors.

Livestock

Livestock is an important production factor on stockbreeding farms and on farms associating crops and livestock.

Some farms are managed for **stockbreeding** purposes and only **fodder** for livestock is grown.

On the other hand, nomadic stock-breeding is only concerned with animal production. No fodder crops are cultivated.

Other farms are managed to produce vegetable and animal products. They are called mixed farms because they associate cropping and raising stock.

A combination of crop and livestock farming, along with the rearing of poultry and small animals usually gives the best results in all areas.

The **constraints** or **restrictions** encountered prevent the farmer from acting freely in accordance with his wishes. He has to take innumerable, highly diversified constraints into consideration. As well as the examples just mentioned, there are the constraints imposed by the **amount of arable land** available, the **quality of the soil**, the quantity and quality of the **seeds**, the **climate**, **disease** in man, crops and cattle, the quality of the **implements** used, and so on. Maybe several production factors are missing at the same time and **if one key factor should prove deficient, say water or seeds, it may be impossible to continue farming.**

There are other constraints, not covered by the production factors described in **table 23**, particularly the constraints of the environment : the **price of products**, the **organization or disorganization of markets**, insufficient **technical knowledge**, the **location of the farm** near to or too far from a road.

Throughout this book, our knowledge of agricultural production factors will be increased, particularly our knowledge of plants and soil. In the sections on soil, many aspects of soil study will be looked at, such as :

- **soil composition** and especially plant nutrients found in the soil itself ;

- land as a **living environment** ;

- **fields** and what they look like on the surface, how they are organized above and below ground ;

- **rural settlements**, all the lands belonging to a farm, a hamlet or a village.

Before discussing these points in detail, our first task will be to find out what is meant by the environment of the farmer, of cultivated plants and of livestock.

The **progressive farmer struggles constantly to overcome the constraints which limit his work.**

But he must be realistic.

Farming is first and foremost concerned with life - the life of men and women, the life of plants and of livestock, the life of the soil because soil contains living things. A lack of respect for life, simply because one has money and can afford chemical fertilizers, may have disastrous effects. **Erosion** in the field of mechanized pineapple cropping **(figure 24)** is due to the farmer's use of powerful agriculture machines and disregard for soil life.

24

Rabat
Algiers
Tunis
TUNISIA
Tripoli
Cairo
MOROCCO
El Aaiún
ALGERIA
LIBYA
EGYPT
WESTERN
SAHARA
MAURITANIA
Nouakchott
MALI
NIGER
CHAD
Khartoum
Dakar
Niamey
SENEGAL
Banjul THE GAMBIA
Bamako
Ouagadougou
N'Djamena
SUDAN
DJIBOUTI
GUINEA
BISSAU
Bissau GUINEA
BURKINA
NIGERIA
ETHIOPIA
Djibouti
Conakry
SIERRA
LEONE
IVORY
COAST
GHANA
TOGO
BENIN
Lagos
CENTRAL
AFRICAN
REPUBLIC
Addis Ababa
Freetown
Monrovia
CAMEROON
Bangui
UGANDA
KENYA
SOMALIA
LIBERIA
Yamoussoukro
Accra
Lomé
Porto-Novo
Malabo
Yaoundé
Kampala
Mogadishu
EQUATORIAL
GUINEA
Libreville
CONGO
ZAIRE
Kigali
Nairobi
GABON
RWANDA
Brazzaville
Bujumbura
BURUNDI
Kinshasa
Dodoma
Luanda
TANZANIA
MALAWI
ANGOLA
Lilongwe
MOZAMBIQUE
ZAMBIA
Lusaka
Harare
NAMIBIA
ZIMBABWE
Antananarivo
BOTSWANA
MADAGASCAR
Windhoek
Gaborone
Pretoria
Maputo
Mbabane
SWAZILAND
Maseru
LESOTHO
SOUTH
AFRICA
Cape Town

AFRICA

● Capital city

—·—· International boundary

0   400   800 Kilometres

Lessons 3 to 5

The environment

Man's environment is made up of everything that surrounds him and constitutes the fabric of his life - material things, plants, climate, dwellings, animals, people.

The environment is subject to change. The seasons come and go, the climate changes, men are born and die.

Plants also have their own environment. Some species can grow in environments quite different from those of other plants. For example, potatoes, maize and beans are successfully cultivated in many parts of the world, in both hot and cold climates. Other plants require special growth conditions. Bananas, for instance, only grow in hot, wet climates. Fonio (hungry rice or acha) prospers in dry tropical zones. The oil palm is found in many regions whereas the date palm, belonging to the same family, only grows in desert soils.

The environment relevant to agriculture will be described under three headings in three separate chapters : The living environment (Lesson 3), The physical environment (Lesson 4), The economic and social environment (Lesson 5).

Lesson 3

The living environment

The living environment of a plant is everything which lives round and on the plant. This environment is called the **biotic environment**.

The biotic environment of the banana is illustrated in **figure 25**. The bananas photographed here are surrounded by big trees, by young umbrella trees, and by tall and bent grasses. They are the habitat of countless **insects, birds, rodents** and **other mammals**. The ground and the felled trees are covered in **mould** and **fungi**. Other organisms, so minute they cannot be seen by the naked eye, penetrate the leaves, the flowers and the tree trunks. They are called **bacteria, yeasts** or **viruses**. Man is present too. He fells the trees, prunes them, gathers the fruit. All these factors make up the living environment of the banana.

The groundnut crop in **figure 26** is on sandy soil. It is obvious at a glance that the biotope of this crop is simpler and much less active than the banana grove environment observed in **figure 25**. Soil life is not so intense. The groundnut plants are growing on their own. Some insects or parasites have attacked the crop. There are few earthworms.

Everything living in the air and in the ground forms the biosphere. All

25

living things depend on others. The living matter or waste of one is food for another. This is the **food chain**. When one living item disappears from the environment, others disappear too because there is no longer food for them. Man deserts his barren lands, livestock leave regions where there is no more grass. Oxpeckers follow the cattle abandoning desert zones, because without cattle there are no flies.

Before examining **figure 27**, which illustrates the biotope of cultivated plants, we need to agree a convention for drawings and symbols to denote biotopic elements such as sun, rain, evaporation, plants, roots, soil, etc. (**figure 28**, page 14).

26

27

The living environment or biotope

This diagram shows all the elements of the living environment on farmlands.

Living things depend on one another. The diagram may be used to examine this interdependence and to ask the question - on whom or on what, does each living thing in the environment depend?

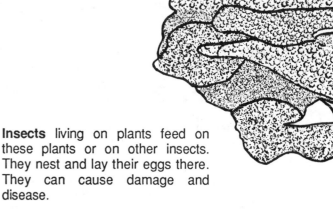

Insects living on plants feed on these plants or on other insects. They nest and lay their eggs there. They can cause damage and disease.

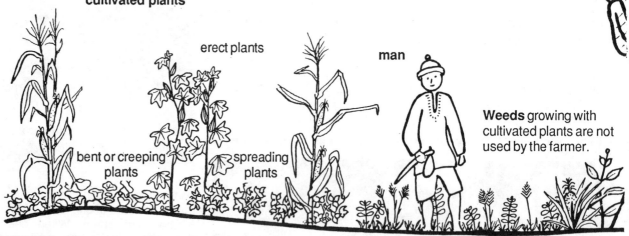

cultivated plants

erect plants

man

bent or creeping plants

spreading plants

Weeds growing with cultivated plants are not used by the farmer.

Some small animals and insects live in the ground, for instance, **worms**, **termites**, **spiders**, **rats**, **ants**.

Organisms invisible to the eye live in waste (litter). They are called **microorganisms**.

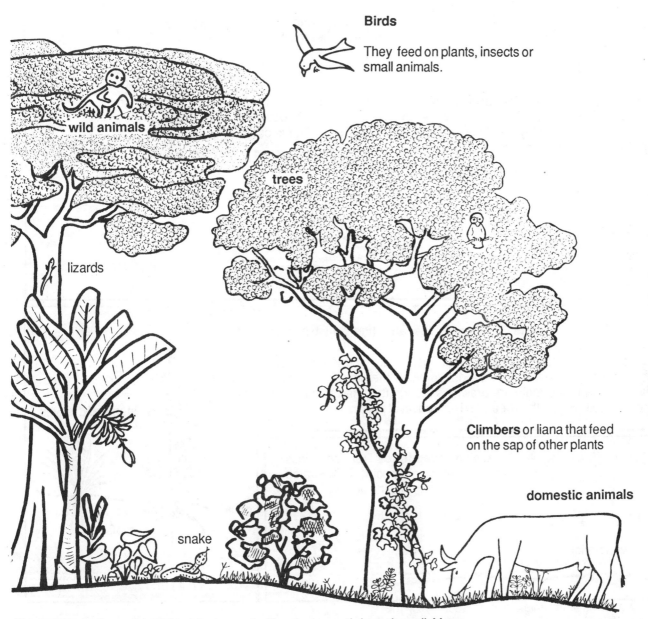

Birds

They feed on plants, insects or small animals.

trees

wild animals

lizards

Climbers or liana that feed on the sap of other plants

domestic animals

snake

Roots take up the water and nutrients needed for plant growth from the soil. Many microorganisms invisible to the naked eye live on the roots and help them to take up food from the soil.

What would lizards do without insects ?
What would hawks do without lizards, mice and small birds ?
What would insects do without plants and animals to feed off ?
What would men do without plants and animals ?
The food of one species depends on the food of others.

This interdependence forms **food chains**.

14

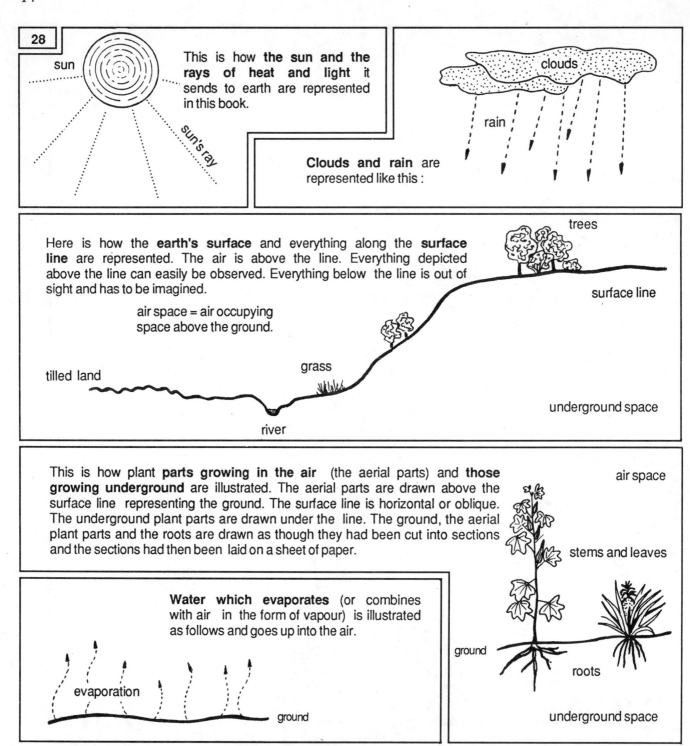

28

sun

sun's ray

This is how **the sun and the rays of heat and light** it sends to earth are represented in this book.

clouds

rain

Clouds and rain are represented like this :

Here is how the **earth's surface** and everything along the **surface line** are represented. The air is above the line. Everything depicted above the line can easily be observed. Everything below the line is out of sight and has to be imagined.

air space = air occupying space above the ground.

trees

surface line

tilled land

grass

underground space

river

This is how plant **parts growing in the air** (the aerial parts) and **those growing underground** are illustrated. The aerial parts are drawn above the surface line representing the ground. The surface line is horizontal or oblique. The underground plant parts are drawn under the line. The ground, the aerial plant parts and the roots are drawn as though they had been cut into sections and the sections had then been laid on a sheet of paper.

air space

stems and leaves

Water which evaporates (or combines with air in the form of vapour) is illustrated as follows and goes up into the air.

evaporation

ground

ground

roots

underground space

Lesson 4

The physical environment

Plants spend their lives above and below ground because they have aerial parts and underground parts. Trunks, stems, leaves and fruit form the aerial parts. They breathe air, transpire water and capture sunlight.

Roots are the underground parts and they feed the plant. They breathe, too, and are fed by waterborne nutrients in the soil.

The **physical** or **abiotic** (non-living) environment is subject to changes because it is composed of the following factors :

- **land forms** depending on the variables **altitude, slope** and **slope exposure** (the way the slope is exposed or orientated to the sun) ;

- **soil composition** depending on the quality of the parent material or bedrock ;

- **climate** which is largely responsible for the way in which plant needs are satisfied.

We are especially interested in studying four plant needs in the physical environment. These needs are light, air, water and heat.

Plants need sunlight

Sunlight varies in intensity. The sun's rays may strike plants directly or be filtered by cloud or by the foliage of taller plants. Daylight duration, also a variable, is a very important factor in plant life.

Leaves absorb light through the action of **chlorophyll**, the substance which gives them a green colour. Chlorophyll is needed to capture **light energy** to make **plant energy**. Food for man uses up plant energy.

Plants need air

Plants breathe air through a large number of little holes in the leaves. These little openings, called **stomata**, open and close depending on light duration (length of day) as well as in response to heat and moisture intensity (see Lesson 23).

Air **temperature** and **humidity** depend on how hot and moist it is. Temperature, expressed in **degrees centigrade,** describes how hot it is. Humidity is the amount of water vapour contained in the air. (**Relative humidity** is the moisture held in the air at a particular temperature. It is measured as a percentage of the total amount of water that could be held in the air at that temperature.)

Air flow can be modified by **wind**, a factor exerting a marked influence on certain regions in the form of, for instance, the **harmattan** which dries up the countryside, or the **monsoon** which brings rain in its wake.

Plants need water

There can be no life without water. Every plant species has specific water requirements. Not only must there be enough water but it must be available when the plant needs it most. (This point will be discussed in Lesson 27.) In many parts of the world, whether there will be an adequate, reliable rainfall is the farmer's greatest anxiety.

Plants transpire water through leaf stomata. The higher the temperature, the more they transpire. Sometimes it is so hot that the amount of water transpired into the air is greater than the amount available to the roots underground and the leaves wilt. (See Lesson 23.)

Plants need heat

Temperature is an important factor of climate. Temperature measures the hotness of the air and ground. Some plant species thrive in high temperatures whereas others prefer cool climates. Air humidity depends on the temperature of the air. The hotter the air, the more water vapour it can absorb.

Figure 29 illustrates the aspects of the physical environment of agriculture.

29

The physical environment

Elements of the physical environment :

- **Land** and **soil**

- **Air** and **wind**

- **Light**

- **Heat** and **cold**

- **Rain**

- **Humidity** and **dryness** of the air

All the elements of the **climate** which form the physical environment are interdependent.

Plants and men breathe the air.

When soil and plant evaporation and transpiration are low, the air becomes dry. There is no cloud formation.

Wind is moving air. It heats or cools the environment. It makes plants transpire, and makes soil water evaporate.

Plants give off some of their water content into the air. This is called **transpiration**.

Sunlight is essential for plant life. The sun's rays heat the earth. The sun imparts **energy** to the biosphere.

sun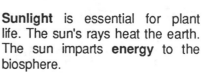

ray of sunlight

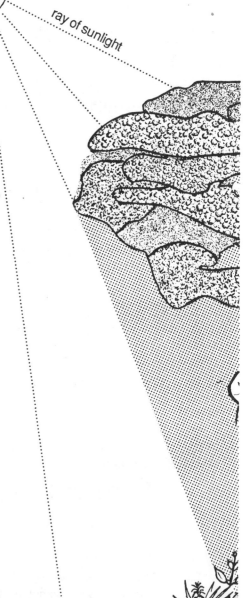

Some plants like plenty of heat and light. They are called **light plants**.

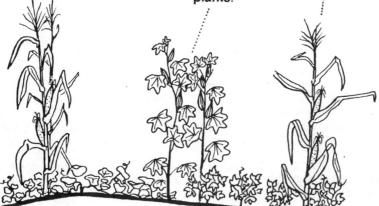

bare ground

Soil water passes into the air in two ways.

1 Plants take up the water which is then evaporated into the air from the soil surface. Water which leaves the soil to humidify the air is called evapotranspired water. This process is called **evapotranspiration**.

2 Soil water is transformed into water vapour - it evaporates into the air. This process is called **evaporation**.

The water vapour evaporated by plants and soil collects to form **clouds**.

Rain falls when clouds are saturated by water and when the surrounding air or atmosphere cools off.

Rain

You cannot see air but you can feel it, when you breathe or when it is windy.

Air contains water in the form of vapour. You can see water vapour or steam escaping into the air above a pot of boiling water, and mist and clouds which form in the sky.

Hot air absorbs a lot of water vapour as is illustrated by the drying effect of the harmattan on bare soil.

Cold air retains less water. Consequently, when hot, damp air gets cold, water drops form and come together to make clouds. When the drops become enlarged, they fall as rain.

For there to be rain, the air must be vapour-saturated and then cool off.

Some plants like shade and cool temperatures. They are called **shade plants**.

ground

When there is no foliage, the ground is heated by direct sunlight. A lot of water evaporates from the neated ground surface. Air under foliage is humid. The ground is not heated by direct sunlight and there is little evaporatioin.

Lesson 5

The economic and social environment

Everything to do with the production, consumption and marketing of goods is part of the economy or economic life of the community. Producing cereals or seeds, crushing seeds to obtain oil, transporting goods, paying labour, selling on the market, preparing food, raising livestock, manufacturing clothes in a factory - all these are part of the economic life.

The term **social life** is used to describe **the organization of communities and relations between their members, for example, relations between young and old in a village, family alliances, getting together for weddings and funerals, relations between men and women, family structures and lineages.**

The economic life and the social life of a community are always closely interwoven. Relations between families illustrate this point. When a marriage takes place, agreements are made about the use of land, which is a very important production factor.

Invitations to **working parties (figure 30)** are another example of the close connection between the economic life and the social life of a community. Farmers join forces to work on each other's land. Working the land is an economic act. But the way the group is formed is a social act. Instead of paying for the work, the host-farmer provides food and drink. He, in turn, will go to work on other farms. This is a form of social organization.

From the farmer's point of view, the main aspects of economic activity are :

- **agricultural production** - its quantity and quality. This covers both vegetable production (plants) and animal production ;

- the **availability of production factors.** These factors were discussed in Lesson 2 ;

- **consumption.** The goods consumed by the family can be produced on the farm (this is the farm consumption mentioned earlier) or they can be purchased from outside the family ;

- **storage (figure 31),** especially of cereals, seeds and tubers. Storage may be required for both long and short periods ;

- when agricultural production is destined for sale to obtain cash income, it must be

30

31

32

marketed. **Marketing** can be handled by traders, by the farmers themselves, or by cooperatives ;

- farm **supplies** also depend on the market, for instance, household goods, such as kerosene, salt, soap, clothes, vessels, radios. Such supplies also include goods for agricultural production, that is, production factors not made on the farm, such as shears, ploughs, fertilizers, pesticides ;

- **seed production (figure 32)** and sale ;

- **transport (figure 38).** Goods must be moved from the fields to the farm, from the farm to the local market, from there to markets further way, or to other countries, and similarly, goods must be moved to the farm and the fields.

All these activities combined form the village economy. Developing these activities means developing the village, the region, the country. However, local economic activity depends on the economic activity of other rural districts and regions throughout the country, on the country's towns and cities, and on foreign countries.

For example, the price of produce such as cotton, coffee, pyrethrin and palm oil are fixed in the U.S.A., Germany, England, France, Japan, and so on. Remember, too, how farms depend on production factors like fuel oil, chemical fertilizers, machinery, whose price is fixed in foreign countries.

Development, however, is more than simply economic progress. The social structure of the village must also be taken into consideration.

- **produce processing** - processing cereals into flour **(figure 33)**, extracting oil, slicing cassava, brewing beer **(figure 34)** ;

- **the work of local craftsmen** - blacksmiths **(figure 35)**, weavers **(figure 36)**, potters, basket makers **(figure 37)**, tanners, bakers, mechanics ;

40 | The economic life of the farm

The resources needed to achieve agricultural production are called

- - - Production factors - - - -

Land is needed.
Does the farmer dispose of land ? Who owns the land ? Does the farmer have to pay for land use ? How much does he have to pay ? Is there a land chief ?

Water is needed.
Is rainfall adequate ? Must water be fetched from off the farm ? Are there irrigation systems ? What is the cost of water ? Who is in charge of water management ?

Farm implements are needed.
Can they be manufactured or found locally ? Are they for sale ? How much do they cost ? Are the implements good ? Is machinery required ?

Labour is needed.
How many workers can the family and each household provide ? Are there enough workers for all the farm work, at all times of the year ? Are workers from outside the family needed ? Are they available ? How much do they cost ? Is the work force jointly responsible or does it have to be paid ? What are the social ties between the members of the work force ?

1

Production factors

2

Agricultural

millet. sorghum groundnut maize

the family

3

Farm consumption

What goods does the family produce and consume on the spot ? Is farm production enough to feed the family and the labour force ? Does the farm produce what the family needs ? Are storage facilities and conservation methods adequate ? Are food processing and preparation well provided for ?

7

■ Consumer expenses

These are the expenses the family has to meet for food, clothing, housing, medical care, information, travel and recreation.

What are these expenses ? How high are consumer prices ? Does the money obtained from the sale of farm produce cover the cost of family consumption ?

■ Circulation of goods

Marketing and supplies 10

Local and external economic environments come together at markets and stalls.

The agricultural produce sold by the farmer must be disposed of, while consumer goods must be brought to the market place to meet customer demand.

Is this buying and selling well organized ? Are transport services efficient ? Are prices fair for farm produce and for consumer goods ? Are the weighing scales in order ? Is the weighing done properly ? Are quality goods available ? Does supply meet demand at all times ?

Are the volumetric measures used in markets standard from market to market ? 9

Plants are needed.

Has the farmer obtained good seeds ? Who produces them ? Does the farmer have to buy them ? How much do they cost ? Can the farmer find the seeds he wants ? Can he buy good seeds ? Who owns the trees ? Who can plant trees ?

Money is needed.

Are the cultural methods cheap or costly ? How much money is needed ? What for ? Do cheaper cultural methods exist ?

The sale of agricultural produce

Can the produce be sold ? To whom is it sold - to other members of the rural community, to tradesmen, to the State, to a cooperative ? How much of the production is sold ? For how much ? Is the product sold on time ? In what form is it sold ? When is it sold ?

Are the products sold in the village, in the country or abroad ? Are they processed before they are sold ?

production

flour 4

 trading

■ **Production expenses**

These are the expenses the farmer must face when not all the production factors are available on the farm.

Can the missing factors be found and brought to the farm ? Are they the right quality ? How much do they cost ? Is the money obtained from produce sales enough to cover the production outlay ?

5

6 Is the money obtained from produce sales enough to cover expenditure which falls into three main categories :

7 **production** expenses
consumer expenses
social and **unavoidable** expenses ?

8

■ How much do **social** and **unavoidable** expenses, for instance, school fees, funerals, medical fees, marriages, taxes and income tax, come to ?

22

Finally, economic activity is also dependent on the roads, railways and ports which form part of the country's **infrastructure (figure 39)**.

Briefly, the economic activity of the labour force on a given farm depends to a large extent on what is happening both within the **rural settlement** and far away from it. The work force living on a self-sufficient farm is obviously less dependent on the outside world than the work force on a farm only growing cash crops to sell.

Figure 40 is an attempt to give an overall picture of economic life on a farm. The figure is in question-form. Factual answers will help the cultivator to get to know his farm better. To understand the figure, the reader must first look at the three illustrations - the photograph of the family on the left ; the photograph of the shop on the right, and the drawing of agricultural produce in the middle. After that, the arrows indicate the right reading order. The family labour force uses the production factors (arrow 1). The production factors are combined to provide agricultural produce (arrow 2). The produce is either consumed on the farm (arrow 3) or traded (arrow 4). The sale of the produce provides money for various expenditures (arrow 5). Arrows 6, 7 and 8 point to the kinds of financial outlay or expenses the family may have to meet, while arrows 9 and 10 show the trade exchanges taking place between the local community and outsiders - other regions and cities, other countries.

Special attention should be paid, in this figure,

37

38

39

Points to remember

Farmers are very dependent people. They depend either on their own land and its potential when they seek self-sufficiency, or on the outside world when their agriculture is geared to cash crops (this is called exchange agriculture).

The modern farm is one which allows the family to achieve its goals, i.e. to produce enough food, to sell the surplus, to buy goods not available on the farm itself, and to preserve the quality of the land and its resources.

Every factor that reduces the farmer's risks helps to modernize agriculture. Having enough to eat, making good use of consumer resources, selling on stable markets - these are all important features of that family well-being which modernization sets out to achieve.

to the network of relationships which make up the economic environment of the farm where the farmer pursues the goals of self-sufficiency, cash income, security and land conservation.

The study of **figure 40** should be followed by a discussion of the text on the opposite side of this page.

Many questions not raised in **figure 40** come to mind in connection with the farm economy.

For instance,

- **What decisions must be made** on a farm? Decisions about land use, labour distribution, cultural methods, produce sales.

- **Who is the decision-maker? Who disposes of the farm produce and the money** - the men, the women, the group, the community?

- **Questions about land rights** - tillage and tenure rights, the mutual obligations of everyone concerned - the old, the young, women, outsiders?

- What about **relations between cultivators and pastoralists**? Do they use the same land? Do they exchange produce? Do they help each other?

- **What wealth is available in and around the community?** Is money borrowed or lent? From whom? To whom? What conditions are attached to lending and borrowing? Who are the people involved: traders, other members of the community?

| 41 | **To observe agriculture fruitfully, we must understand its goals, means and limitations** |

What the farmer wants to do (his goals):

- satisfy his own needs;
- dispose of cash income;
- make a secure living and ensure good relations with the family and the community;
- preserve the resources of his land.

What the farmer can do with the resources at his disposal on the farm. These resources are called production factors:

- soil;
- water;
- soil nutrients;
- plants;
- labour input;
- implements;
- money;
- livestock.

Getting to know about agriculture means studying the way in which the farmer combines the production factors at his disposal.

What the farmer cannot do (reasons for his limitations):

- production factors are inadequate;
- nature bars the way;
- economic factors hold him back;
- the land tenure system stops him;
- his community offers resistance.

We must also understand the living environment.

The living environment of the plant includes:

- other plants;
- other living beings;
- man.

The economic and social environment comprises the economic life surrounding the farm:

- local aspects (in the vicinity of the farm);
- external aspects - national and international.

The physical environment includes all aspects of climate:

- light;
- temperature;
- humidity;

and the land, especially:

- land forms (configuration);
- soil composition.

organized relations between rural inhabitants:

- the way families are organized;
- the way age groups are organized;
- relations between social groups (men, women, young people).

24

■ Another very important question concerns the hazards or **dangers** confronting farmers, the **risks** he has to take. These hazards are linked to unpredictable weather conditions, to unforeseen damage caused by crickets and birds, to disease, to the fall in market prices, etc. They may also be linked to the way infrastructures and trade are managed. Firms may not be able to supply the goods required. Credit to buy machinery may not be available or may be too onerous when roads are impassable.

■ **The rural exodus** (young people leaving the land) **is another aspect of village economy.** The younger generation can find neither the work nor the money it wants. Migration from the land undermines the economic and social life of rural communities. Thus, asking questions about the rural exodus is one way of getting to understand village economies.

Many of the young are disillusioned when they reach the cities. They find no work, high living costs and none of the support they got from family and village life.

42

Many questions may be asked about the economic environment, and, when the opportunity arises, it is worthwhile asking them in village circles

We shall now attempt to summarize in the form of a diagram all the elements of agriculture discussed in the first five lessons. All these elements vary from one place to another and from one time to another. The reader may not always understand exactly how things happen, but he will have grasped one important fact - that agriculture results from the combined, changing action of all these elements. They form the spider's web mentioned in the introduction to the book. They figure in **table 41**, page 23, which presents and classifies the entire contents of the book.

43

Notes

donkey-drawn cultivator

44

Lessons 6 to 9

🔲🔲🔲🔲🔲🔲🔲🔲🔲🔲🔲🔲🔲🔲🔲🔲🔲🔲🔲🔲🔲🔲🔲🔲🔲🔲🔲

Fields change with the seasons

In the first lesson, the visit to the fields took place at a particular point in their life cycle, but fields undergo change depending on the season and the characteristics of the varieties of crops cultivated.

To understand agriculture properly, we need to be able to **see how fields change as time goes by**. Here are two examples of changing fields. The first comes from an arid zone with only one short rainy season (Lesson 6). The second comes from a wet forest zone with two rainy seasons (Lesson 7).

Lesson 6   🔲🔲🔲🔲🔲🔲🔲🔲🔲🔲🔲🔲🔲🔲🔲🔲🔲🔲🔲🔲🔲🔲🔲🔲🔲🔲🔲🔲🔲🔲

The story of a field in a dry region

This field is situated in the Sine Saloum region of Senegal. The land is **flat,** the soils are **sandy.** Rainfall is between 500 and 700 mm per annum, in the months of July, August, September and October.

The cultivated areas around the villages are dotted with trees left there by the farmers. The trees are mostly baobab, locust bean, shea butter, *Acacia albida* and some thorny legume varieties **(figures 42 and 43). Trees vary in number and quality from place to place**, depending on the villagers' outlook. Some farmers take little interest in the role played by trees and think only of the profit made from selling wood. Others react differently. They recognize the importance of trees and make sure that trees are planted and respected.

The number of trees growing in the fields also depends on the amount of rainwater penetrating the soil during the rainy season. In order to live through the dry season, the trees have to share the water stored in the soil. If water supplies are plentiful, there may be plenty of trees. If water is in short supply, the number of trees is limited.

At the beginning of July, the farmer ploughs the field lightly with a donkey-drawn cultivator **(figure 44)**. Tilling can start before the rains because the sandy soil is soft and therefore easy to work.

45

seeder

46

Next, he sows the millet using a seeder. The millet seeds buried in the ground **(figure 45)** will sprout with the first rains.

Figure 46 shows the millet growing. The sprouts at the foot of the tree are developing faster than elsewhere in the field because it is cooler and moister there. The ears of millet will be harvested after three or four months. The dry stalks left behind will be flattened to the ground **(figure 47)**.

Straight after harvesting, the waste is carefully buried. This is good practice because it feeds the soil and maintains moisture. Elsewhere in the field, the farmer leaves the surface residue which will be eaten up by termites and finally return to the soil.

47

Groundnut likes clean soil because the plant must be able to stick the young pods into the ground. Hence the importance of weeding the field two or three times during the cultural season to get rid of unwanted weeds. In November, the pods are ripe and the crop can be harvested **(figure 49)**.

49

groundnut haulms

50

Shortly before the next rainy season, the farmer comes back to clear the field of all the waste left behind so that he can sow groundnut.

The waste from the previous season's crop is piled into little heaps and burnt **(figure 48)**. **Ashes enrich the soil** wherever they fall. Then the farmer prepares the seedbed and waits for the rains.

burnt waste

48

The ground lies bare throughout the dry season because the natural growth which would have followed the groundnut crop has been kept down by hoeing. Only the trees provide a little shade. The soil is barer than usual because the groundnut shells have been heaped up to be carried away for forage. **Figure 50** shows the stacks ready to be removed and sold in Dakar.

After two cultural seasons, the land is exhausted because it has borne seed crops (millet and groundnut), fodder (groundnut haulms), wood, legume pods and other tree produce. This yield from the land has been removed and along with it, all the soil nutrients it consumed. So the field will lie **fallow** for two years. Lesson 35 deals with fallow land.

This is the order in which the crops succeed each other from one year to the next.

- In the first year, millet is grown.
- In the second year, groundnut is grown.
- In the third and fourth year, the land is left fallow.
- In the fifth year, millet is grown again.

Such an ordered succession of crops (or crops and lying fallow) is called rotation because the same crops are grown in the same fields after a lapse of several years.

Rotation and **cropping pattern** go hand in hand. In the case under discussion, the farmer has four plots of land which bear the same crops in a rotative or repetitive manner. The succession is such that the cropping pattern in any one season always includes a field of millet, a field of groundnut, and two fields left fallow.

When the farmer chooses the crop which will follow the harvest on a plot, he is taking a decision about rotation. When he decides on which plots are for cultivation and which will lie fallow, he is choosing a cropping pattern. Rotation takes place on one plot over a cycle of several years, whereas a cropping pattern concerns the state of all the plots in the same year **(figure 51)**.

Certain plant species, such as trees, are always present. Trees and their usefulness will be studied in Lessons 16 and 17. They preserve life in the aerial environment from one season to the next, especially *Acacia albida* which stays green during the dry season. Its leaves and pods provide precious fodder for livestock.

---

51 | **Cropping pattern and rotation**

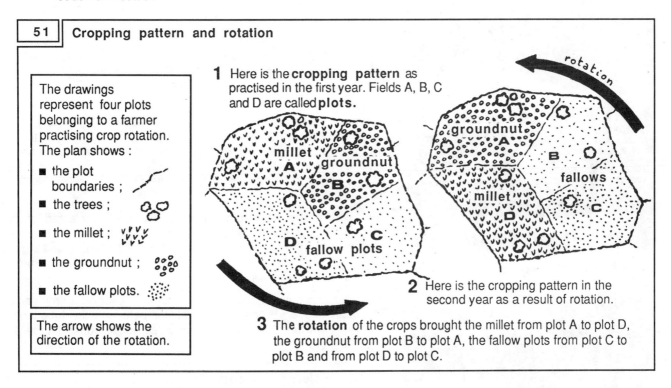

The drawings represent four plots belonging to a farmer practising crop rotation. The plan shows :

- the plot boundaries ;
- the trees ;
- the millet ;
- the groundnut ;
- the fallow plots.

The arrow shows the direction of the rotation.

1 Here is the **cropping pattern** as practised in the first year. Fields A, B, C and D are called **plots.**

2 Here is the cropping pattern in the second year as a result of rotation.

3 The **rotation** of the crops brought the millet from plot A to plot D, the groundnut from plot B to plot A, the fallow plots from plot C to plot B and from plot D to plot C.

In this particular part of Senegal, agriculture is faced with the following difficulties and limitations :

- the sandy soil is of poor quality ;
- rainfall is too low and too irregular ;
- the weather is too hot and the harmattan is strong at certain times of the year. Soil and plants are desiccated by evaporation ;
- **man demands a big yield from the land and does not always nourish it in return.** There are too many human and animal consumers compared with the soil capacity and labour distribution. The farmers are providing for the family, townsfolk, domestic animals in town and village, and people in other lands who eat groundnuts ;
- pests, such as mice, birds, insects, bacteria, attack the crops.

As a result of these difficulties, **great risks are attached to this farming.** Production rarely meets expectations after the labour input has been taken into account.

Lesson 7

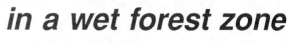

The story of a field
in a wet forest zone

52

53   cucumber

In East Cameroon there are **two rainy seasons in the year**. The story of this field begins when the farmer marks out a fertile plot of forest land. He fells most of the trees during the dry season **(figure 52)** and leaves them to dry out for many weeks before burning them. This is the **slash** and **burn** technique. Some farmers burn twice in succession at a few weeks' interval, thus eliminating all the waste which hampers work in the field. However, fire is a double-edged sword, proving both useful and dangerous.

The farmer only sets fire to the field we are studying once. He then lets nature finish the work begun by fire, by sowing a variety of cucumber whose leaves choke the young forest plants starting to grow again **(figure 53)**. Cucumber is a **cleaning plant**.

Here and there, banana suckers are planted and the shoots show above the cucumber leaves. Cucumber is harvested in the dry season about four months after sowing, or roughly six months after the field was first cleared. The ground is still full of tree trunks and big branches, but the waste can be heaped up fairly easily and the plot cleared because fire and termites have already hastened the decomposition of the wood.

Just before the next rainy season, the field is cleared of trash and closely spaced maize is sown **(figure 54)**. Later on, the farmer hoes between the rows of maize to get rid of the young forest plants which are cramping crop growth. Only the banana shoots are left between the maize stalks.

The maize is harvested **at the end of the second rainy season**. The trash and anything which might compete with the groundnut to follow, is piled up. Groundnut needs clean, loose soil to let the pods enter the ground.

How to clear forest without squandering natural soil fertility

In forest zones, clearing is an operation with marked effects on soil fertility. Clearing must be adapted to fit the farmer's plan. Groundnut cannot be treated in the same way as bananas. Groundnut needs clean soil, while banana will make do with partial ground clearance.

*The best method of clearance advances step by step. The work done by nature is always better than the work done by fire or machinery. If the farmer clears the land ruthlessly, if he burns and removes all the vegetation, soil fertility will deteriorate sharply during the next cultural season. It is better to let the residue from living organisms rot slowly on the ground. This residue is an important source of nutrients and water for the soil. Only large trees and branches which really get in the farmer's way should be burnt down. If the soil is very rich in organic material, it is a good idea to cultivate a **pioneer plant** suited to this type of soil, such as the cucumber grown by the Cameroon farmer.*

Land clearance requires the right combination of fire, plants, implements, and machines (if there are any).

54

55

Groundnut is sown densely, but is not the only crop in the field. The banana shoots, sown among the cucumber during the first cultural season, are a year old. New shoots now show round the young trees. Here and there, cocoyam shoots, from corms planted along with the groundnut, are visible, as well as maize stalks propagated from grain that fell to the ground during the previous season.

So the cultural association consists mainly of banana, groundnut, cocoyam and maize. **Figure 55** shows what the field looks like before the groundnut in the foreground is harvested.

On close inspection, the cultural events of the life of the field can be detected easily. **In the current season, the third since the ground was cleared, the farmer has divided the field into four plots** each of which will evolve in its own way. The plots are outlined in **figure 56**.

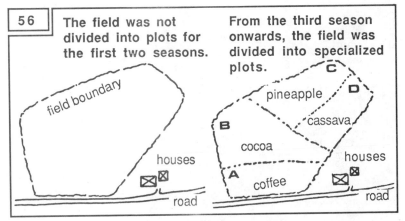

56 | The field was not divided into plots for the first two seasons. | From the third season onwards, the field was divided into specialized plots.

One of the plots (A) will be used for coffee, another for cocoa (B), the third for pineapple (C), and the fourth for cassava (D).

Plots A and B, used for seasonal crops during the first two seasons, will slowly be taken over by perennial (permanent) crops which will remain permanently on the cleared land. Plots C and D, on the other hand, will go on being tilled for non-perennial or seasonal crops.

However, the life cycle of the species cultivated after the groundnut, **from the fourth cultural season onwards**, is longer. Instead of four to five months for cucumber, maize and groundnut, it will vary from six to twenty-four months, or more, for sweet potato, sugar cane, pineapple and cassava.

Plots A and B will be turned into stands of trees.

The young coffee and cocoa shrubs were transplanted when the groundnut was weeded for the first time. It rained heavily and

57

the young trees responded well, benefitting from the canopy provided by taller plants - the cocoyam, the banana or the large forest trees left standing.

In **figure 57**, the coffee plant is doing well on the ground cleared after the groundnut harvest. Note the groundnuts laid out to dry on the tree trunk to the right and the heaps of trash which will soon be spread round the young shrubs. A tobacco plant growing by chance among the groundnut somehow escaped the hoe. A few seasons later, cocoyam, sweet potato and vegetables will be thriving under the coffee trees **(figure 58)**.

During the coming seasons, the young trees will prosper in the shade of large forest trees and bananas, now more and more numerous. The banana grove will slowly die out with age and the coffee trees will take over the whole patch, barely leaving room for low-growing herbaceous plants. The tall forest trees spared by the farmer will still be there.

The cocoa in **plot B** will develop like the coffee in **plot A,** only it will end up by choking practically all the low-growing plants. When the field is about five or six years old, the trees will reach peak production **(figure 59)**.

Plot C will be used for growing pineapple **(figure 60)**. The groundnut has been harvested and is drying on the ground. Pineapple suckers were planted during the groundnut season. More suckers will be planted when the rains start as the crop is too widely spaced **(figure 60)**.

The pineapple plants will soon spread and cover the whole patch, along with a wide variety of vegetables, such as black nightshade, okra, tomato, eggplant.

After three or four years, the pineapple plantation will be exhausted. Many wild plants will regrow while the ground lies fallow for some years, until all the associated wild species have restored soil fertility and allow cultivation once more.

Plot D will only grow produce for farm consumption. Cassava will predominate until the ground is left fallow and the soil fertility is restored.

When the crop is lifted, at the end of the **third** or at the beginning of the **fourth cultural season**, the groundnut roots are interplanted with many useful varieties - cocoyam, tobacco, sugar cane, gourds, oil palm, vegetables, cassava. The time of sowing the various plants may be spaced out to stagger the picking period for each crop **(figure 61)**.

58 — coffee, cocoyam, sweet potato, vegetables

59 — cocoa, cocoa pods

60 — pineapple, unshelled groundnut

In this case, the farmer is especially interested in cultivating cassava. He does not wait for the rains, but plants stem-cuttings, three by three, all over the field, intercropping with the groundnut.

During the fourth cultural season, bananas, cocoyam, gourds, tobacco and vegetables are picked while the cassava tubers are still enlarging.

From the fifth season, the cassava spreads over the whole plot, only leaving room for the taller bananas and the oil palms, which the farmer tends carefully. Cassava is a greedy, **unsociable** plant when the tubers are enlarging. When planted densely, cassava eliminates associated crops, a process which lasts while the leaves are active and the tubers developing **(figures 62 and 63)**.

When the tubers are ripe, the cassava leaves begin to wilt and the roots stop taking up water and nutrients from the soil. Fallow plants then take over **(figure 64)** and the farmer does not stand in their way because he knows that the land must be rested. The cassava tubers that have not been lifted remain buried in the fallow plot for months and are dug

32

up as required until supplies run out, or until they rot. The regeneration of wild species will not stop oil palm growth. These trees will bear fruit some years later when the land is cultivated again.

So, on plot D, the cultivation of produce for farm consumption will last three to four years from the time of the first clearing.

Plots C and D will lie fallow for several years. The wild plants, both herbaceous plants and trees, will regenerate the soil by taking up, from the various soil strata, everything needed to make large quantities of organic matter which will later be restored to the soil, will rot and end up by revitalising it.

What is a fallow plot ?

A fallow plot is a piece of land left to rest after cultivation. Natural vegetation begins to thrive again, regenerated by the mass of seeds, trash, cuttings, bulbs, roots which survived the cropping period. The speed of vegetal regrowth is directly related to the length of time the land was cultivated and the level of soil exhaustion.

The plants found on a fallow plot are those of the bordering zones, whether grass or forest land. The value of a fallow plot, indicated by the variety in the plants growing there, dwindles when the land has been overcultivated, because the amount of forest seeds in the soil is inadequate. Hence, in an overcultivated forest clearing, it is not unusual to observe a fallow plot composed solely of grassland species, because their seeds are propagated more easily.

In drier regions, fallow plots contain mostly herbaceous plants and brushwood. If there is overcultivation and periods of drought, the variety in the wild plants is limited and the land may look like desert.

Points to remember about this forest clearing

After cutting down the trees, the farmer cleaned the land, first by setting fire to it, then by growing cucumber. Termites, insects, fungi and many kinds of organisms invisible to the naked eye all helped to do the work.

Some cultivated plants, like groundnut and maize, need clean fields, whereas others, like cucumber, cocoyam, and banana can tolerate residues.

Not all the plots in the field were managed in the same way. From the third season onwards, the cultivator knew what he was going to do with each plot, and gradually organized the field in keeping with his plan.

On plots used for seasonal crops, rotation of crops and lying fallow is practised. Cropping for three years will be followed by a fallow period lasting from five to ten years or more. So, the soil is exploited for three years by seasonal crops and fertility is restored in the succeeding five to ten years.

Rotation in this field

association of cultivated plants

for a three-year period

forest fallow lasting from five to ten years

What can be learned from the story of this field ?

The story is typical of a farm pursuing **several aims** simultaneously and practising a wide range of agricultural activities. All the plots cultivated grew produce for **farm consumption**. Two plots (A and B) became perennial stands to produce **market crops**.

There are two striking features about this field - firstly, the sequence of crops and harvests, and secondly, the characteristics of the associated crops.

The sequence of crops and harvests

We observed that in this field many plants were associated and sown one after the other. **However, at each cultural season, certain crops were dominant and farm work was specialized to suit the needs of these main cultures,** cucumber, maize, groundnut, cassava, etc.

Table 65 shows, season by season, the evolution of the cultural associations (second column) and of the dominant crops (third column).

The associated plants and their characteristics

The field contains **long-lived and short-lived plants.** Some are seasonal, some last from one season to the next, others remain permanently in the field. What differentiates the farming practised in wet and dry zones is the fact that, **in arid zones, fewer plants remain in the field, year in year out.**

65 | **The evolution of the cultural associations and of the dominant crops**

In this column, the **succession of cultural seasons** following the forest clearing is indicated.	*In this column*, the **associated crops** in the field are indicated. Some crops are harvested in the season in which they were sown, others are lifted later. The letters, A, B, C, D, indicate on which plot the crop is sited.	*In this column*, the **produce harvested** is indicated. The dominant crops are in **heavy print**, the secondary crops in light print. The letters A, B, C, D, indicate the plot on which each crop is sited. Where there is no letter, it means that the crop is grown simultaneously on the four plots.
lst season	cucumber banana suckers	**cucumber**
2nd season	maize banana	**maize**
3rd season	groundnut coffee (A) banana cocoa (B) tobacco pineapple (C) maize vegetables oil palm cocoyam	**groundnut** banana, vegetables, tobacco, maize
4th season	banana coffee (A) cocoyam cocoa (B) tobacco pineapple (C) vegetables cassava (D) sweet potato oil palm	**banana, cocoyam** gourds, sugar cane, sweet potato, tobacco, vegetables
5th season	banana coffee (A) cocoyam cocoa (B) sweet potato pineapple (C) oil palm cassava (D)	**banana, cocoyam** sweet potato, vegetables, pineapple
6th season	banana coffee (A) cocoyam cocoa (B) vegetables pineapple (C) oil palm cassava (D)	**banana, cocoyam** cassava (D), pineapple (C), vegetables
7th season	banana coffee (A) cocoyam cocoa (B) vegetables pineapple (C) oil palm cassava (D)	**banana, cocoyam, cassava** (D)
Following seasons	Banana plants, cocoyams and vegetables disappear gradually to make room for coffee (plot A), cocoa (plot B), pineapple (plot C), cassava (plot D). The trees will begin to yield fruit in the 8th season and will go on flourishing in the field. But pineapple and cassava will slowly be overtaken by fallow growth which will be dominant by the 9th or 10th season.	

Some of the plants in the field are sown from seed, some from cuttings, some are transplanted. The differences between the plants means that farm operations must be adapted to suit each plant.

The cultural work is done partly by the farmer with the help of implements and partly by the plants themselves. Encouraged by the farmer, the cucumber cleans the ground for the maize which, in turn, creates good growth conditions for the groundnut.

Some plants like company, some do not. Some overrun other plants in the field. The capacity of plants to associate or not often depends on their age.

The most sociable plants from those studied are banana, cocoyam, sugar cane, tobacco, tomatoes and vegetables. They can be found growing in the field for several consecutive seasons. They are often self-propagating.

Other species are sometimes **sociable** but not at all times, for example, maize, groundnut, pineapple, oil palm. Groundnut, for instance, does not tolerate close associated crops during the period of fruit enlargement.

Cucumber, cassava and cocoa are asocial plants, at least at certain times in their lives. Fully grown cucumber, for example, is so vigorous that it can overrun forest regrowth. Cassava tolerates companion crops in the early stages of its development, but later eliminates them almost entirely as it approaches maturity. It does this in many ways - it smothers smaller plants under its foliage by depriving them of light, it is very greedy and does not share soil nutrients with other species. Cocoa becomes asocial with age. If the plantation is dense, herbaceous growth disappears gradually unless the farmer protects it.

The farmer used cucumber to clean the field after burning **(figure 53)**. It is not quite true to say that cucumber does not tolerate companion crops. It depends on how and when cucumber is planted.

In **figure 66**, cucumber has spread over the whole field of maize, but the maize stalks are tall enough to avoid being choked by the dense cucumber foliage.

Every plant has a specific role to play in crop association. This role depends on the stage of growth or maturity it has reached.

A range of goals

The way this field is farmed allows the cultivator to pursue several goals at the same time :

- **to obtain a selection of produce** for farm consumption and for the market ;

- to build up stands of trees when seasonal crops are phased out ;

- **to ensure safe, regular income** by diversifying the produce and staggering the harvests ;

- **to protect the land** from the bad effects of the climate by letting large quantities of organic matter find their way back into the soil to rot.

Points to remember about this field

- *Climate permitting, many plants can be associated in the same field.*

- *The farmer sows the plants in the order determined by the characteristics and needs of each species.*

- *Every cultural season, the associated plants which are ready for harvest and those to be harvested later are recognizable.*

- *When the seedbed preparation and the sowing are done well, the plants help the cultivator, as happened with the cucumber. Every plant changes the environment to some extent and benefits, in turn, from the changes brought about by its companions.*

- *Natural forest plants stay in the field, because all the trees have not been cut down and because many seeds of forest species remain in the field and will grow again if the farmer no longer controls them.*

Lesson 8

The rotation of crops and fallow land

When forest or grass land is cleared, the soil is still covered by large quantities of plant and wild animal residues. This residue, combined with the quality of the sand and clay, makes the soil fertile.

While the land is under cultivation, the plants gradually use up the fertility of the soil. So, when the farmer has cropped the land for several consecutive seasons, he lets it revert to fallow for many years. By doing so, he is practising the **rotation of crops and fallow lands**.

Why is this rotation necessary ?

Every plant exploits a particular soil layer (Lessons 31 and 32), by taking from it all the nutrients it requires. When the crops are harvested, the farmer removes the agricultural produce containing the nutrients extracted from the soil. So, in fact, **the soil layer exploited by the cultivated plants is impoverished by the harvesting of crops**. This soil degradation is made manifest by the increasingly lower yields which result from the same crops being grown for several consecutive years on the same land.

Figure 67 shows the yield pattern in three fields on land in Zaire where the same plants were cropped over a period of five seasons.

67

Finger millet was sown for five consecutive seasons on plot 1. Here is the number of baskets harvested.

first season	second season	third season	fourth season	fifth season

Groundnut was cropped on plot 2 for five consecutive seasons. Here is the number of bags lifted.

first season	second season	third season	fourth season	fifth season

Cotton was grown on plot 3 for five consecutive seasons. Here is the size of the piles picked.

first season	second season	third season	fourth season	fifth season

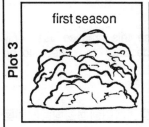    

The yield is high for the season immediately after forest clearance. Then it slowly dwindles. From the third season onwards, the yield does not pay for the labour input, and the soil is exhausted by the crop. Moreover, when a plant is grown in the same field for many consecutive years, the pests and diseases to which it is prone, cause more and more damage, resulting in poor yields, in both quantity and quality.

When two or more crops are grown in sequence in the same field, each crop uses up the fertility of the soil in its own particular way. Different plants grow to different depths and use up different nutrients.

Consequently, the yield of a newly cleared field, which only supplies nutrients from its own natural fertility resources, is greater when different crops are grown in sequence. Also, losses from disease are smaller because plant-related pests have less time to establish themselves firmly.

To make a comparison with the first three fields in **figure 67**, here is the yield pattern in a field in the same region of Zaire where crops are grown in sequence each cultural season **(figure 68)**. In the sixth year, rotation starts all over again with finger millet. Yields for each crop are satisfactory but they will be much lower if the same crop sequence is practised after the fifth season.

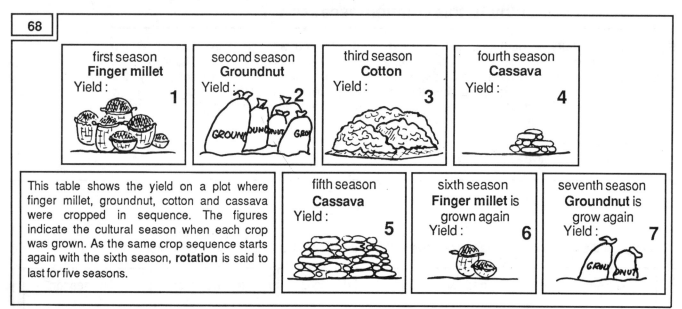

68

| first season **Finger millet** Yield : 1 | second season **Groundnut** Yield : 2 | third season **Cotton** Yield : 3 | fourth season **Cassava** Yield : 4 |

This table shows the yield on a plot where finger millet, groundnut, cotton and cassava were cropped in sequence. The figures indicate the cultural season when each crop was grown. As the same crop sequence starts again with the sixth season, **rotation** is said to last for five seasons.

| fifth season **Cassava** Yield : 5 | sixth season **Finger millet** is grown again Yield : 6 | seventh season **Groundnut** is grow again Yield : 7 |

Cultivated plants exhaust the natural fertility of the soil after some years. So, one way or another, fertility must be restored to allow long-term cultivation of the land.

Why does the soil become exhausted ?

- Firstly, **man removes the harvest produce from the field and the produce contains nutrients taken from the soil.** So the farmer must only remove the produce that can be consumed. **All unnecessary removal is a loss to natural fertility.** Hence **the importance of putting back all crop residues** to let them rot in the field.

- Secondly, **a one-crop field does not produce enough trash and organic matter** to maintain soil fertility.

  When many plants are associated in the same field, more waste and organic matter are available. In fact, the more associated species there are, the better they transform and improve the soil.

- Thirdly, season after season, crops, rain, sun and microorganisms in the soil (Lesson 33) combine to change the soil with the result that **the rainwater penetrating the ground carries away with it large quantities of the nutrients found in the upper soil layers.**

  Leaving the land fallow is a cheap way of restoring soil fertility, because this practice uses only production factors available on the farm itself. However, it is often inadequate and needs to be supplemented.

69

Lesson 9

Legume crops

Plants of the family *Leguminosae* are often present in the fields considered so far in this book.

They are valuable plants for two reasons :

- they provide excellent nutritious food products ;
- they improve and sustain soil fertility.

Soil improvement by *Leguminosae* is carried out by the little bulbs or **nodules** attached to the roots. **Figure 69** shows the stalk and roots of a bean plantlet, with the root nodules clearly visible.

Legumes are recognized by their **leaves**, their fruit in the form of **pods** and their **flowers**. The presence of **nodules** on the roots is another way of identifying them.

70   locust bean

71

72   Black wattle

The leaves are composed of many leaflets (figure 73).

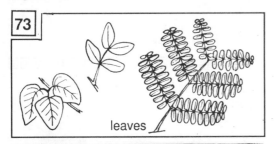

leaves

cassia

The flowers, like those of the groundnut and bean, are often shaped like butterflies (**figure 71**).

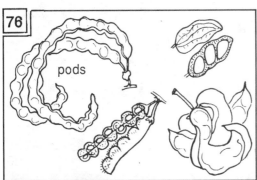

pods

Acacia albida

Sudan gum arabic

Pueraria phaseoloides
(Tropical kudzu)

Or, they may look like tiny globules or little coloured tails, like those of the sensitive *Mimosa* which closes when touched. There are other kinds of legume flowers with a variety of shapes

The **fruit** is a **pod**, more or less elongated, **which splits open lengthwise to shed the seeds (figure 76).**

Common bean

79

While the root nodules on most leguminous plants can easily be observed, they may not be visible when the soil is very rich or when the nodules are attached to the roots of large trees. The presence of nodules, in itself, is not evidence of effectiveness. On cutting open, effective nodules appear pale pinkish, which shows that nitrogen fixation is actually occurring.

Figures 70 to 79 illustrate plants from the family *Leguminosae*.

Remember

Farmers and scientists agree on the very great value of legumes in agriculture.

One way towards better farming is to look for the most effective associated cropping of Leguminosae and other useful plants.

The following members of the *Leguminosae* are cultivated for their beans :

groundnut, beans, cowpea, the green pea, the pigeon pea, the soya bean, *Voandzeia* (Bambarra groundnut or earthpea), varieties of *Dolichos* (Lablab bean).

There are also many species used to improve the soil or to provide shade. Scientific names are : *Pueraria, Centrosema, Mucuna, Derris, Calopogonium, Desmodium, Stylosanthes, Canavalia.* They are herbaceous plants.

Among the trees of the family *Leguminosae* used in agriculture are the locust bean *(Parkia)*, several Acacias including *Acacia albida, Leucaena* .

Other trees are used for their timber.

These species have been mentioned because they are of use in agriculture, but the list is not exhaustive. There are, in fact, many thousands of *leguminosae* species in tropical Africa.

Lessons 10 to 15

Some features of rural settlements

What ground have been covered so far in our study of tropical agriculture ? Lesson 1 was spent visiting the fields. Lesson 2 discussed the farmer's goals and the means of achieving them. It was here that the term **rural settlement** was used for the first time to mean **all the land, cultivable or not, of a farm, a hamlet or a village**. Lessons 3, 4 and 5 went into three aspects of the farming environment - the biotope or living environment, the physical environment and the economic and social environment. The study of the economic and social environment introduced many facets of rural life and explained how rural people are dependent on the outside world.

Next we approached the study of fields in very different climatic regions. Of the many ideas that emerged, two are particularly useful - the practice of crop and fallow rotation (Lessons 6, 7, 8) and the role of legumes in tropical agriculture (Lesson 9).

However, many other features of rural settlements also have to be taken into account, more specifically those related to the **nature of the countryside**. Four lessons will deal with the subject under the headings **mapping rural settlements** (Lesson 10), **land relief** (Lesson 11), **water resources** (Lesson 12), and **land allotments** (Lesson 13), with special reference to the main problems confronting country people, particularly the farmers. Lessons 14 and 15 present two types of rural settlements.

Lesson 10

Mapping rural settlements

Some attempt must be made to picture a rural settlement in order to describe accurately its geographical traits and the life of the inhabitants. **Drawing such a picture is one way of representing the basic features of a rural settlement, of locating the inhabitants, the fields, the roads and the rivers, and of fixing distances between houses, fields, forests, schools, dispensaries, etc.** This picture, called a **map**, provides an immediate overall view of a whole range of features of the rural settlement, of the village, the farm, the region or the country.

Here is a map of **Rwanda (figure 80)**. The outline of the country is obtained by following the frontiers on the ground. The main roads, rivers and towns are indicated, as well as prefecture boundaries.

Figure 81 is a map of Kabuya subprefecture in Rwanda. This part of the country is shown in greater detail by marking secondary roads and trails, villages, mountains, streams, rivers and lakes. Lines, dots and other little

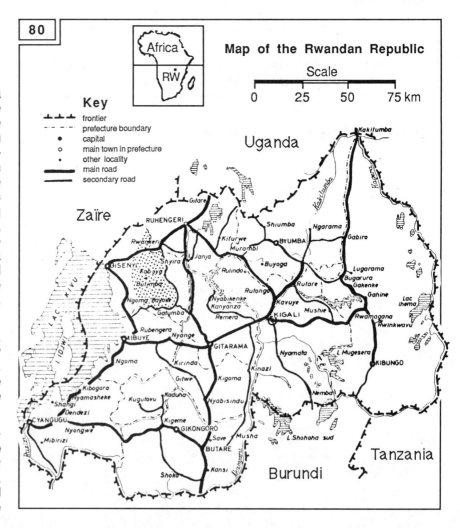

80

Map of the Rwandan Republic

Scale

0   25   50   75 km

Key

━┿━┿━ frontier
─·─·─ prefecture boundary
● capital
○ main town in prefecture
○ other locality
━━━ main road
━━ secondary road

Africa
RW

Uganda
Zaïre
Tanzania
Burundi

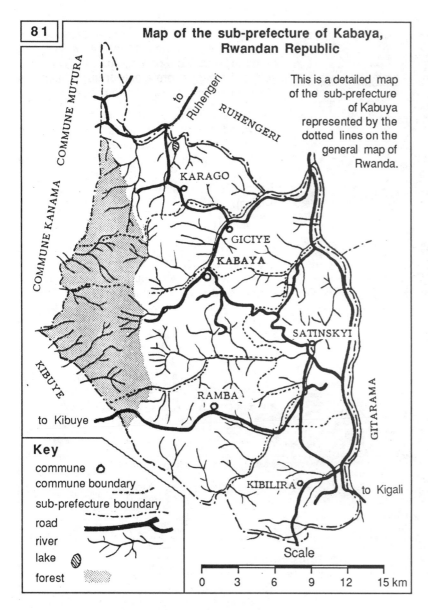

81

Map of the sub-prefecture of Kabaya, Rwandan Republic

This is a detailed map of the sub-prefecture of Kabuya represented by the dotted lines on the general map of Rwanda.

COMMUNE MUTURA

COMMUNE KANAMA

to Ruhengeri

RUHENGERI

KARAGO

GICIYE

KABAYA

SATINSKYI

GITARAMA

KIBUYE

RAMBA

to Kibuye

KIBILIRA

to Kigali

Key

commune ⭕

commune boundary

sub-prefecture boundary

road

river

lake

forest

Scale

| 0 | 3 | 6 | 9 | 12 | 15 km |

drawings are used to locate everything that had to be shown.

Both maps have a **scale** to make them intelligible. A scale allows the map reader to calculate the distance between the places marked on the map. The scale of the Kabuya map is given at the bottom right, and the scale of the map of Rwanda is shown at the top. The meaning of a scale is explained at the end of this lesson.

The art of observing and representing our own rural settlements

There are three different, yet complementary, ways of representing on a sheet of paper everything that can be observed in a rural settlement :

■ by taking a photograph or making a drawing ;

■ by making a transect ;

■ by making a map, i.e. a drawing representing everything on the land as it might be seen from above by a bird in flight.

Specialists in map-making find that **aerial photographs** make mapping very much easier. **Figures 82, 83 and 84** are aerial photographs. **Figure 82** shows a river basin taken from a high altitude. **Figures 83 and 84** show two inhabited areas near the same river.

As an aeroplane is obviously needed for aerial photography, the technique falls outside the scope of this book. But those who are landbound can still use other procedures.

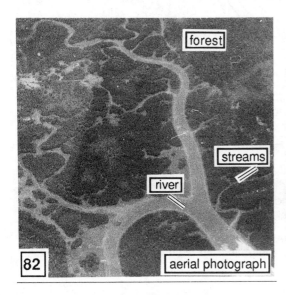

82 forest · streams · river · aerial photograph

83 river · canoes · houses · palms · aerial photograph

42

84 | aerial photograph

The first thing to do is to reproduce the settlement just as the person sees it. This can be done by taking a photograph or drawing the scene on paper, preferably from a height to make the picture as complete as possible.

Here is a view of a small hillside settlement shown in plenty of detail **(figure 85)**. The house with surrounding trees and the bean fields can be made out. In the forefront, there are herbaceous plants and bushes growing on the banks of a stream (not included in the picture).

The dotted line is explained below.

Figure 86, beneath the photograph, is a **drawing** by hand of the same scene. The details are missing but all the important elements are there - the contour of the hill

shaded in with long strokes, the house, the trees, the fields.

When no camera is available, a drawing will do very well.

If the observer walks round the settlement, he can make drawings representing it **from different angles** and note all its distinctive features - types of field, springs, ponds, odd or unusual trees, roads, walls, enclosure.

However, these pictures are not enough. The settlement has still to be measured by walking over it in straight lines and taking note of everything observed on the way. Such straight lines, combined with careful observation of every detail including distances, slopes, vegetation, etc., are called **transects**. Several transects can be made for the same settlement.

Looking at **figure 85**, we are going to take the path marked by the dotted line. The departure point, **A**, is behind the hill to the left. The destination, **B**, is situated at the front right, just after the path crosses the stream **(figure 87)**.

Keeping to the route traced by the dotted line, the following landmarks are observed - first, herbaceous fallow and a tree. Then the upward slope gets steeper and we cross a field of maize and cocoyam, and a banana grove. Near the

85

86 | **Drawing reproducing the landscape**

trees
house
palm tree
fallow land
fallow land
field of beans
bushes
brushwood

Simplified drawings reproduce only the most important features of the landscape : hill contours, houses, wooded areas, field and fallow sites, etc.

summit, our path crosses a road. We go through trees, past the house and down the slope. There is more fallow land for a few metres, then a big tree marks the fallow boundary. Further down the slope, a large field of beans stretches right up to the tall grass growing along the banks of the stream.

We counted out steps as we went along. From departure point **A** to the house was 120, and from the house to the stream was 70. We also counted out the number of steps separating the main landscape features on the way - trees, road, house, bushes, stream.

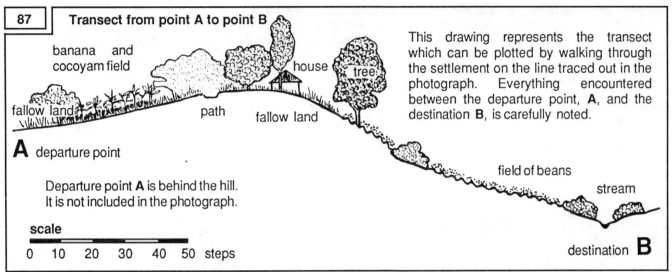

87 **Transect from point A to point B**

banana and cocoyam field

fallow land

path

fallow land

house

tree

A departure point

This drawing represents the transect which can be plotted by walking through the settlement on the line traced out in the photograph. Everything encountered between the departure point, **A**, and the destination **B**, is carefully noted.

field of beans

stream

Departure point **A** is behind the hill. It is not included in the photograph.

destination **B**

scale

0 10 20 30 40 50 steps

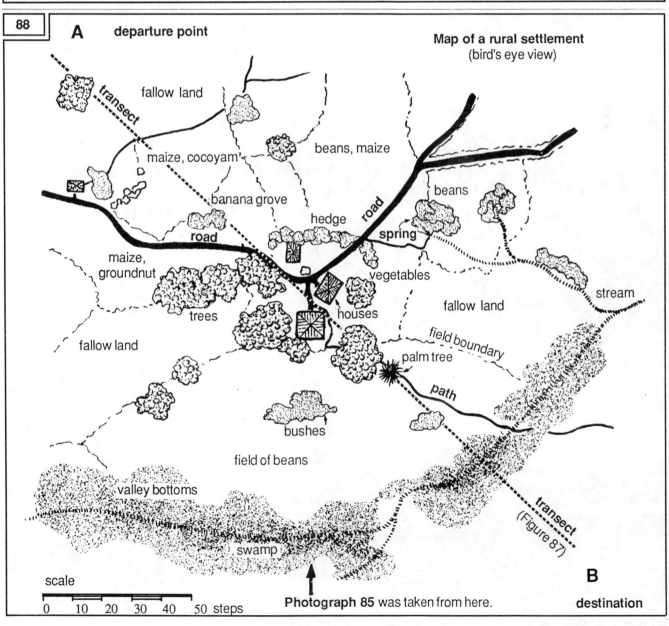

88 **A** departure point

Map of a rural settlement
(bird's eye view)

transect

fallow land

maize, cocoyam

beans, maize

banana grove

hedge

road

beans

spring

road

maize, groundnut

vegetables

trees

houses

fallow land

fallow land

field boundary

palm tree

path

bushes

field of beans

stream

transect
(Figure 87)

valley bottoms

swamp

B

scale

0 10 20 30 40 50 steps

Photograph 85 was taken from here.

destination

The **map of a rural settlement** can be traced fairly accurately using drawings and transects **(figure 88)**. Roads and paths are plotted, houses marked in, fields, valley bottoms and springs are located, and anything else which may help us to become better acquainted with the settlement is also shown

A **scale** is needed to make the map in **figure 88** accurate. Suppose we have measured the distances in steps, we could say, for example, that every ten steps taken on the way are represented by one centimetre on the map. The scale chosen must be marked on the map so that anyone using it can read it properly.

89

90

Lesson 11

Land relief

Relief is the term used to describe land forms mountains, tablelands, plains, valleys, and so on. Land incline, exposure to sun, the location of gullies and valleys, the shape and depth of valleys, etc., depend on the relief of the land.

Here is a settlement on level ground in Senegal **(figure 89)** showing **land relief typical of a plain**. The rivers and valley bottoms are shallow, the slopes only slightly inclined, with very little water runoff.

In contrast, **figure 90** shows a **mountainous relief** on Santiago Island, Cape Verde. This type of relief is **pronounced**. Very steep slopes cannot be cultivated, the rocks are visible and all the water falling on the steep inclines is swept away in the torrents.

Here is a **hillside settlement** in Rwanda **(figure 91)**. The land relief is less pronounced. The slopes are fairly gentle and the rocks do not show on the surface.

91

Agricultural land falls into four categories where land relief is concerned

92

- **Plateaus** are high level regions, situated on the top of sloping ground. They are never flooded because rainwater penetrates the soil or flows down and away to lower ground.

- **Land on slopes** situated on the flanks of valleys have the highest erosion risks.

- **Plains** situated near rivers are sometimes flooded but there is no permanent water stagnation.

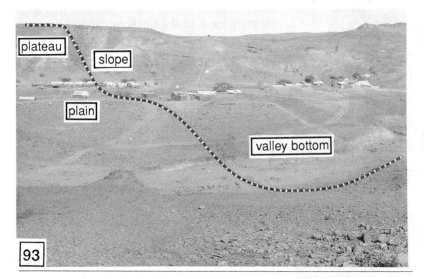

93

- **Low-points or valley bottoms** is land near springs, streams and rivers where water-flow is more sluggish. Here the silt and clay, swept down by water runoff, accumulate. Valley bottoms are often flooded and swampy.

Plateaus and lands on slopes are generally impoverished by water and wind erosion. The soil is often sandier and stonier because the runoff water and wind carry away the finest particles (grains of clay and silt) which contain the best plant nutrients.

Valley bottom land is always moister because of the water runoff from adjoining valleys. The soil contains more clay, and supplies of plant nutrients are greater than on sandy ground. Nevertheless, valley bottoms are often harder to farm because the soil is heavy whereas

sandy soils are light. Unless there is some system of water control, these soils are only suited to plants with water tolerant roots.

In **figure 92**, the plateau, slopes and plain stand out. The plateau is arid and barren whereas the plain retains more moisture. Notice the trees and cultivated plots. The slopes are rocky and without topsoil, i.e. there are no cultivable lands on these slopes.

Figure 93 is a photograph of mountainous country in the Sahel during the dry season. The plateau, the slopes, the plain where the village is located, and the valley bottom can be seen. As the amount of rainfall is negligible, the valley bottom is completely dried up. Even when it does rain, the flow is so torrential that water does not penetrate the soil.

94

95

96

97

The same relief elements are shown in **figure 94,** in a more enclosed valley. There is no actual valley bottom, but the course of a quiet stream can be seen.

Figure 95 shows a typical plain where the land is flat or on only a slight slope. Water does not stagnate ; it flows off.

Figures 96 and 97 show two types of valley bottom swamps. The first one **(figure 96)** is subject to flooding during the rainy season. In the dry season, the soil retains quite a lot of moisture and can be cultivated here and there, provided that a little water is available in case it is needed. Catchpits, or shallow holes, in the ground can provide water for irrigation purposes.

Figure 97 illustrates a permanently flooded valley bottom. Notice the vegetation, typical of swampy land. In the background can be seen the plain, which is never flooded.

Notes

Lesson 12

Water resources in rural settlements

Land without water turns into desert. On the other hand, when there is too much water, the land may be flooded. One way or another, through desertification or flooding, land can become uninhabitable and unproductive. That is why the study of **water resources** plays an important part in our understanding of the agriculture and life of a locality or region. Emphasis has been placed on water resources in this lesson and also in Lesson 21, which deals with water movements.

> ## The water resources in rural settlements and their uses

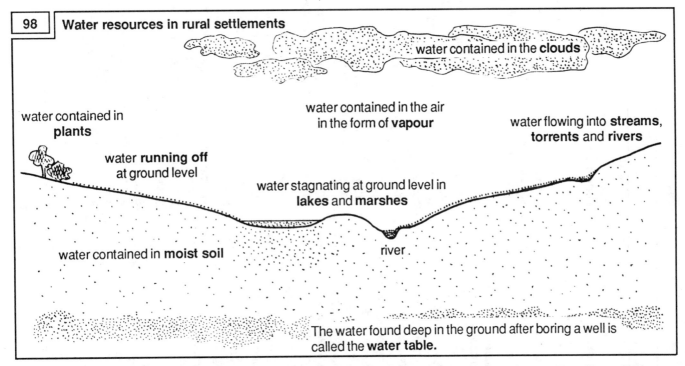

98 Water resources in rural settlements

water contained in the **clouds**

water contained in the air in the form of **vapour**

water flowing into **streams, torrents** and **rivers**

water contained in **plants**

water **running off** at ground level

water stagnating at ground level in **lakes** and **marshes**

water contained in **moist soil**

river

The water found deep in the ground after boring a well is called the **water table.**

All the water resources of the settlement, whether permanently or only occasionally available, are sketched in **figure 98**.

Table 99 explains and completes the drawing. The column on the left gives the list of **water resources** according to the movement of the water. For it should be remembered that:

- some water simply passes through the land. If it is left untapped, it is of little use to the inhabitants;

- some water is stored on the land itself and is at the disposal of the inhabitants, if they know how to use it;

- there is water in the air in the form of water vapour.

A good understanding of water resources and the movement of water is important in order to make plans for water utilization.

The middle column gives **the possible uses of water** on land, while the right-hand column sets out the **factors affecting the utilization of water resources.**

100

48

99 | Water resources on rural settlements

Factors affecting the utilization of water resources in rural settlements

Climate has a marked influence on the water resources of rural settlements. Man cannot change the rainfall, be it heavy or light. Rains may be regular one year, unreliable the next. Winds vary - strong winds desiccate the land.

The *physical features* of the land must be taken into account. Water runs off bare, unprotected slopes, for example. It flows into rivers and is lost to the soil. On the other hand, there might be little runoff and water will penetrate the soil, if the land has good vegetal cover and erosion barriers.

Fire usage - uncontrolled fires are bad for water resources.

Water is used for *agriculture*. Some plants demand more water than others. Some cultural practices call for large quantities of water compared with other methods which save water.

The *number of people and animals* living in the settlement can influence the use of the water resources.

Water management affects the use of resources. Water may be so scarce that the support of plant and animal life is difficult, in which case the farmer uses the water sparingly. But if there is plenty of water for farming and for human and animal consumption, it is used without forethought.

Education. Educators must inculcate a responsible attitude towards water resource. They are the people responsible for teaching the younger generation all about water - its importance, economic use, proper management, etc.

Social structures. Water is owned collectively and concerns everyone living in the settlement. Social structures sometimes affect water usage. They can, for example, prevent landowners and farm tenants from agreeing on ways of keeping water on the land.

The uses of water resources in rural settlements

Water is a *production factor in farming*. This water is called *agricultural water.*

Domestic water is needed for human consumption - for drinking, cooking, washing.

Water is needed *for animals.*

Water is a *health factor.* Dirty water transmits disease.

Water is needed for *processing* some agricultural produce - to pulp coffee cherries, to wash cassava, heat palm nuts before extracting the oil, and so on. Water is also needed for *trades and crafts* - to freeze fish, to temper iron in the forge, etc.

Water is the life-giving and supporting environment of *fish.*

Water is often a *commodity,* especially in big towns and cities.

Water can *produce energy,* it can generate electricity, power a mill, and so on.

Water can be used for *transportation* - canoes, boats, goods (timber etc.).

The water found on rural settlements or flowing through them

Water that flows through the land without stopping :

runoff water that flows away from the settlement ;

water flowing through in streams and rivers.

Water that remains in store in the settlement :

surface water in lakes, ponds, marshes, dams, cisterns, etc. ;

some *underground water* is near the surface and ensures *soil moisture.* Some penetrates the ground and flows down to the *water tables.* Water from these tables sometimes reappears on the surface, for instance, in springs ;

water contained in *plants* and in all living matter. Lush vegetation is an important store of water for the settlement.

Water that is contained in the air in the form of water vapour and clouds

Living matter and organic residues are an important source of moisture

101 aloe

Where there is water there is life. Water in large quantities is found in the plants themselves and in all living matter in the settlement. It is common knowledge that water can be extracted by pressing the leaves, fruits, stems and roots of plants.

Some plants are capable of retaining water by storing it in the stems and leaves. Here are four examples of water-retentive plants.

Spurge *(Euphorbia)* has thick leaves and stems which release a milky juice when broken. This type of plant with fleshy stems, full of water, is seen in **figure 100**.

Aloe **(figure 101)** is used to make anti-erosive strips. The plant grows in regions where it hardly ever rains. It collects even the slightest amount of moisture and stores it in its pointed leaves.

Like all the members of the cactus family, prickly pear **(figure 102)** is well adapted for storing water. In some arid regions it is planted as a reserve forage for stock to fall back on when other sources of food are exhausted.

dried herbs

purslane

103

prickly pear

102

104 purslane

Purslane is seen in **figures 103 and 104**. Its tiny green leaves are full of water and

50

contrast with the dried herbage on the left **(figure 103)**. Some varieties, with big red flowers, are decorative garden plants, others are grown as vegetables. The varieties photographed are self-propagating. All purslane plants can store water and maintain good soil moisture. The small fleshy leaves of one variety are clearly visible in **figure 104**.

Lesson 13

Land allotment

All the land in rural settlements is divided into plots. These may be cultivated, inhabited, flooded or left fallow, wooded, and so on. When a map showing the shape and size of the plots is drawn on paper, it is called a **land allotment map**. If it is a village allotment,

| 106 | Land allotment |

Roads, paths, trees and hedges, as well as the boundaries of the cultivated plots, can be seen.

106

| 107 |

This diagram shows only the **boundaries** of the plots and the roads and paths.

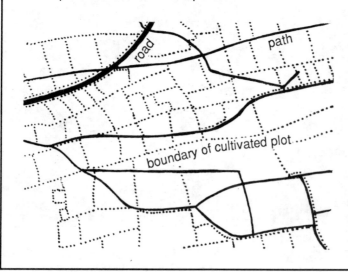

road

path

boundary of cultivated plot

Agricultural water and drinking water - two different qualities

105

There are different qualities of water - dirty ditch water, clear water in a stream, silty water from a torrent, saline water from a rice field, soft rainwater.

The quality of the water is important for farming but even more so when water is for human use - drinking, cooking, washing.

Water in its natural state contains germs, worms and larvae which cause disease when they enter the human or animal body.

*The man in **figure 105** is drinking unfiltered water from a swampy pool dug out to provide water for crops, and polluted by people and livestock. The man may well catch waterborne diseases such as parasitic worms or fevers, and will consequently not be able to work.*

If the means are available, it is best to develop a spring or construct a proper well. Clean drinking water will keep farm workers in good health. Providing and drinking clean water is a way of sharing in agricultural development.

the fields are distinguishable and the farmers know exactly where their own plots are situated.

A **cadastral plan** is a land register or cadastre made from an accurate map of the plots and then registered by the authorities.

The two figures represent land allotments.

Figure 106 is an aerial photograph of an allotment. The field and plot boundaries are visible. They are often marked by hedges and trees. **Figure 107** is the allotment map drawn

51

from the photograph giving field and plot shapes and sizes.

Figure 108 is an aerial photograph of a rice allotment. The earth banks demarcating the parcels of land and the drainage ditches are both visible.

Ideally, all local inhabitants should be interested in the allotment plan as one way of studying land problems.

In some parts of Africa, land is freely available to would-be cultivators who, with the permission of the land authority, simply set about clearing and cultivating a parcel of land.

The expression 'land authority' in this book is used to mean a chief, clan, family or families, or government - whoever decides how the land will be allocated. It sometimes also stipulates how the land will be used.

Figure 110 shows a sparsely-populated **village in forest land**. The villagers have settled by the roadside and go into the forest to open up fields and temporary farmsteads.

In front of the dwellings along the road are large spaces, devoid of vegetation (**figure 109**). Alongside the houses, there are tiny gardens growing tobacco, gourds, a few cotton and castor shrubs, leafy vegetables, medicinal herbs, etc. These plots are very small and more often than not are trampled by domestic animals on the loose.

Close by, there are coffee and cocoa plantations manured by domestic waste and excreta. Notice how hard it is to distinguish the

108

109

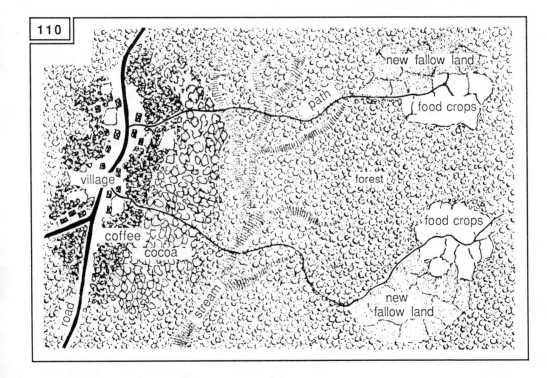
110

allotment plan. Only the farmers themselves are able to follow the demarcation lines.

Virgin forest lies behind the coffee and cocoa plantations. It is communally owned by all the villagers and, as yet, there are no cleared plots.

If you follow the paths in the forest and cross several little streams, after a little while, you come to cultivated food crop clearings. The plots, managed by women, are delimitated during the growing season but the boundaries disappear completely when the land goes back to fallow. This shows that the allotment in these clearings has not been fixed permanently.

Reasons why food crops are cultivated some distance from the village

- The cocoa and coffee **plantations** are near the houses to make for better surveillance. So perennial cultures are on lands close by.

- Seasonal crops cannot be grown near the village, because **small animals** (goats, sheep, pigs) are allowed to **wander**.

- **Cultural practices** are such that new forest must be cleared every year , so the cultivators choose the best forest lands even though they are far from the village.

- The roads in this region run alongside the streams which tend to spread out into large swamps. The valley bottoms are hard to cultivate, so the marshes must be crossed to reach flat ground better suited to the local crops.

111

The location of the land parcels is a source of **many problems** for the inhabitants of this settlement.

- The fields are **a long way** from the village and so involve hard, heavy transportation on bad forest trails.

- **Seasonal food crops are badly guarded and often devastated** by wild forest animals.

- **Moving the fields** frequently to new sites inside the forest, means that proper trails to facilitate transport are not cut.

It is interesting to note that in this village no plots are cultivated collectively. The land belongs to the village collectivity, but the plots are cultivated on an individual basis. Food crop plots are simply grouped together by the two big village families.

The parcelization of a **savanna hamlet** is shown in **figure 112**. The allotment plan is very different from the one in the preceding village, mainly because the lands cultivated collectively and those cultivated individually are clearly demarcated.

The hamlet is set back from the road and divided into three areas, or neighbourhoods, corresponding to the three families on the holding. One of these areas with its enclosures can be seen in **figure 111**. A study of the parcelization reveals **five kinds of parcels or land lots :**

- small individual gardens near the dwellings ;

- **big collective fields** where the whole workforce of the village must work on a communal basis ;

- **big family fields** controlled by the area or family heads ;

- **individual fields ;**

- **state lands.**

112

road
estate
pasture
individual fields
gardens
family field
quarter
family field
family field
collective field
path
pasture
tree
wood

Relations between cultivators and pastoralists are a serious problem. During the dry season, herds graze in the big fields, but during the rainy season, when they are not guarded properly, cattle sometimes break into cultivated plots and consume and trample the crops.

Fencing off the fields is one answer to the problem but it is not enough in itself because the cattle must be fed. So, as well as fencing, forage cropping is also needed.

In this particular case, it is worth examining the allotment plan in relation to the respective needs of cultivators and pastoralists.

Clashes can be more easily avoided if both parties agree to plant forage crops which will guarantee supplies of animal feed at all times. When agreement is reached, planting forage trees, hedges and grass strips can go on hand in hand with enclosure fencing.

Figure 113 reveals **a hillside settlement** where the land is fragmented into small, irregular parcels. The land is densely populated, which explains why it has been broken up into ever smaller plots under the pressure of successive waves of new young farmers. All land has been distributed on a permanent basis and collective rights no longer exist.

In **figure 113**, there is a clear distinction between the plots given over to bananas and those with seasonal crops. A landslide caused by erosion can be seen in the middle of the photograph.

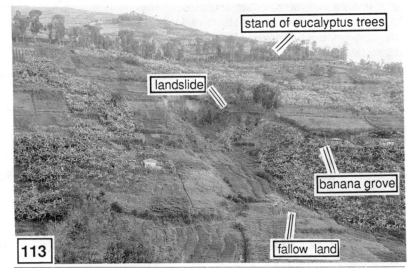

The farmers in such a settlement are not in a position to increase the size of their plots as they could have done in the past. If they want to increase family output, the only way they can improve the productivity of their plots is to use more intensive farming practices.

A farmer may have plots scattered over quite a distance. This fragmentation of land impedes farm work because a lot of time is wasted moving from one parcel of land to the next. Consequently, some village authorities, with the consent of the families concerned, **relocate** or redistribute the plots and apply **land consolidation reforms** to regroup

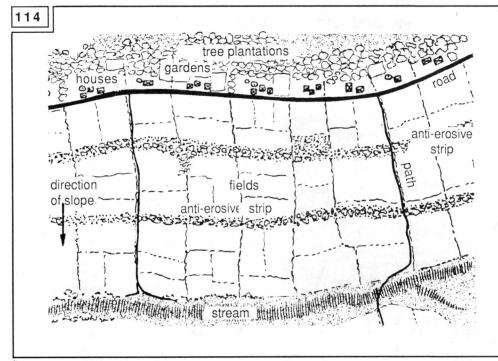

each family's land. Land consolidation reforms call for protracted discussions which must be handled by people with an intimate knowledge of local land-tenure and of the problems facing every family.

Figure 114 is our final example of parcelization.

Each farmer received a long strip of land crossing the road and perpendicular to it. Every farm strip runs down to the stream. So, all the farmers have a combination of three types of land - plateau, slope and valley bottoms.

When the land was distributed, two long stands of tall vegetation were sown between the road and the stream to form permanent anti-erosion barriers which may not be touched.

However, within the limits of his own land strip, the farmer may split up the land and cultivate it as he sees fit.

Here is how the strips are cultivated from top to bottom. First, perennial plantations (coffee and fruit trees), then vegetable gardens by the houses and the road, next, plots of seasonal crops, separated here and there by the herbaceous strips, and finally the valley bottom where some farmers grow more vegetables.

On this settlement, the parcelization has been fixed definitively. Also the farmers may not sell or distribute their land in bits and pieces. This restriction prevents fragmentation of farms in the long run, but the actual division of fields on a given farm can be changed.

Points to remember

The examples chosen show that the basic problems a village has can be discovered by examining its parcelization plan. This is a consequence of the history of the inhabitants. Indeed, the way parcelization has evolved depends on :

- *how the first families occupied the land from the outset ;*

- *how the land was then divided ;*

- *how the land was used by cultivators and stockbreeders ;*

- *how authority was exercised to harmonize collective and individual land rights ;*

- *the number of inhabitants living on the land, i.e.* **land occupation density** *;*

- *the characteristics of the living and physical environments ;*

- *the farming systems used, i.e. the way the cultivators use the production factors - shifting cultivation (moving from one season to the next), fixed or sedentary cultures, cultures associated with stock, etc.*

Lesson 14
A rural settlement near Bobo-Dioulasso (Burkina Faso)

Village and temporary farmstead

We have seen that the characteristics of a settlement are the result of natural factors (biotic and abiotic environments) on the one hand, and of human factors (economic and social environments) on the other.

115

This lesson shows how a village settlement has evolved near Bobo-Dioulasso in Burkina Faso. There is one good, though somewhat variable, rainy season per year. We will look at the **evolution** of the settlement and of the countryside brought about by the **work of man**.

This village, that we have called Fouma, is situated by a river and has a population of over a hundred families living in adobe houses **(figure 115)**. The people of Fouma have a strong community spirit, so the houses are very close to each other. In the past, the people often had to fight for their lives. This explains why the village is **compact**, in contrast to other regions where the houses are **dispersed** over several hamlets.

There are few fields in the immediate vicinity of Fouma. The ground is eroded by the passage of man and animals **(figure 115)**.

Round the village there are some **small gardens** where cereals and vegetables are grown.

When people first settled in the village, they began tilling the land near their homes. By the sweat of their brows, they acquired for themselves and their descendents, family and individual **land rights** over communal land. So, the oldest families enjoy the right to farm the lands near the village, while newcomers have to go three or four kilometres to be able to exercise land rights over communal property. A visit to one of these new families and their fields reveals how their work has transformed the landscape of the Fouma settlement.

Here, four kilometres from the village, is the **temporary homestead** of Yamou, his wife and five children **(figure 116)**, where they live, in the midst of their fields, during the rainy season. This practice avoids the long trips every day between village and fields.

Unlike the village house made of adobe, the temporary farmstead is made of African fan palm.

What is the point of a temporary homestead ?

- It is situated in the fields which can then be easily guarded from the farmstead.

- It avoids the necessity of long treks, and reduces **transportation** time from village to fields.

- The temporary homestead is a simple, light structure. It can easily be **moved** when crop location is changed to make way for a fallow season.

What are the disadvantages of life on temporary farmsteads ?

- The family is **far from the village.**

- Children on the site are far from **school** and have no playmates.

- Clean **water** is not always available.

- **Health care** is precarious.

How the family transforms the landscape

With the permission of the land chief, Yamou chooses the site of his temporary farmstead near secondary forest, on land which has been allowed to lie fallow for ten years or more **(figure 117)**.

Yamou's first task is to **clear the land.** As this is strenuous work, he invites about ten men of his own age to help him, in return for food and drink while the work lasts. Later on, he himself will answer requests for help from his farming

56

friends. The clearing and removal of brush is a lively, merry occasion. As a result of the invitation, it is done quickly, in time for the planting season in June.

Certain **trees** considered useful, such as the African locust bean, the shea butter tree and the fan palm, are left standing. Their presence is proof that the cleared land had been cultivated in the past. The wood is sorted and whatever is saleable is tied in bundles and left by the road. The remainder is burnt on the site (**figure 118**).

Wood burnt on the spot releases the mineral salts it contains in the form of ashes (Lesson 18) The ashes are then scattered over the fields thus making the mineral salts available to the crops.

After bush clearing and burning, the sandy soil is hoed for the earthpea crop which, in this region, tolerates freshly cleared soil and is used as a pioneer plant. This means that, unlike many other plants, the earthpea thrives on land rich in organic residue.

Figure 119 shows up the earthpea plantlets. Note the tufts of herbs and leaves here and there on the sandy surface. Close observation reveals the little wavy lines caused by rainsplash on the bare ground.

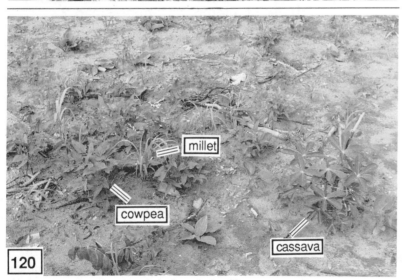

Several crops, mainly millet and cowpea, are associated in the second year but sweet cassava cuttings have also been planted here and there (**figure 120**).

This crop association will remain unchanged for some years. In the field in **figure 121**, it has lasted five years. The trees have grown and the cassava has developed in scattered clumps. The fan palm in the forefront is used for matting and roofing.

After six to eight years, Yamou gives up intercropping millet and cowpea because soil fertility has declined too much. He then proceeds to weed the plot thoroughly and mound up the soil to grow groundnut or earthpea. Groundnut grows well on poor soil (without organic matter) and earthpea adapts to similar conditions. In face, earthpea is an **adaptable** plant - it can grow on soil rich in organic matter and also on poor, already exhausted soil (**figure 122**).

The following year, fonio replaces the groundnut or the earthpea (**figure 123**). The trees have matured noticeably.

Fonio, (hungry rice or acha), is the last crop before the land is allowed to revert to fallow for several years. The duration of the fallow period depends on the amount of land at Yamou's disposal. If he has plenty of land, he will let the field rest for many years. But if land is short, the fallow period will also be short, and Yamou will come back to start clearing after only three, five, or eight years.

Figure 124 shows the fallow plot two years later, after the fonio has been harvested.

earthpea

122

shea butter tree

fonio

123

fan palm

shea butter tree

124

Customary land tenure

Land rights are always very complex and, in most cases, are only fully understood by the inhabitants of the settlement involved.

However, some land tenure concepts more or less common to villages in black Africa may be expressed as follows :

- *The land belongs to the community. Land ownership is **collective**.*

- ***Authority over the land** is vested in one or more eminent people whose duty is to preserve collective rights.*

- *The **produce** of the land belongs, nevertheless, to the individual **worker or group of workers**.*

- *Permission to use the land is obtained beforehand by **formally acknowledging the land authority**. This is done by offering a gift to the land chief, whereby the worker acknowledges the collective to whom the land belongs.*

- *The farmer who has cultivated the land **has the right to go on doing so,** provided he recognizes the land authority and collective ownership. The right of use usually applies to an area which is in proportion to the needs of the family.*

- *Authority over the land is vested in one or more eminent people who recognize the **land use rights of every cultivator**. However, the land authority can impose **rules** to regulate land use rules about fallow land, about respecting certain trees, about pastoral fires, water resources, and so on.*

- ***Trees are the property of the planter.** A newcomer is not allowed to plant trees, because they would give him permanent occupancy rights on land only temporarily allocated to him.*

The land authority is tending to disappear in parts of Africa today. More and more, ownership is vested in individuals, private companies and the State.

One thing is certain when considering the future of agriculture : land, uncontrolled by a competent authority, established in the vicinity, is land condemned to die.

Agricultural work has influenced and changed the landscape after a few years. When the initial forest in **figure 117** is compared to the transformed landscape in **figure 125**, we can see that, by his actions and of his own decisions, man can change the settlement to his advantage, even if he has no significant means at his disposal.

The changes helping to modernize agriculture are those whose effects are controlled and freely decided upon by man, as opposed to changes imposed by nature and beyond the effect of human intervention.

How man can exercise control over this settlement

Using only manual tools, Yamou has managed to transform the physical landscape with the help of his family and friends. When it rains enough and at the right time, his crop yields are satisfactory. He alters crop patterns to keep in step with changing soil fertility. Yet some elements, notably the trees and the soil, are beyond his control.

The trees

Yamou knows the value of trees. So, during clearing, he decided to save some of his trees. But the trees were planted **haphazardly** in the first place, leaving some land treeless and some overburdened. Consequently, there is plenty of shade in some parts, while the ground is completely exposed to the sun's rays in others. Also, the roots are so dense in some areas that they compete with cultivated crops for soil nutrients.

Yamou has little control over the trees making up the highest layer of vegetation in his fields. This situation arises largely from the fact that a fallow period is essential when the soil has been exhausted by cultivation.

The haphazard siting of the trees has another disadvantage - it makes mechanization difficult to plan. Cultivators using agricultural machinery tend to eliminate trees and stumps rather than integrate them.

Yet, the management of trees for agricultural purposes is within everybody's ability and is an excellent way of improving and modernizing agricultural practices.

This can be checked by leaving the Fouma settlement, for the moment, and visiting a field cultivated in a natural palm grove (**figure 126**). The old palms and other trees were planted at random. Now they will gradually be replaced by fruit trees, planted in rows and protected by the palm fronds. The arrangement in rows will facilitate the use of

125

126

young fruit trees

palms

127

citrus trees

machinery, should it become available to the farmer.

Figure 127 shows land with seasonal crops and citrus - oranges, lemons, grapefruit - also planted in rows. The ploughed furrows can be clearly seen.

To the front of the land planted with citrus, notice fields bare of vegetation, and, in the background, trees scattered here and there.

The soil

Yamou has not much control over soil evolution or gradual soil exhaustion for which two factors are mainly responsible. Firstly, cultivated plants take up the soil nutrients, and the plant produce is taken from the field, with the consequent loss of soil nutrients. Secondly, rainwater infiltrates the cultivable soil layers and drains away the nutrients to deeper levels.

The techniques that allow the farmer to exercise control over the evolution of soil fertility will be studied at a later stage, but one essential point, incorporated into the agricultural practices of the region, can be stressed here - tree management and the evolution of soil fertility are closely linked. The customary chiefs have always understood the relationship between these two factors. They have made and apply rules forbidding the felling of certain trees. Unfortunately trees are not always respected by those farmers who prefer immediate return on their land to the long-term preservation of land fertility.

Notes

A worthwhile product on the Fouma settlement

The caterpillars, seen on the shea butter trees in figure 128, are valued for making relishes. Like meat, eggs and fish, they are rich in protein. They can be dried or grilled, or used plain in sauces. Every year, in the month of July, they swarm over the shea butter trees and attack the leaves (figure 129). Collected and dried, they can be a useful source of income.

128

129

Caterpillars, and similar products such as larvae, termites, locusts and crickets, are found in many regions. They have tended to be incidental by-products but perhaps, some day, people will manage them more efficiently.

Lesson 15 🔲🔲🔲🔲🔲🔲🔲🔲🔲🔲🔲🔲🔲🔲🔲🔲🔲🔲🔲🔲🔲🔲🔲🔲🔲🔲🔲🔲🔲🔲🔲🔲🔲🔲🔲🔲

Land management in Serer country (Senegal)

The Serer people occupy a vast region to the south-east of Dakar. The land is densely populated. Serer agriculture, relying only on rainfall for water, is called **rainfed** agriculture, as opposed to **irrigated** agriculture. This lesson will deal with Serer land management in the past and what it can teach us today.

The foundations of the economic and social structures in Serer country

---

Land tenure

The first Serer who reached this land traced a wide circle round the site where he wanted to settle. This circle, made on horseback, fixed the collective ownership of the land once and for all. By this act of demarcation, the ancestor gave all his descendants the right to till the delimited region and to harvest the fruit of their work. This is known as the **right of the hoof** because the country where the Serer people exercise land rights was delimited by the horse's hooves.

All the descendants of the first settler have preserved the right to cultivate the land delimited by the right of the hoof, but since then, every Serer has acquired, by virtue of his work, permanent hereditary **land rights** through family lineage. The Serer who, with the permission of the land chief, has tilled a plot of land with his family or as an individual, has the right to remain there and to enjoy the fruits of his labour. When a family splits up, family rights to land are shared out.

There is a great difference between **ownership rights** and **land use rights**. Traditionally, the community has collective ownership of the land. All the members of the community have the inalienable right to work the land, provided they respect the rules laid down by the land authority who has the duty to assign farmland to every family or every worker. Land use rights, whether family or individual, last for as long as the **beneficiaries** cultivate the plots distributed to them. Conversely, should the land be abandoned, land rights are forfeited, and the plots return to the collective pool where they await reallocation.

The right to work the soil is conveyed by individual or community (family) **labour**. Regular cultivation confirms the dual right to harvest the produce of the land and to go on using the same land.

This concept of land tenure is peculiar to black Africa. In Europe, North Africa and America, the land usually belongs to one individual owner who has absolute rights of disposal with the power to engage or dismiss farming tenants. In black Africa, everyone enjoys the undisputed right to cultivate the land of his family ancestors.

130

Obligations connected with the land

The ancient Serers knew perfectly well that the land rights of every member of the community went hand in hand with certain obligations, such as the obligation to share in community tasks so as to guarantee collective security and the obligation to respect rules designed to protect group ownership of the land.

The worker who wants to farm must first recognize the collective ownership of the land. If he is a member of the community to whom the land belongs, he has to share in collective tasks - tilling communal fields, contributing to grain reserves, etc. If he is an outsider, he has to recognize communal authority, for instance, by offering a gift such as a chicken.

The land authority

In the past, traditional authority over land was strong and it determined how best collective ownership might be preserved for the future. The land was deemed a precious possession requiring careful management.

The people responsible would therefore take decisions concerning land and land use at regular intervals For instance, every year they decided which fields were to be left fallow to permit soil restoration. All individual and collective cultural rights were abrogated for the mandatory period. Similarly, the felling of certain trees was prohibited, and *Acacia albida*, known to benefit the land, was given special protection.

Today, the land authority has been undermined and in some parts, no longer exists. People can do what they like - burn, clear, sell, build and so on - without the slightest respect for the land and the customary rules which used to preserve it.

131

132

133

Outside Serer settlements
What happens when the land authority has ceased to exist

These figures illustrate the damage caused when forest lands are unscrupulously burnt and exploited.

Charcoal makers enter the forest to chop down the trees and make charcoal. The large pile of wood in **figure 131** will soon be set alight to produce charcoal. All the big trees on the site are cut down and the men depart leaving behind a degraded brushwood forest.

Little by little, the forest is transformed by felling and fires **(figures 132 and 133)**. Indeed bush fires are often started on purpose, by charcoal burners, to make clearings and facilitate penetration. A forest may be destroyed to obtain a few truckloads of charcoal, and no trees planted in replacement. This happens mostly in areas where authority over the use of land has fallen into abeyance and regulations about fires and forest exploitation are no longer applied.

134

gully

135

*Every time we see cartloads of wood leave the land, as in **figure 136**, we should ask these questions : If the towns use up all our wood, what will become of our forests and trees and the benefits we derive from them ? Are we sure that this rampant exploitation will not ruin our lands ?*

136

After burning, erosion sets in. This too is hard to combat if there is no land authority able to organize the local inhabitants. **Figures 134 and 135** illustrate the aftermath of indiscriminate fires.

When a land authority is present, improvement schemes bigger than anything the individual farmer might attempt, can be undertaken as shown in **figures 137, 138 and 139**. **Figure 137** shows hillside works carried out by Cape Verde peasants.

Figure 138 shows the men from a village in Burkina Faso making earth banks to protect against erosion.

Figure 139 illustrates a valley bottom developed solely for irrigated rice after lengthy discussions among village leaders, the inhabitants, State officials and technicians. In this instance, the role played by the land authority had to be complemented by the water authority responsible for sharing out the available water resources.

The role of a land authority may be summed up as follows :

■ it makes sure that no inhabitants, young or old, are left without farmland ;

■ in the interest of the community, it sees that the rules to prevent degradation of the living and physical environments are obeyed. More particularly, it draws up the regulations about fires, tree conservation, erosion control, etc ;

■ it regulates cohabitation between farmers and stockbreeders, bearing their respective interests in mind.

The unanimous acknowledgement of the fact that the land is a collective possession and must not be exploited ruthlessly, is an important step towards village modernization. An effective land authority is one which can communicate with all the beneficiaries and work out with them the rules and schemes needed to protect and improve the soil.

An efficient land authority, recognized by all the interested parties, is an important factor in modernizing agriculture.

The management of Serer settlements in practice

Three very important elements characterize the traditional management of Serer settlements as devised by the elders of the tribe. They are :

■ the **allocation of fields, rotation and cropping patterns** ;

137

138

139

- **the presence of trees** in the fields and their permanent association with seasonal crops ;
- the association of **livestock and agriculture**.

The allocation of fields, rotation of crops and cropping pattern

The settlement is usually divided into five parts or plots **(figure 140)**. The plot encircling the houses is called the **pombod** in Serer. It is reserved for bulrush millet (variety *souna*) and cowpea. The same crop association is practised every year because the plot is fertilized by animal manure spread on the land during the dry season. In the pombod, cotton may be mixed with vegetables, such as calabash, cassava and okra, grown for family use.

The pombod is a **safety zone worked collectively. The cereals harvested are stored by the headman against hard times.**

Because **livestock and agriculture are associated** and the soil is amply fertilized by animal manure, bulrush millet and cowpea can be grown all the time in the pombod with no intervening fallow years.

The zone beyond the pombod is set aside for **large fields** where families and individuals may cultivate freely provided fallow rules are respected.

The large field zone is divided into four plots on the basis of the cropping pattern applied. Two plots, called **tos**, lie fallow. Cereals, for instance, sorghum and millet (variety *sanio*), are cropped on the third plot. The fourth plot is occupied by groundnut, a cash crop introduced in the last hundred years and grown in pure stands as directed by official instructors. These four plots are surrounded by thorn tree fencing to keep out cattle. However, this tradition of making fences to stop cattle from wandering is dying out with the gradual disappearance of thorn trees.

So, rotation in the large fields is as follows : a cereal crop or cereals associated with cowpea, followed by groundnut, and then fallow for two years. This type of rotation is usual because the cereals grown first, i.e. the first crop of the rotation, like soil enriched by fallow waste and manure, while groundnut is more suited to a well-cleaned, poorer soil.

The presence of trees

Many kinds of trees are to be found all over Serer settlements. Each one is useful in its own way, but the elders were so struck by one particular tree, *Acacia albida*, that severe penalties were meted out

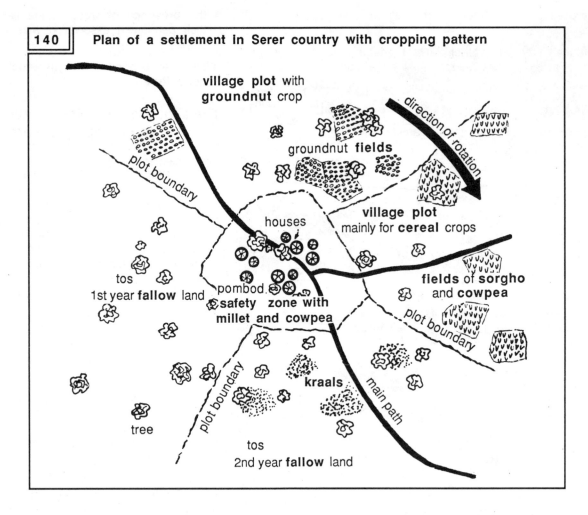

140   **Plan of a settlement in Serer country with cropping pattern**

village plot with
groundnut crop

direction of rotation

groundnut **fields**

village plot
mainly for **cereal** crops

houses

fields of **sorgho**
and **cowpea**

tos
1st year **fallow** land

pombod.
**safety zone with
millet and cowpea**

plot boundary

plot boundary

kraals

main path

tree

tos
2nd year **fallow** land

to anyone daring to destroy it **(figure 141)**. In fact, by deliberatley associating this tree with cultivated crops, the elders proved that every kind of tree should be studied from several points of view.

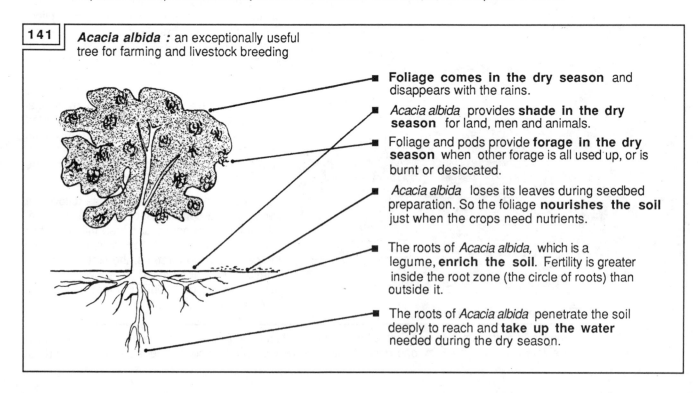

141   *Acacia albida :* an exceptionally useful tree for farming and livestock breeding

- **Foliage comes in the dry season** and disappears with the rains.

- *Acacia albida* provides **shade in the dry season** for land, men and animals.

- Foliage and pods provide **forage in the dry season** when other forage is all used up, or is burnt or desiccated.

- *Acacia albida* loses its leaves during seedbed preparation. So the foliage **nourishes the soil** just when the crops need nutrients.

- The roots of *Acacia albida,* which is a legume, **enrich the soil**. Fertility is greater inside the root zone (the circle of roots) than outside it.

- The roots of *Acacia albida* penetrate the soil deeply to reach and **take up the water** needed during the dry season.

- What does the tree produce (fruit, seeds, leaves, flowers, bark, wood) ? How do people and livestock use this produce ?

- How does the tree help other plants ? (shade, yield of vegetal waste good for the soil, improvement of the soil because it has deep roots, increase or decrease in soil moisture, etc.)

- In what way is the tree useful when associated with cultivated or wild plants ?

These and similar questions must be asked before planting a tree or cutting one down.

Stockbreeding in association with farming

A large herd of cattle was the sign of social eminence and wealth among Serers in the past. The herd guarantees prestige and security but as well as that, it provides the manure that the land requires (figure 142).

During the rainy season when the fields are under cultivation, the cattle graze on fallow land, and wandering is checked by cowherds and thornbush fencing. But every evening the farmers take home their own cattle to let them graze on the fields that are to be cropped the following year. The animals are hitched to pickets that are moved every few days so that several plots may be manured in turn.

A big cattle owner may lend or hire his herd to a farmer who has no cattle, since manure is essential for cultivation.

kraal

manure

142

The fields are cleared after the harvest. The cattle are let loose to eat the trash and haulms lying on the ground. It is the dry season. Every day the cattle are brought back to the pombod, the central zone round the houses, so that this area may be fertilized at night. Consequently, millet and cowpea may be cultivated here all the time and a fallow period dispensed with.

In the heart of the dry season, forage becomes scarce. That is also the time when the green leaves and pods of *Acacia albida* are gathered. So, thanks to this tree, the cattle are able to stay at home during the dry season and do not need to travel far in search of food.

Basic lessons handed down by the Serer elders

The Serer elders evolved an efficient, balanced agricultural system which stood the test of time for hundreds of years. The elements of the system were :

- a land authority empowered to discuss and negotiate with all the land users and draw up plans for land management ;

- a farming system based on the judicious association of seasonal crops, trees and stock. The system could be built upon to introduce improved farming practices ;

- the use of plants which nourished the soil and benefited crops.

Notes

They lived for a long time on the same land because they maintained the right balance between all these farming components :

- cultivated fields and fallow land ;

- the quality and number of trees in relation to cultivated crops ;

- trees and cattle ;

- cattle and crops ;

- the combination of community life and the activities of the individual ;

- individual land use rights and collective obligations ;

- the safety of the individual along with the safety of the community.

As the system stands, there is nothing to prevent the use of machinery if trees are spaced in a way which allows easy access. Mineral fertilizers could be used, seed varieties with better yields in companion cropping introduced, crops could be watered where possible, fallow land improved, forage crops encouraged, etc.

But over the last twenty or thirty years, the increase in population and the marked preference for cash crops, **especially groundnut**, have had several consequences :

- **the area reserved for fallow has shrunk** and fallow duration shortened, so soil fertility is no longer restored ;

- less fallow space means **less space for cattle** ;

- fewer cattle produce **less manure** for the fields, and particularly for the pombod or safety zone, with a consequent drop in crop yields ;

- cash crops are a most attractive source of income but they now take more and more room and **the traditonal rotation of crops is no longer practised**. What is taken off the land for export is less and less counterbalanced by what is put into the land (Lesson 40) ;

- **modernization brings in machinery.** But machines are handicapped by the presence of trees **which are then cut down** to make way for the machines. The absence of trees leads, in turn, to a scarcity of forage during the dry season and to changes in the climate.

Summing up

There are many lessons to be learnt about modernizing agriculture in tropical Africa. Agriculture is not just a question of machines, fertilizers, pesticides and improved varieties. Modernizing agriculture means achieving ***a better balance*** *between all the production factors :*

- *balance between machinery, trees and plants ;*

- *balance between the plants fit for human consumption and forage crops for animal feed ;*

- *balance between the space reserved for crops and for pasture ;*

- *balance between what is taken from the land in the form of agricultural produce and what is put into the land in the form of fertilizers ;*

- *balance between the crops grown for farm consumption and cash crops ;*

- *balance between the produce stored and the produce marketed ;*

- *balance between the production factors available on the settlement itself and the off-farm factors which must be purchased ;*

- *balance between income and expenditure ;*

- *balance between the farmer's immediate return and what he is attempting to set aside for the future ;*

- *balance between human population density and sustainable land-use practices.*

Lessons 16 and 17

The need for trees

Lessons 16 and 17 deal with the usefulness of trees in the landscape and in the fields. Lesson 16 discusses the general usefulness of trees, while Lesson 17 considers the effect of trees on climate and their contribution to the fight against erosion.

143

Why trees are useful in the landscape

Trees can be arranged in many ways on a cultivated settlement. They may be scattered here and there, planted in clusters, or form hedges (**figures 143, 144 and 145**).

The presence of trees in or around fields is beneficial for four reasons :

■ **trees mitigate the effects of climate** on cultivated land ;

■ **they transform the soil** ;

■ **they provide worthwhile produce** such as wood, forage, vegetables, seeds, medicinal herbs ;

■ **they establish land use rights** and the tree planter can dispose of the produce.

144

Trees mitigate the effects of climate

Figure 29 in Lesson 4 considered climate and its component parts : air and wind, rain, moisture and drought, light, heat and cold. How do trees affect these components ?

Trees act as windbreaks

Strong wind can break plants and even dislodge them. Wind can wither plants. The more air that comes into contact with plants, the more they dry up (Lesson 23).

When soil is bare, strong winds can carry away soil particles. This is called **aeolian or wind erosion**. Trees, on the other hand, can withstand wind and diminish its effects on other plants. So they protect plants and save them from being damaged or desiccated, thereby reducing the risks of wind erosion.

145

Trees provide protection from heavy rain

Raindrops are the main agent of erosion. Gathering speed as they fall from great heights, they strike the little clods of earth, bursting them apart unless stones or vegetative cover afford protection.

Figure 146 illustrates the effect of raindrop impact on soil. The exposed soil has been eroded but the soil protected by small stones has escaped the shattering action of the rain droplets.

small stones

eroded earth

little mound sheltered from raindrop impact

146

In the same way, thanks to their foliage, trees act like umbrellas, absorbing the shock of the raindrops and making them lose their force before they reach the ground. In **figure 147**, young cassava leaves afford good cover for the soil.

Trees provide protection from the energy of the sun's rays (radiant energy)

Trees provide shade and stop the soil from overheating. Temperatures that are too high are dangerous because they dry up the soil and kill soil life.

Trees, on the other hand, need radiant energy in their leaves in order to live and thrive.

Land with a tree canopy is in the shade at least part of the day and remains cooler. Indeed, the effect of trees on soil temperature is significant even in changing shade conditions **(figure 148)**.

Shade is always beneficial for tropical soils and shade management, whenever feasible, will always contribute to soil fertility. It is particularly important for some plantation crops, such as cocoa.

147

148

Trees preserve soil life and transform it

The effect of trees on soil life and the consequences for associated crops can only be fully grasped by finding out at what level annual crops take up nutrients and the kind of nutrients they need. Some aspects of the question are mentioned here and these will be discussed in detail later on in the book.

- Most seasonal or annual crops take up nutrients in the shallow soil layers. As root penetration is rarely more than 50 cm, it follows that nutrients for seasonal plants must be available in these superficial layers.

- There are two kinds of plant nutrients - non-living nutrients called **mineral nutrients**, originating in stone, sand, clay and air (Lesson 40), and nutrients from **organic** sources, i.e. from all kinds of **decomposed** living plant and animal matter (Lesson 33).

- For seasonal crops, **fertile** soil means soil with the right mixture of mineral and organic nutrients.

These three points will help us to interpret the illustrations that follow.

The sandy soil in **figure 149** has been stripped bare after the groundnut harvest. The little hollows caused by the impact of the last rains are noticeable. This soil will remain uncovered right through the dry season. The sun's rays will overheat the field and remove any remaining moisture. The first rains will cause severe damage.

Rain and sun have also turned the bare soil in **figure 150** into barren land. The sparse herbage will soon die too. This is a type of **laterized soil** with red stones which make cultivation impossible (see Lesson 19).

The rich layer of decomposing debris in **figure 151** was photographed in a forest plantation. The litter gives protective soil cover and, by depositing mineral and organic nutrients, it nourishes the soil and activates soil life.

sand

149

laterite

150

litter

151

The role of litter in soil life is confirmed by simply scratching the soil and observing the great number of living things - insects, worms, spiders, fungi, little plants, seeds, mosses and so on - all contributing actively to soil transformation.

The level of natural soil fertility depends very much on tree life and on the activity of many small animals and microorganisms living in the litter of leaves and other residues covering the ground.

Trees are a source of many useful consumer products

On our own and on nearby farms, which trees are most commonly associated with seasonal crops and what are they used for ?

Trees meet human requirements in a surprising number of ways.

Take the fan palm as an example **(figure 152)**. It grows in many regions and is often found on sandy soil. Village elders knew how valuable it was and the many uses to which it could be put. Few trees are so multipurpose :

- its fruit can be eaten like coconut ;
- the sap can be made into palm wine or drunk fresh ;
- the leaves are used for matting, roofing and basketwork ;
- the young shoots are eaten as vegetables **(figure 153)** ;
- the wood is rot-proof and is ideal for building **(figure 154)**, for making bridges and other constructions. In the past, it was hollowed out to channel water. In some areas, it provides more or less termite-proof fencing stakes.

152

young fan palm shoots eaten as vegetables

153

154

155

156

In cultivated fields, the fan palm does not get in the way of cultivated crops because its roots do not extend laterally and the foliage gives little shade especially when the tree is trimmed.

The old fan palms growing here and there on the land in **figure 155** have been topped. In many parts of Africa, fan palms have been mismanaged, chopped down and not replaced, and their rightful role on farmlands has not been respected.

Today, young farmers in Casamance (Senegal) understand why fan palms used to be highly prized and have decided to restore the tree's ancestral value. They take care of the best shoots during cultural operations and see that they are not damaged **(figure 156)**. Up till now, these trees have been seeded broadcast, but methodical spacing will soon be introduced and trees interplanted on cultivated lands.

The fan palm is an outstanding example of a tree which the farmer can use in many different ways. But in the attempt to modernize agriculture, many other equally useful trees should also be considered. Other palms, *Acacia albida*, locust bean, shea butter, baobab, and all kinds of fruit and timber trees can all be grown profitably.

Summary

Where there are trees, there is always life. Where the climate makes farming difficult, trees can be used to help, if the farmer knows how to use them.

Trees and perennial plants, i.e. plants lasting year after year, enable the cultivator to overcome the rigours of the climate to some extent. Cutting down trees or neglecting them is a great mistake and may endanger the life of the fields.

Looking after trees, keeping them in good condition and managing them properly is a sure way of promoting intensive, modern agriculture.

In tropical climates, farming methods associating trees and annual cultivated crops is one way of ensuring the future of agriculture.

Lesson 17

🔲🔲🔲🔲🔲🔲🔲🔲🔲🔲🔲🔲🔲🔲🔲🔲🔲🔲🔲🔲🔲🔲🔲🔲🔲🔲🔲🔲🔲🔲🔲🔲🔲🔲🔲🔲🔲

Trees act as windbreaks

Although **wind erosion** is not always as noticeable as **rain erosion**, it is considerable in some parts of Africa and acts by carrying away the finest soil particles.

This explains the dust storms which sometimes blow over deserts or in the Sahel region.

In savanna with dry herbaceous cover, strong winds often cause dust whirls.

In the same way, after the passage of fire over bare fields, the wind picks up the ashes and deposits them far away. The wavy appearance of sandy soil subjected to wind erosion stands out clearly in

figure 157, picturing farmlands swept by the harmattan in the dry season.

In regions where wind is not slowed down by vegetation, dunes or hills are formed by the accumulation of sand particles deposited by the wind.

When land is bare, the wind carries off the finest soil particles, espeally clay and loam containing many plant nutrients.

Figure 158 illustrates the basic principles of wind erosion. To understand the drawings in this lesson, it should be remembered that wind is the movement of a considerable mass of air, at high or low velocity. Wind may be compared to clouds moving in the sky or to water flowing on the ground. In the drawings, wind is represented by arrows.

sandy soil beaten by the wind

little ridges heaped up by the wind

157

---

158 | **Wind erosion is called Aeolian erosion**

This table shows what happens to the **denuded lands** on the left and to the **sheltered** lands on the right. The lands with no cover lose their fertile particles which accumulate on the lands to the right. The arrows indicate the direction and the force of the wind.

Strong wind carries away the finest soil particles, e.g. clay and loam dusts, and leaves behind the coarser particles like grains of sand, gravel, small stones, rocks.

The wind moves in an eddy as it carries away the clay and loam containing the **most useful mineral nutrients for plant growth.**

After blowing over an obstacle, **the wind slows down and deposits the clay and loam dust** which accumulates in sheltered places.

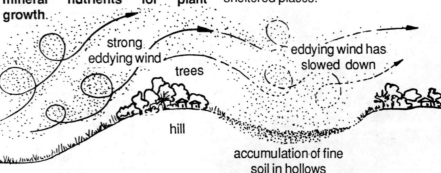

strong eddying wind

trees

eddying wind has slowed down

strong wind

hill

bare soil

accumulation of fine soil in hollows

If the farmer is not careful, the wind may well remove large quantities of fertile soil from his land, leaving behind nothing but sand and stone. Taking steps to **slow down the wind and cover the soil** is the most effective way of fighting erosion and preventing the dangerous removal of soil.

72

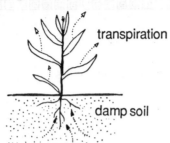

159 **The effect of the wind on plants**

transpiration

wind · transpiration

damp soil

water

desiccated soil

1 The air is calm
The plant pumps the soil water up through its roots, and then **transpires** water through the **stomata** or tiny holes near the leaf surface.

2 The air is in motion
Wind blowing past takes up and removes **all the water vapour it can get.** This forces the plant to pump more and more water from the soil, which eventually dries up.

3 Wilting
When the **soil is parched**, the plant, deprived of water, closes the stomata and stops breathing. The leaves begin to wilt.

Wind speeds up the wilting process in plants. It lessens respiratory capacity, reduces the supply of nutrients available to cultivated plants and results in poorer crop yields.

Well-sited hedges, trees and windbreaks break the force of the wind and contribute to improved agricultural production.

Wind not only carries away soil particles and damages plants, it also removes soil and plant moisture. Wind action on plants is illustrated in **figure 159**.

How do trees break the force of the wind ?

The aerial photograph **(figure 160)** shows countryside dotted with trees and divided into plots bordered by hedges. The trees and hedges form a succession of obstacles which lessen the force of the wind.

The way trees modify the effects of wind is examined closely in **figures 161 to 165**.

160 aerial photograph — road, plots of land, hedges, trees

166

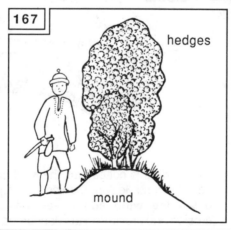

167 hedges / mound

How exactly are hedges drawn in this lesson ? **Figure 166** shows a hedge bounding pastures. The hedge is composed of many kinds of trees. The drawing on the left **(figure 167)** shows how this type of hedge or hedgerow is represented in the illustrations that follow.

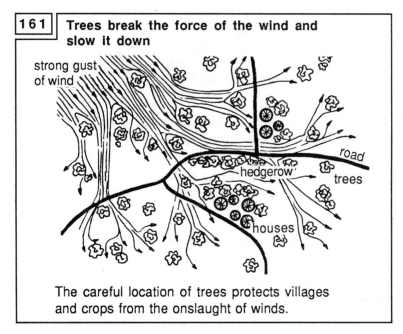

161 | Trees break the force of the wind and slow it down

strong gust of wind

road

hedgerow

trees

houses

The careful location of trees protects villages and crops from the onslaught of winds.

The drawing **(figure 161)** shows land dotted here and there with hedges and trees, rather like the countryside in **figure 160**. Land with these physical features is called a **grove**. Note the houses, trees and hedges round the fields.

A great mass of air coming from the left suddenly meets the grove head on. The air is represented by a series of arrows.

If there were no trees, the huge air mass would blow straight over the field drying up the vegetation on the surface of the ground. But because there are trees, the air mass is jostled and, to pursue its course, has to slow down as it whirls through the trees. The large mass of air is broken up into smaller and smaller masses that bump into one another and blend together. Retarding air flow limits the drying effect on the soil.

What does the wind route look like, not from a plane but observed by a person on the ground ?

Air and water are elements that both flow in a similar way, especially when they encounter obstacles. Water flow can be observed easily, but air flow is invisible. If it is hard to understand the air flow figures, we can think of water and how it flows into streams or over fields.

Wind blows over cultivated land with no tree cover (figure 162)

The arrows indicate a dry wind blowing across the land at high speed. It snatches large quantities of water from the cultivated plants and these transpire all the more because the weather is hot.

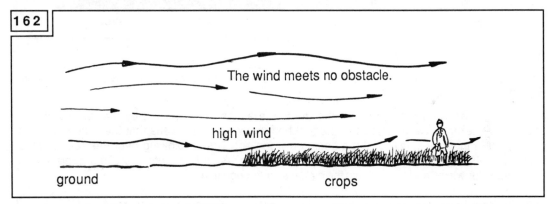

162

The wind meets no obstacle.

high wind

ground

crops

On reaching the land, the wind encounters the foliage of a row of trees with bare trunks (figure 163)

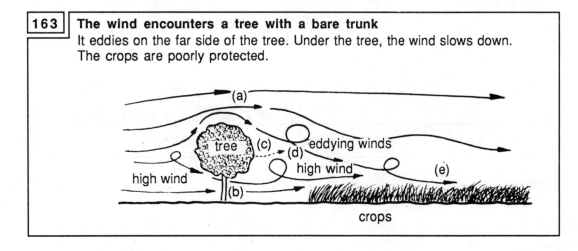

163 | The wind encounters a tree with a bare trunk
It eddies on the far side of the tree. Under the tree, the wind slows down. The crops are poorly protected.

(a)

tree (c) (d) eddying winds

high wind high wind (e)

(b)

crops

74

The air is split into three parts : one part, (a) in the figure, blows over the tree tops. The second part (b) blows under the foliage. The third part (c) blows through the leaves. On the far side of the foliage, there is a calm zone where wind velocity is very low (this light wind is shown by a dotted arrow). Next, there is a zone where the wind moves in eddies (d). Finally, the air regains its original speed over the cultivated crops (e) and dries them up. A row of trees with bare trunks gives poor protection to crops.

The field is protected by trees and hedgerows (figure 164)

In this instance, there is just one hedgerow beyond the line of trees. As in **figure 163**, the trees first split up the air, making it eddy and then slow down.

The hedgerow also splits up the air. The mass of air blowing under the tree foliage hits the hedgerow and is sent swirling upwards (f). Some blows through the hedgerow, slowing down considerably on the way (g).

So, these crops, compared to those in **figure 162**, are much better protected from the drying effects of the wind. The wind only picks up speed near the end of the cultural zone where it skims once more over the cultivated plants (h).

| 164 | **The wind encounters a tree with a bare trunk, then a hedgerow.** The tree and the hedgerow cause the wind to eddy and then slow down. The crops are protected over a certain distance. |

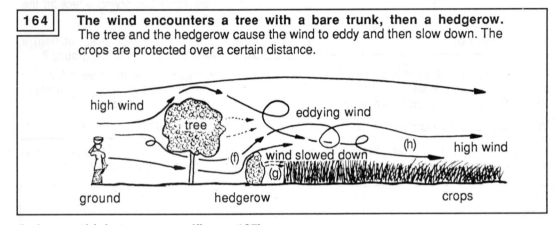

A denser thicket or copse (figure 165)

Here the hedgerows are properly spaced and there are more of them. The big trees have leafy branches up the trunk.

The force of the wind blowing under the treetop is almost all absorbed by the foliage on the trunk and, after that, by the leaves of the first hedgerow (i). This means that the wind is slowed down a lot in the lowest layers of the aerial environment. The air blowing over the treetop goes whirling on its way and comes near the ground. But the second hedgerow forces a large part of the air mass upwards and slows down the air passing through it (j).

| 165 | **The wind encounters first a tree with a branched trunk, then a succession of hedges.** It is slowed down considerably. The crops are well protected. |

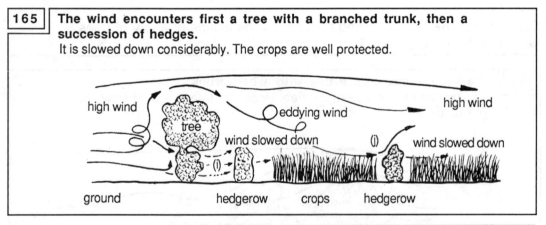

Notes

The composition of a good hedge

Hedges must always be adapted to suit local conditions. They may have to be tall in order to act as effective windbreaks for the crops they are protecting, or they may need to be lower to avoid giving too much shade.

It would be useful to compare the hedges in **figures 168 and 169**. The first hedge **(figure 168)** is composed of different species, for instance, banana, legume trees, shrubs and herbaceous plants like tannia but the second hedge **(figure 169)** only contains *Azadirachta indica* (Neem). Because the base of this hedgerow is bare, it is a poor windbreak. The first hedge is effective as a windbreak and useful in many ways.

thick hedge composed of different kinds of plants

168

Azadirachta indica (Neem)

A thin hedge with only one kind of tree.

169

The second hedge is ineffective and only good for firewood.

Figure 170 represents a good hedgerow seen from the front. Here are some of the species it contains :

■ shrubs with foliage about a man's height ;

■ small trees or large shrubs of fast-growing varieties that provide poles, especially for

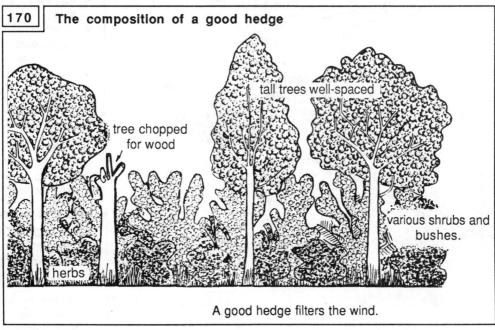

170　**The composition of a good hedge**

tall trees well-spaced

tree chopped for wood

various shrubs and bushes.

herbs

A good hedge filters the wind.

construction work. Care must be taken to cut them at regular intervals. This part of the vegetation, called the copse, may grow to a height of four or five metres ;

- lastly, taller trees planted eight to ten metres apart. The crowns may grow to a fair height.

A good hedge slows down the wind but does not block its passage. In other words, a hedge must be permeable to wind.

Other functions of hedges

Hedges form very useful living environments

The term **ecosystem** means the communities of living beings, dependent on one another for their existence, that can be found in an environment (Lesson 3), and this term can be applied to hedgerows. Here are examples of these interdependencies.

Small rodents gnaw wood and seeds. By scattering their excreta on the ground, they help to make humus. In this, lots of bacteria and other microorganisms provide food for earthworms. The earthworms in their turn are active in the soil, busily helping the growth of root systems, particularly in cultivated plants. The plants themselves are eaten by insects, some of which cause disease. These insects are eaten by birds, lizards and snakes, and these may be victims of birds of prey.

Living beings cannot live without each other. Suppose, for instance, that an insecticide killed off all insects. Where would lizards, snakes and insect-eating birds find food ? If all birds were banished from our fields, who would attack the insects decimating the crops ?

A well-managed hedge with the right plants is a good way of checking the damage done to crops by the multiplication of insects, small mammals and birds.

Hedges and their produce

Hedges can provide valuable produce. If the varieties of trees, shrubs and herbs composing hedgerows are chosen properly, the yield may include :

- firewood ;
- poles for building ;
- timber for planks ;
- fruit for family consumption or for sale ;
- honey ;
- vegetables, condiments, medication, forage ;
- dyes, glues, ropes.

In addition, many small edible animals, caterpillars, worms and small game, live in hedges.

Hedges, properly planted with useful varieties instead of being allowed to run wild, can be a source of extra income.

Hedges have an effect on the soil

- **They enrich the soil.**

  Well-composed hedges are able to enrich the land where they grow. Their roots take up water and nutrients from deep in the soil while the leaves fall onto the

171

big tree

shrub

crops

bush

crops

shallow roots

deep roots

fields. To some extent, growing hedges has the same effect as allowing land to lie fallow.

Because the ground under hedges retains more moisture and is more active than the ground under crops, it becomes the habitat of microorganisms, worms, fungi, insects and useful small animals (Lesson 33).

- **They limit land erosion** which can be caused by water runoff (Lesson 22), as well as by wind..

- The roots of hedge plants develop abundantly and penetrate deeply. Where rooting occurs, water penetrates the soil more easily. **Hedges force the water to enter the soil and prevent it from flowing off the land** (Lesson 34).

- Water channelled into the ground by hedgerows is found lower down in the watertables and in springs (Lesson 22).

Planting hedgerows is a satisfactory way of gradually replacing the use of fallow land, and is also a way of controlling forest plantations. Hedgerows are 'linear forests' because they are planted in rows on rural settlements. **In this way, as with natural forests, man can make use of hedges to his own advantage and exercise control over them at the same time.**

Hedges and fields

Trees and hedges growing in fields may disturb cultivated plants growing close by. Roots from the plants in a hedge may compete with cultivated plants for available water, or foliage may be too shady..

As well as the adverse effect of trees on plants near them, the effect on the field as a whole should be noted. Plants near hedges may not grow so tall because of root competition but they are taller where winds are not so strong and shorter where winds pick up speed again **(figure 172)**.

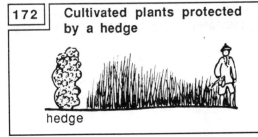

172 | **Cultivated plants protected by a hedge**

hedge

small eucalyptus wood

sparse litter

173

row of eucalyptus trees

irrigated banana

174 | ditch

Some trees dry up the surface soil. Consequently, the association of hedges and seasonal crops in regions short of water **(figure 173)** should be avoided. On the other hand, trees greedy for water sometimes drain overmoist soils and are beneficial to crops.

Note that when tree or hedge roots disturb cultivated plants too much, a ditch can be cut to stop the tree roots from spreading through the superficial soil layers **(figure 174)**.

Wind affects the soil in various ways and can cause a drop in crop yields. If the farmer uses trees advantageously, he will minimize the harmful effects of wind and will harvest useful produce.

Notes

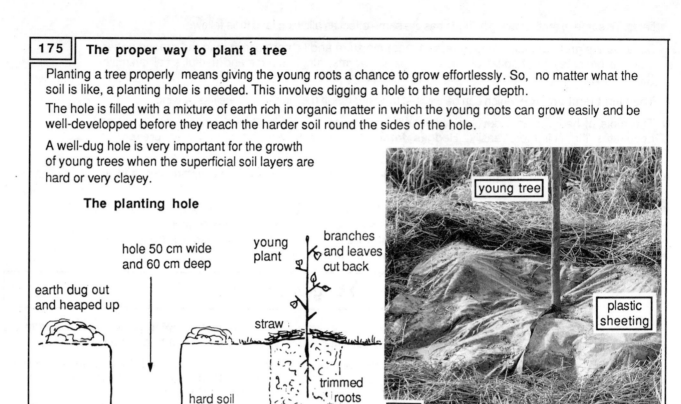

175 | The proper way to plant a tree

Planting a tree properly means giving the young roots a chance to grow effortlessly. So, no matter what the soil is like, a planting hole is needed. This involves digging a hole to the required depth.

The hole is filled with a mixture of earth rich in organic matter in which the young roots can grow easily and be well-developed before they reach the harder soil round the sides of the hole.

A well-dug hole is very important for the growth of young trees when the superficial soil layers are hard or very clayey.

The planting hole

earth dug out and heaped up

hole 50 cm wide and 60 cm deep

hard soil

young plant

branches and leaves cut back

straw

trimmed roots

young tree

plastic sheeting

175

Lessons 18 to 22

Soil and water

Rural settlements and fields have been studied from various angles in this book but the question remains, 'What happens in the soil itself ?' Examining the soil on the surface and in depth will provide the answer. An attempt should also be made to understand how soil and water are closely linked, and how water transforms the soil.

The next five lessons deal with :

- ■ Soil composition (Lesson 18) ;
- ■ Soil organization or structure (Lesson 19) ;
- ■ Soil layers (Lesson 20) ;
- ■ Water movement on and in soil (Lesson 21) ;
- ■ How water movement can be altered to benefit agriculture (Lesson 22).

Studying the soil means digging it up, and if the student cannot dig deeply enough, he has to imagine what is happening deep below the surface.

Digging and examining the soil

Sometimes the soil need only be scratched with the hand or with a knife. This is how the roots of the tossa jute seedling (**figure 176**) and of the yam (**figure 177**) were uncovered.

177   young yam tubers   roots

176   roots

But the best way to learn about soil and what lives in it is to dig **profiles** (vertical sections) deeper than the height of a man. First choose a place where the soil is sure to be in a natural state, i.e. not overworked by man or by machinery.

One of the sides of the hole must be cut really straight and then scraped with a knife so that everything in the soil can be observed easily (**figure 178**).

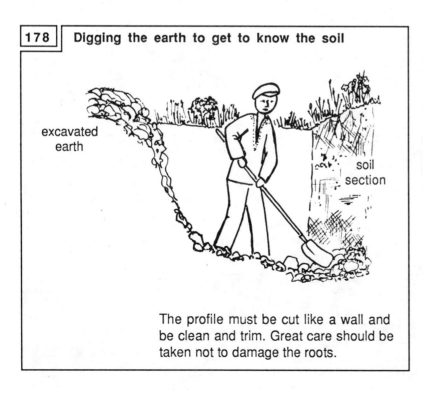

178   **Digging the earth to get to know the soil**

excavated earth

soil section

The profile must be cut like a wall and be clean and trim. Great care should be taken not to damage the roots.

179   roots   soil profile in a field of maize

Figure 179 shows a soil section in a field of maize. The roots are free of earth and are clearly visible. A close look at the picture reveals the different soil layers.

Lesson 18

Soil composition

Soil with living roots is composed of a mixture of all kinds of substances each with its own properties. These soil components can be **solid, liquid** or **gaseous**. When they are not visible to the naked eye, they are examined with the help of a magnifying glass.

Part of the soil is solid

The solid part of the soil comes from two substances. One substance derives from rocks, and its components are called the **mineral elements** or **mineral fraction**. The other substance originates in living beings, and its components are called **organic matter** or the **organic fraction**.

The mineral fraction

This fraction separates into finer and finer components.

Small stones, gravel **(figure 180)** and sand **(figure 181)** are formed of coarse particles, whereas silt and clay **(figure 182)** are formed of very fine particles.

Mineral salts are found in small stones, gravel, sand, silt and clay. They are taken up by the roots and nourish the plant. Mineral salts are often invisible just like kitchen salt when it is mixed with water.

Silt and clay contain the most mineral salts. Consequently, **silty** and **clayey** soils always contribute more to plant growth than sandy soils. **Ashes** are also a form of mineral salts.

The organic fraction

This fraction is derived from the dead and decaying matter of everything living above and below the ground. This means the waste matter of plants, animals living on or under the ground, worms, termites, larvae, fungi, mosses, all of which are visible. There are many more living beings, invisible to the human eye. We know they exist because the effects of their presence appear in rot and mould, in the diseases found in plant life, animals and man. These living beings are called **microorganisms** and can only be seen through a **microscope**.

Faeces and the remains of all these living beings rot in the soil and break down into **humus**. Humus can be detected by its black or dark brown colour, especially in places where the soil is covered by a good layer of

180   gravel

181   small stones   sand

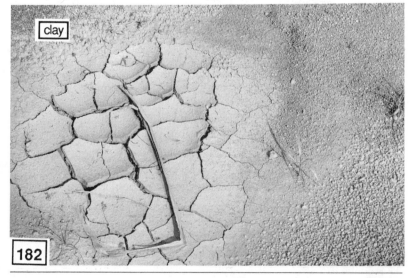

182   clay

dead leaves or household waste. Humus must not be mistaken for ash. Like silt and clay, organic matter, including humus, is an excellent supplier and holder of plant nutrients.

Part of the soil is gaseous

Air is the gas that fills soil pores if these are free of water. The air comes from outside or originates in the soil when plants and living beings breathe.

Some microorganisms cannot live in the soil if there is not enough air while others can live without air (Lesson 33).

Part of the soil is liquid

Rainwater infiltrates through the soil particles into the cracks in the soil. Some of the water, called **free water**, merely passes through on its way into the earth. But some of the water clings to the solid grains, especially to the grains of silt, clay and organic matter, and is termed **available capillary water**.

The water settling on the solid grains and the water flowing deep into the earth **dissolves** the mineral salts. Once the salts are dissolved, the roots can take them up to nourish the plant. This is why plant life without water is impossible.

Farming depends on the reserves of water in the soil and these reserves depend, in turn, on four factors :

- the amount of water falling on the ground and penetrating the soil ;

- the amount of water retained by the soil particles ;

- the amount of water taken up by plant life ;

- the amount of water lost through evaporation caused by heat.

Rainwater can penetrate the soil and create groundwater reserves at varying depths. Some groundwater is retained by the earth itself in the form of moisture, while some flows freely between the soil particles (Lesson 21).

This process is apparent when heavy rain falls on dry ground. The ground absorbs the first raindrops, so that even if the earth is squeezed between the fingers, no water oozes from it.

Later, the rain is not absorbed by the soil particles to the same extent. At this stage, drops of water can be pressed from a clod of earth. This free water, that has not been retained by the particles of clay, silt and organic matter, now travels through the slender soil ducts to form deep water reserves called **groundwater**. The upper limit of this is called the **water table**.

Free water penetrating the soil is useful because it increases the groundwater reserves, but as it is below the root zone, it does not contribute much to plant growth. Available capillary water is much more valuable because it stays near the growing roots.

Water fixation is influenced by the type of soil and varies depending on whether the soil is clayey or a mixture of stones and sand. To understand why, the composition of these soil parts must be carefully examined under a microscope because they are not visible to the naked eye.

Commercial fertilizers and ashes are mineral salts

Mineral salts sold commercially are chemical fertilizers processed in granules of varying coarseness (figure 183), or as spherical prills which do not stick together easily and so spread more easily than granules. When spread on the ground, they are dissolved by rainwater and carried into the soil and down to the plant roots.

183 | granules of chemical fertilizers

Ashes are a good soil fertilizer. The young men in figure 184 are well aware of the fact. They collect household ashes in barrels for use on their vegetable plots. Ashes can also be spread in thin layers between old leaves and crop residues to make compost.

ashes

184

Sand and clay are two very different substances

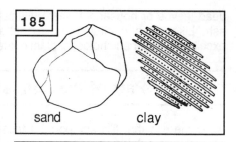

185

sand   clay

Sand is like stone in that each grain is one piece, just like a pebble.

Here is a drawing **(figure 185)** of a grain of sand under a microscope. The soil water flows over the surface of the grain but not into it.

Clay is the opposite. It drinks in the water and can stock it. A clay particle is composed of many leaflets that can only be seen through a powerful microscope. Clay can be likened to a book, the pages of which are superimposed. The water and mineral salts are able to slip between the clay leaflets.

If we take a stone, wet it and leave it in the sun, it will dry quickly. But if we take a book, leave it for a moment or two in water and then spread it out in the sun beside the stone, it will remain wet for a long time. The pages of the book trap the water. The same thing happens with clay. Its thin leaflets, lying one on top of the other, lock the soil water in.

Some clay soils are known to swell when wet. This is because the clay grains absorb water between the leaflets. Then, in dry weather, the water escapes by slow evaporation and the clay shrinks, causing slits or cracks in the soil as seen in **figure 182.**

When mineral salts are dissolved in water, they are carried along with it between the leaflets of clay particles and can stick to them. That is why a clayey soil feeds plants better than a sandy soil, which retains hardly any mineral salts.

This is also the reason why, in dry weather, clayey land at the bottom of a hill remains moist longer than land in the plains or on slopes.

What is the difference between a sandy soil and a clay soil ?

When clay is damp, it can be handled easily and is used for modelling objects. Pottery is made from clay. Sand is the opposite. It crumbles away at once and is no good for modelling.

There are many soil mixtures ranging from the all-sand to the all-clay soil. If the mixture contains more sand than clay, it is referred to as a sandy-clay soil. On the other hand, if the mixture contains much clay and little sand, it is called a clay-sand soil.

The magnifying glass and the microscope : two instruments to help see objects that are invisible to the naked eye

*Here are two magnifying glasses **(figure 186)**. The larger of the two, (a), makes a small object look three times larger, while the smaller glass, (b), magnifies the object eight times. Reading lenses are the commonest form of magnifying glass but their magnifying power is usually very limited.*

*A microscope is an instrument for examining much smaller objects, invisible to the naked eye and through a magnifying glass. A microscope like the one in **figure 187** magnifies objects between 80 and 500 times. Other, much more powerful, microscopes also exist.*

leaves as seen through a magnifying glass

magnifying glasses

(a)

(b)

186

microscope

the viewer

glass slide on which the tiny object to be viewed is placed

the mirror lights up the object to be viewed from underneath

187

Lesson 19

Soil structure

The **structure** of a soil means the way in which its parts fit together. Different soil structures will be discussed and illustrated in this lesson to consider which are best for farming.

Sandy soil is light. The grains of sand are piled randomly on top of one another. Air and water can easily slip into the gaps between the grains, and roots can penetrate without any difficulty. Yet, because there is no clay, this type of soil is often poor unless it contains sizeable amounts of organic matter and humus.

Figure 188 shows sandy soil covered with plant and groundnut waste.
Figure 189 shows the way the grains of sand are packed in this soil, as if seen under a microscope.

Figures 190 and 191 represent **a clay soil.** The grains are much smaller than those of a sandy soil and are formed of leaflets as was explained in Lesson 18.

This type of soil is rich in nutrients for plant life, but the roots find it hard to penetrate because there is hardly any room between the clay grains. This is called **a heavy soil.**

When sand and clay are mixed, water and roots can penetrate the soil more easily. Soil, enriched by clay and loosened by sand, can be cultivated without any difficulty (**figures 192 and 193**). This type of soil is more suited to farming than soil which is all sand or all clay, and it is even better when mixed with organic matter and humus.

188

groundnut shells

light sandy soil

190

heavy clay soil

193

soil aggregates

The **best soil structure for farming is one in which sand, silt, clay, humus and organic matter are thoroughly mixed** and together form aggregates between which water, air and roots can move freely. These are **loamy soils (figure 194)**.

In a **loamy soil**, roots find all the nutrients they need because water and mineral salts are present in generous quantities in the clay and the humus. When there is too much water, it flows away between the aggregates (little lumps) and makes way for air **(figure 195)**.

194 cloddy soil rich in humus / roots

---

196   Rootlets and root hairs in soil

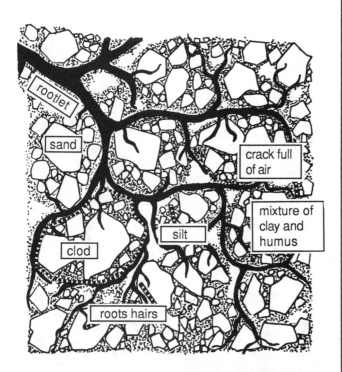

rootlet / sand / crack full of air / mixture of clay and humus / silt / clod / roots hairs

The rootlets make their way through the cracks in the soil. The root hairs feed on nutrients and water from the mixture of clay and humus that surrounds the grains of sand and silt.

Good soil (figure 196) calls for

- earth containing sand and clay ;

- a generous quantity of living matter rotting on top of the soil ;

- plenty of strong, active roots and organisms living in the soil.

By hoeing and ploughing, the farmer makes sure that organic residues are carefully mixed with the soil, thus stimulating soil life. Later on in the book, Lessons 36 and 37 explain how soil structure can be improved.

Some soil structures are bad for agriculture

Here are three kinds of soil unfit for cultivation because of bad structure :

The pebbly soil in **figure 197** is formed of **laterite** stones. This soil has been completely eroded by water and deteriorated by sun. Nothing grows here anymore. Heat has rotted the humus, and rainwater has swept the mineral salts deep into the ground. This soil is dead (Lesson 20).

The soil rich in clay **(figure 198)** is the sort you might find at the bottom of an eroded slope. The fine clay grains have been washed to the valley bottom by the water pouring down the slope (water runoff) while the coarser grains of sand remain on the plateau.

This clayey soil contains many mineral salts and is rich, but its structure is bad. Roots cannot grow in it. To make this soil cultivable, it would have to be mixed with organic waste and much sand.

The soil profile **(figure 199)** is composed of a mixture of **small stones**, **sand and clay**, with a certain amount of humus.

Yet, this soil structure is also bad because water is present all the time. This prevents the soil from breathing and chokes the roots.

To be of any use for farming, this kind of soil must be **drained**, i.e. the water must be given an outlet so that it can flow away. This means breaking up the earth and leaving plenty of cracks on the surface.

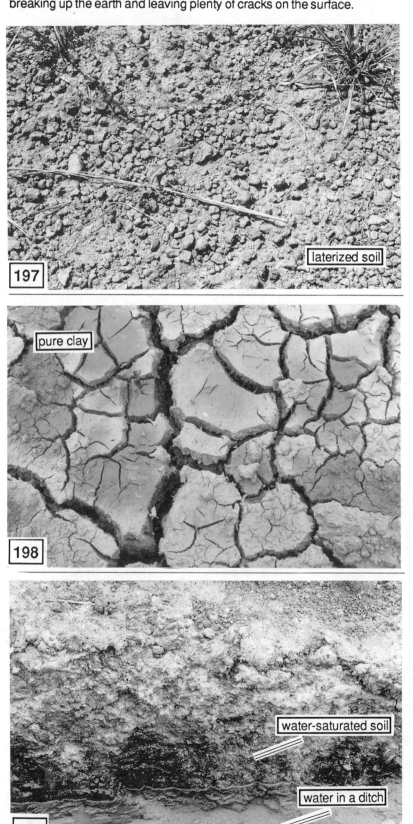

197 laterized soil

198 pure clay

199 water-saturated soil

water in a ditch

Lesson 20

🁢🁢🁢🁢🁢🁢🁢🁢🁢🁢🁢🁢🁢🁢🁢🁢🁢

Soil horizons

Lesson 18, the first in this group of lessons, deals with soil profiles. It explains how, by digging a straight-sided hole, several soil **layers** called **horizons** of different colours can be distinguished. **Their composition and structures vary.** In some, stone predominates, in others, sand and clay.

In **figure 200**, the soil horizons are clearly visible. The layer nearest the surface is made of a mixture of small stones and sand. It is full of roots, a little humus, and ashes caused by regular bush fires. The layers lower down are formed mostly of sand and clay and are compacted.

If we look closely at the picture, we notice that just under the level of the stony soil, there is quite a lot of clay, but lower down there is mostly sand.

So in this soil **profile** or soil section, four layers or **horizons**, are distinguishable.

In **figure 201**, there are three different layers, but here the stony layer is about 50 cm below the surface. Close examination reveals five soil layers which are reproduced in **figure 202**.

200

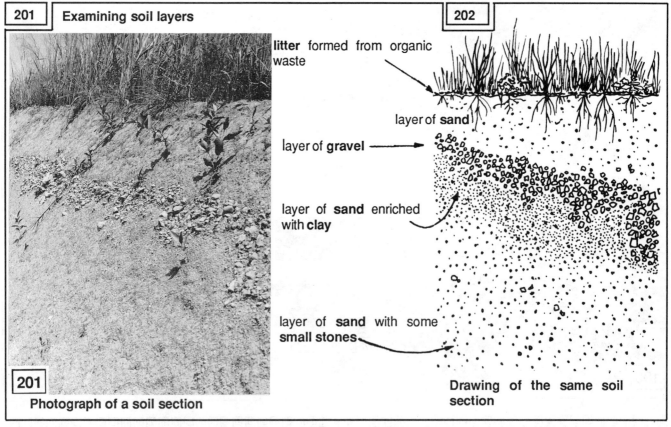

201 **Examining soil layers**

202

litter formed from organic waste

layer of **sand**

layer of **gravel**

layer of **sand** enriched with **clay**

layer of **sand** with some **small stones**

201
Photograph of a soil section

Drawing of the same soil section

Every soil horizon has its own level of fertility

Because clay and humus nourish plants better than sand and stone, it is understandable that, in a given soil, **fertility** varies from one horizon to the next, depending on soil composition.

However, other factors also affect soil fertility and make it vary from layer to layer. They include :

- water and air circulation which is affected by soil structure ;
- the circulation of mineral salts which is affected by the acidity of the soil ;
- the decomposition rate of organic residues.

The fertility of each soil horizon is its ability now and in the long-term future to produce useful crop yields.

Deep soils, shallow soils

Sometimes even a very deep hole may not strike a layer of rock or a water table. This happens in **deep soil**, and it means that the roots of plants and trees can develop far below the soil surface without meeting any major obstacles.

What is the difference between the three soil profiles in **figures 203, 204 and 205** ? In **figure 203**, the soil lies on rocks. Roots can only grow 50 to 80 cm deep. Here and there, stones break through the surface. This type of soil is **superficial** or **shallow**.

Figures 204 and 210 reveal another type of shallow soil found in valley bottoms. A layer of impermeable clay keeps the water near the surface of the soil and stops the roots from penetrating to any depth.

The opposite is shown in **figures 205 and 206** where the soil is deep and roots grow to a depth of several metres without meeting any obstacle.

Water movement in the soil is different in each case. In the first example, water infiltrates the top soil, then flows freely away through large cracks in the rock. In the second example, the water barely infiltrates the soil. It either stagnates on the surface or flows sidewards into streams. In the third example, the water slowly percolates into the earth.

In each case, water movement influences the composition of the soil and its structure.

shallow soil

parent rock

203

deep soil

205

The parent material or parent rock

The **parent material** of soil is the rock from which the soil comes. This parent material is clearly visible in **figure 203**. Climate and the

shallow soil

impermeable clay

water

204

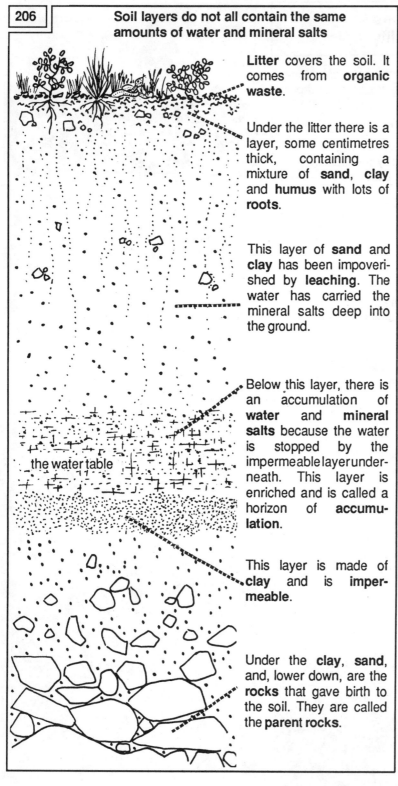

206

Soil layers do not all contain the same amounts of water and mineral salts

Litter covers the soil. It comes from **organic waste**.

Under the litter there is a layer, some centimetres thick, containing a mixture of **sand**, **clay** and **humus** with lots of **roots**.

This layer of **sand** and **clay** has been impoverished by **leaching**. The water has carried the mineral salts deep into the ground.

Below this layer, there is an accumulation of **water** and **mineral salts** because the water is stopped by the impermeable layer underneath. This layer is enriched and is called a horizon of **accumulation**.

the water table

This layer is made of **clay** and is **impermeable**.

Under the **clay**, **sand**, and, lower down, are the **rocks** that gave birth to the soil. They are called the **parent rocks**.

roots of plants together break down and disintegrate the rock. This process makes the surface soil seen in the same figure.

The parent material of soil is not always found just below the soil itself. The parent rock of soil in a valley bottom made of earth deposited there by water runoff is found at the top of the slopes.

The earth carried downstream by water runoff is called **alluvial earth** or **alluvium**.

The interaction of water and soil

What does water do when it infiltrates the soil ? Where does it go ? How and why does it change the soil ? The answers to these questions may be found by examining the way water and mineral salts travel through the soil.

Not all soil layers are **permeable**. A permeable layer is a layer which lets water through. An impermeable layer is a layer that water cannot get through.

- A layer of small stones is very permeable, water flows through it easily.
- A layer of coarse sand is permeable.
- A layer of sand and clay mixed is not so permeable.
- A layer of clay is impermeable.

When water falls on sandy or stony soil, the water flows away between the soil particles because these soils are permeable. However, the water may meet an impermeable clay layer further down. When this happens, the water stops flowing and forms a reserve of water in the ground.

If rainfall is high, the water table level increases. If rainfall is low, the water table gets less. So, the depth of the water table depends on the depth of the impermeable layers of soil.

Pumped water is drawn from the **water table**, as diggers of wells know.

When water flows through soil, it dissolves the mineral salts and carries them away. This is called **mineral leaching**. But the process is a variable one. It may dissolve large quantities of mineral salts near the surface and get rid of them deep in the soil at quite a distance from where the salts were dissolved.

Whether the leaching process carries the salts away from a soil layer or deposits them on a soil layer, the water changes the composition of the soil horizons, as can be seen in **figure 206** which represents deep soil.

The rainwater infiltrating this soil carries the mineral salts of the superficial soil horizons to deeper horizons and especially to the soil horizon where the water table lies. **As a result, the horizons near the soil surface are deprived of their mineral salts. They are leached. But, at the same time, the deeper soil layers are enriched because mineral salts accumulate there.**

Now, bearing in mind that most cultivated seasonal crops use only the top soil layers for their growth, we can understand why soil leaching is so harmful for farming. The mineral particles most valuable for soil fertility are removed. That is why it is always wise to combine plants using the surface soil horizons with others whose roots tap the mineral salts in the layers where they accumulate.

For the same reason, growing trees along with seasonal crops is worthwhile because trees usually root deeper than herbaceous crops. They bring the mineral salts up from the deeper layers to the surface, in the form of waste such as leaves, stalks, branches.

Here is a second type of deep soil found on plateaus or on slopes (figure 207).

The parent rock is on sloping ground. Above it, there is a layer of impermeable clay also on a slope. The water infiltrating the soil first flows vertically and then sideways. So the mineral salts are leached away with the water off the field and maybe even right off the settlement. In this type of soil, we find impoverished or leached horizons and no enriched horizons at all.

In this particular case, soil leaching can be fought by ensuring that lots of roots take up the mineral salts in the top soil layers before the water has removed them.

Figure 207 shows why spring water and water in valley bottoms is sometimes salty. The water has collected the mineral salts from the soil layers through which it flowed.

In the shallow soil illustrated in **figure 208**, the rainwater passes quickly through the sand and clay, then flows freely away through the cracks in the parent rock. The amount of soil water available for the roots is negligible because there is only a thin layer of clay and sand.

Mountain farmers often make little ridges to ensure more soil for plant growth and to store up a little more moisture in the soil mass **(figure 209).**

Farming is always hard work on this kind of soil because of the risk of drought. When evaporation is high, water escapes into the

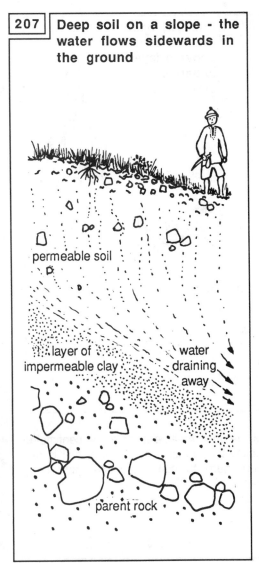

207 Deep soil on a slope - the water flows sideways in the ground

permeable soil

layer of impermeable clay

water draining away

parent rock

208 Shallow soil on sloping rock

ground

parent rock

mounds

direction of slope

209

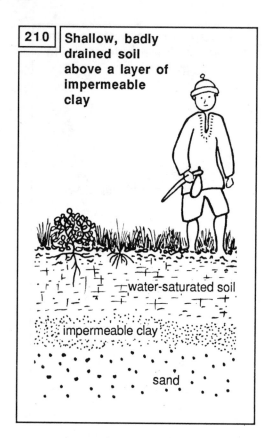

210 Shallow, badly drained soil above a layer of impermeable clay

water-saturated soil

impermeable clay

sand

211 laterite hard pan

atmosphere, but there is no store of water under the soil surface to make it moist again.

Unlike the soil in **figure 209**, the shallow soil in **figure 210** is unlikely to suffer from drought. It is water-saturated all the time because of the impermeable layer a few centimetres below its surface. This type of soil is also very hard to cultivate since most crops cannot live with their roots permanently in water.

Soil laterization is disastrous for farming

Laterite is red rock usually weathered into small red pebbles **(figure 212)** or into larger stones or hard pan **(figure 211)**. **Soil laterization** is typical of degraded soils in tropical climates. It comes from very intense leaching of the upper soil layers by heavy rains and hot sun.

When soil is laterized, as shown in **figures 211 and 212**, it is useless for farming. It has lost all the mineral elements needed for cultivated plants, becoming at times a kind of rock that roots cannot penetrate.

Laterite generally occurs when the soil has been badly tended. Such poor husbandry may include uncontrolled bushfires **(figure 213)**, land clearance and untimely farming practices with no regard for the environment, and intensive cropping which exploits the soil without restoring it.

212 laterite hard pan

213 burnt trash

laterite

plain

To combat laterization, the farmer must at all costs avoid clearing the ground too drastically, and do all he can to cover it with cultivated or uncultivated plants. Life in the soil fights laterization.

Points to remember

The study of the soil is a complicated science and only some aspects of it are covered in these lessons. We need to remember that :

■ Soil is composed of different mineral and organic fractions.

■ These fractions are assembled in such a way as to give the soil its structure.

■ The mineral salts found in the different soil fractions nourish the plants. Soil fertility depends on the amount of mineral salts found in the soil fractions, as well as other factors. The fertility of soil is a measure of its suitability to support crops.

■ Earth is composed of layers (horizons) lying one on top of the other. Each layer has its own composition, permeability and fertility.

■ Water flows more or less freely through the soil horizons. When water meets impermeable layers, it can form groundwater reserves with upper limits called water tables.

■ Water dissolves mineral salts and removes them. Water movement through soil layers is the cause of this leaching, but water movement can also cause the accumulation of mineral salts or bring them closer to the roots.

Lesson 21

Water movements

---

214 | **Main water movements in the ground**

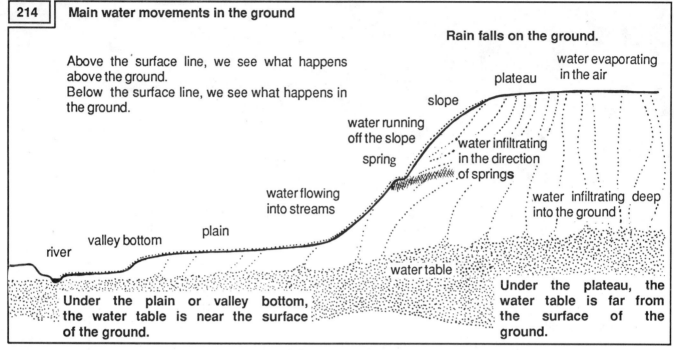

Above the surface line, we see what happens above the ground.
Below the surface line, we see what happens in the ground.

Rain falls on the ground.

water evaporating in the air

plateau

slope

water running off the slope

spring

water infiltrating in the direction of springs

water flowing into streams

water infiltrating deep into the ground

plain

valley bottom

river

water table

Under the plain or valley bottom, the water table is near the surface of the ground.

Under the plateau, the water table is far from the surface of the ground.

---

215 | **The soil structure is damaged by the beating of raindrops (rain splash)**

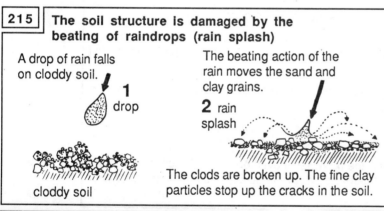

A drop of rain falls on cloddy soil.

1 drop

The beating action of the rain moves the sand and clay grains.

2 rain splash

cloddy soil

The clods are broken up. The fine clay particles stop up the cracks in the soil.

Water travels a great deal and rarely on its own. Water is also the greatest transporter in the world. Water moves rocks in torrents, tears up sand, sweeps away clay and silt, carries off the mineral salts in the soil. **Figure 214** is a study of water movements and how they affect the soil.

Rain falls on the ground

Raindrops fall from a height. They beat the ground and the leaves in their path. If they hit the ground directly, they move the grains of sand, clay and humus. They break down the little clods of earth and sort them into fine and coarse parts **(figure 215)**.

Water barely infiltrates a sorted soil because the fine particles clog up the cracks between the coarser grains. In other words, the top soil layer stops water penetration **(figure 216)**.

Rainwater falling on the ground can move in three different ways :

- it can run off ;

- it can penetrate the ground ;

- it can evaporate.

216

217

sheet erosion in
a field of millet

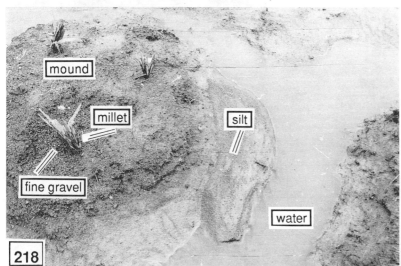

218

mound

millet

silt

fine gravel

water

220

gully
erosion

Rainwater runs off sloping ground

On gently sloping ground, the water begins by flowing on the surface. This is called **runoff**. At first the water runs off **in sheets**, it flows all over the ground (**figure 217**). After that, it collects in the drains.

Water running off in sheets has not much energy or force. It only removes the finer, lighter soil particles, leaving sand and gravel behind.

The mound in **figure 218** was photographed during a heavy downpour. The top of the mound is composed mostly of sand and small gravel, the bottom is covered in silt and clay. The water flowing away in the furrow is tinged with the red clay it is sweeping along.

If the watercourse is not barred, the water moves faster and faster with increasing force. Streams form with swift waters that tear away sand and uproot young plants (**figure 219**).

219

If nothing is done about it, water runoff carries away everything in its path and forms **gullies** (**figure 220**).

Gully water is lost to the rural settlement. Instead, it fills up rivers, lakes and seas. The loss to the farmer is all the more serious when rainfall is low and irregular. An even more serious loss is the large quantity of fertile soil which leaves the land for good.

Figure 225 illustrates the huge quantities of soil which may be lost through runoff.

Nothing can grow on the barren land in **figure 221**. It has dried up because rainwater flowed away into gullies instead of penetrating the soil.

The land cannot be cultivated. The fine particles of clay and silt have been removed by runoff and burning sunshine has beaten down on the bare, unprotected soil. Gullies make the terrain difficult.

Erosion by water is the worst calamity which can befall a rural settlement.

221

Water runoff and erosion endanger agriculture

Mountainous, deforested land is shown in **figure 222**. There is fallow land in the foreground with **wild plants** growing up through the maize stalks left from the previous harvest.

In the background, long ridges have just been tilled on the hillside. Most of them run with the slope though a few are perpendicular to the hill and cut across the downward furrows.

But what is actually happening on the **ridges** and in the **furrows** ?

If we look at the potato plant growing on the side of the ridge in **figure 223**, we see that rain splash and water runoff have stripped the roots,

222

223

224

and the potato is bare of soil. If there is more rain, the plant will become dislodged.

A few metres further down, furrow drains that will turn into gullies are apparent. When it rains, the water running off the ridges flows faster and faster into the furrows instead of penetrating the ground **(figure 224)**. The rain removes large quantities of earth by a process called **pluvial** or **rain erosion**.

The future of agriculture on the hill in **figure 222** is in the balance, because the soil is running off the slopes. When the water streaming downwards has carried off all the loose earth, nothing will remain but a hard pan of stones unfit for cultivation. The topsoil and stones will clutter fertile land below.

Why is the soil running off these fields ?

Firstly, because it is **beaten by the raindrops** that break up the clods so that the flowing water carries off the finer particles.

Secondly, because these fields are on a steep slope and **water runoff moves quickly**.

Thirdly and perhaps worst of all, nothing is being done to slow down the **runoff and force the water to penetrate the ground**.

Reducing rain splash, slowing down runoff along the slopes are the two main steps needed to save the fields for farming.

How much earth is carried away by erosion ?

Agriculturalists in the Ivory Coast compared the erosion of bare lands with the erosion of lands covered with vegetation. All the lands studied lay on 7% gradients. (A 7% gradient means that, as you go down the hill, the ground is 7 metres lower for every 100 metres covered.) Over a whole year, water runoff at the bottom of every slope was collected. The specialists then separated the earth and water and weighed the two to find out the difference between the fields. **Figure 225** illustrates the result of their work.

225	**Land erosion**

(adapted from E. Roose, 1973)

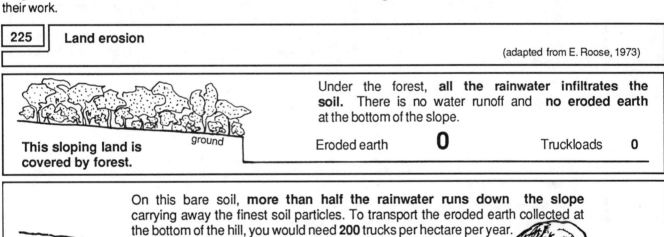

Under the forest, **all the rainwater infiltrates the soil.** There is no water runoff and **no eroded earth** at the bottom of the slope.

This sloping land is covered by forest.

Eroded earth **0**   Truckloads **0**

On this bare soil, **more than half the rainwater runs down the slope** carrying away the finest soil particles. To transport the eroded earth collected at the bottom of the hill, you would need **200** trucks per hectare per year.

This land is denuded (stripped of vegetative cover).

200

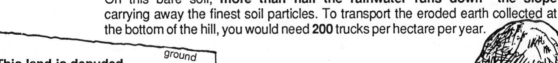

Water runoff on the denuded areas, with its earth content, **is stopped** by the anti-erosive strips. Little water runs to the bottom of the slope. To transport the eroded earth collected at the bottom of the hill, **1** truck only would be enough.

This land is managed in anti-erosive strips planted with various associated plants.

1

These results relate to a fairly gentle slope with a gradient of only 7%. But on steeper slopes the amounts of water and earth flowing off bare fields can be far greater.

To sum up :

■ **Good soil cover,** such as that provided by forest or grass fallow, stops erosion.

■ **A bare field runs a high risk** of erosion because much of the rainwater leaves the field and

rocks denuded by erosion

226

carries away large quantities of earth.

■ On cultivated slopes, the ground must be cleared at the beginning of every growing season and is consequently exposed to the risk of rain erosion. **Anti-erosive strips of trees and associated plants stop water runoff before it endangers the land.**

If no steps are taken to halt the loss of rainwater and soil on the rural settlement in **figure 222**, the farmers will abandon their land because all that will remain will be stone and rock. This is what happened to the land photographed in **figure 226** - water runoff has carried away almost all the soil.

It is disastrous for farming when a settlement loses its water and soil.

Water infiltrates the soil

When soil is permeable and water runoff insignificant, water seeps down between the grains of sand and clay. It may also swell the clay particles as explained in Lesson 18. Infiltrating water may travel a short distance or it may travel far. Water travels a short distance when it remains near the soil surface. It travels far when it penetrates deep into the ground. Everything depends on the obstacles it meets in the soil layers. Infiltrating water removes the mineral salts found in the upper layers to the deeper soil layers.

These two aspects of water infiltration have to be examined carefully.

Infiltrating water on its way through the ground

Infiltrating water meets obstacles. **Figure 227**, representing a hill section, illustrates this point.

On the left, the infiltrating water travels to a great depth (a). It accumulates on the aquifer lying on a layer of impermeable clay (b). If a well is dug (c), the aquifer water will flow slowly towards the well. The water moves sideways.

In the middle of **figure 227**, under the slope, there is a layer of rock and clay preventing deep infiltration (d). The water flows to the right and emerges at the spring.

This figure demonstrates why wells and bores often get their water supplies from quite a distance away.

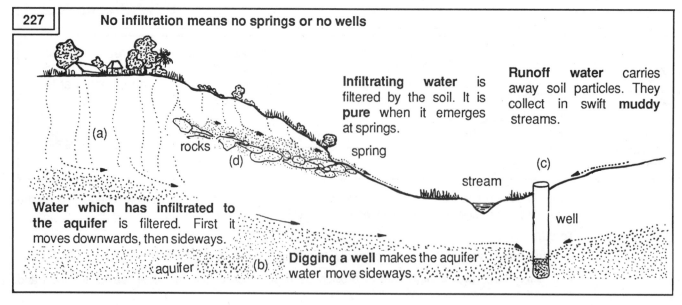

227   No infiltration means no springs or no wells

Infiltrating **water** is filtered by the soil. It is **pure** when it emerges at springs.

Runoff water carries away soil particles. They collect in swift **muddy** streams.

(a)

rocks

(d)

spring

(c)

stream

well

Water which has infiltrated to the aquifer is filtered. First it moves downwards, then sideways.

aquifer   (b)   **Digging a well** makes the aquifer water move sideways.

Infiltrating water carries the mineral salts along with it

Water penetrating to a depth carries away the mineral salts which are the wealth of the soil for farming. The mineral salts accumulate deep in the soil. The superficial layers in which the plants should find nutrients are leached (washed) and there is accumulation of minerals in the deep layers.

A leached soil is poor in plant nutrients, so everything must be done to combat leaching. For instance, the activity and variety of roots in the top layers can be increased in order to capture the mineral salts before they are carried away by the water. The soil can also be enriched by humus that retains the mineral salts. Mineral salts in the form of commercial fertilizers can also be carried away by leaching.

Soil evaporation and plant transpiration

Water in the soil, especially infiltrating water, is used by plants. They need the water to draw on the mineral salts that are then transformed into organic matter with the help of air and light.

Plant water may be transpired, that is, the leaves and stalks send the water back into the air in the form of droplets or vapour. The stomata (Lesson 23) allow this process to take place.

Soil water can also evaporate directly without passing through plants. This happens when water leaves the soil and combines with air.

The process of water leaving the ground in the form of water vapour, either directly or through plants, is called **evapotranspiration**.

Evapotranspiration can be observed in many ways :

■ the soil is dried up by wind and heat ;

■ plants wither when they have transpired a lot and when there is not enough water in the soil (Lesson 23) ;

■ **mist** lies over the fields and forest in the morning. As the air is still cool, water evapotranspired during the night is found in the air in the form of tiny, visible droplets. Later in the day when the sun has warmed the air, the droplets become smaller and smaller, and finally become invisible ;

■ **dew** covers the leaves in the morning. Water transpired by the plants in the cool night forms drops on the leaves before falling to the ground or evaporating in warmer air during the day ;

■ sometimes a light salty **crust** forms on the soil surface as the sun gets warmer and the water evaporates. As it rises to the surface, water brings the mineral salts with it and then abandons them as it evaporates. The dried-up salts then form the salty crust.

Water in valley bottoms

valley bottom drained

banana grove

228

In valley bottoms, water is always present on, or a few centimetres below, the surface. Marshlands are lands where water running off plateaus or nearby hills accumulates and stagnates.

As runoff water contains a lot of clay, silt and mineral salts, valley bottoms and marshes are often quite fertile though not easy to cultivate. This type of land is only suitable for plants whose roots can live in water, such as some varieties of rice. Most other kinds of cultivated plants do not adapt to this kind of land. So, marshy land on rural settlements must be drained and dried out before it can be cultivated.

Draining land means letting the water saturating the soil flow away.

229

Here are two examples of drainage. In **figure 228**, a large stretch of marshy land has been drained. Long outlet rills channel the water away to the stream in the middle of the photograph.

Figure 229 shows in more detail how a valley bottom was drained by forming strips of cultivated land. The strips were planted with sweet potato and maize. Rice was sown in the channels between the strips **(figure 230)**.

maize

maize and sweet potato

230

rice

Flooding

Flooding is a scourge in some countries. Floods are often due to deforestation. When hills and mountains are deforested, rainwater begins to run off the soil instead of penetrating it. The water then flows into the valleys and accumulates there instead of recharging the water tables.

This is why valley flooding often goes hand in hand with drought on plateau lands and on slopes.

One way of fighting drought and flooding is to halt runoff and force rainwater to infiltrate the ground.

Notes

Lesson 22

Altering water movements for the benefit of farming

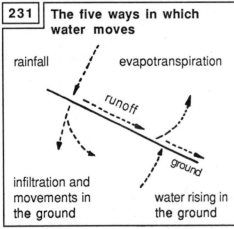

231 | **The five ways in which water moves**

rainfall

evapotranspiration

runoff

ground

infiltration and movements in the ground

water rising in the ground

Figure 231 lists the five ways in which water moves. The farmer should understand these water movements and control them as far as possible. Admittedly he cannot decide the amount and distribution of rainfall on his farm but he can exercise control over runoff. When runoff is under control, infiltration and evapotranspiration are improved. This lesson will study different ways of :

- **fighting rainsplash and incipient runoff by providing vegetative cover ;**

- **using plants to slow down runoff ;**

- **fighting runoff by mechanical means ;**

- **promoting rainwater infiltration in the soil ;**

- **reducing evapotranspiration.**

The role of plants

The role played by plants in protecting the soil and fighting erosion is described in **figure 232**.

---

232 | **Cultivated plants protect the soil from erosion, especially when seeded early**

Leaves absorb the shock of the falling raindrops and protect the ground from the erosion these drops would have caused.

Living plants and their residues store up plenty of water. The more abundant the supply of living matter, the more drought-resistant the environment will become.

Water evaporated from the soil into the air is useless. **Water transpired by plants helps to form living matter.**

Under the foliage, the rainwater can penetrate easily. The rate of evaporation is low because the ground is protected from the sun's rays.

Abundant roots make water penetration easier.

Water stored in plants remains available to the farmer. Soil water evaporated by direct sunlight is completely wasted.

The farmer who **associates plants wisely** saves water and can use it advantageously to produce the living matter of his harvests.

Rainsplash and soil degradation

How can soil degradation be prevented?

Soil with good vegetative cover is not beaten by rainsplash, but, on bare ground, direct rainsplash destroys the topsoil structure and leads to runoff (Lesson 21).

Figure 233 proves this point. Note the steep slope in the foreground, with gullies and landslides in two places. In the background, on the same slope, the ground is protected by forestation.

In **figure 234**, the farm lying on a sharp incline is protected from the risk of erosion and runoff by the small wood situated behind it upslope. To the right is a landslide caused by runoff. The farmer and his family are shielded by the wood planted for the purpose of guarding against such landslides.

234

235

236

The effectiveness of forestation in preventing runoff and erosion is demonstrated again in **figure 235**, where runoff resulted in a big landslide (centre). Up the slope, above the landslide, the fields are quite bare ; there is not a single tree in sight.

But go up to the wood, top left, and come straight down the slope from there. No landslide is to be seen although farming practices are the same as those above the landslide.

We may conclude that the cover afforded by a forest or planted trees prevents erosion and runoff. However, even thinner vegetal cover can be as effective. **Figures 236 and 237** bear out this point. In **figure 236**, soil on a slope is beaten by rainsplash, though, under the bean leaves, it is not quite so bad. In the same field, only a few metres away, the leaves of the sweet potato crop provide a perfect umbrella against rainsplash.

237

Careful intercropping **(figure 238)** ensures soil cover as good as that afforded by tree plantations.

During the sowing season the farmer should, in his own interest, strike a balance between :

■ **the cultivated plant or plants cropped mainly for their produce ;**

238

239

240

- **the cultivated plant or plants able to protect the soil quickly so that the beating action of the first heavy rains of the season, always the most dangerous, can be minimized.**

Beans and maize are intercropped in the field in **figure 239**. The maize plantlets give very little soil cover whereas the young bean leaves already protect the soil well.

In another field **(figure 240)**, full soil cover and protection against erosion and runoff is provided by a forage plant.

Living plants are not alone in supplying soil cover against erosion. Plant materials such as straw, branches and dead leaves also do a good job. The cabbage plantlet in **figure 241** is bedded down with straw, a cover almost as efficient as the leafy plants in **figure 240**.

241

Three kinds of vegetal cover protect the soil in this field **(figure 242)**. They are maize, cotton shrubs and straw spread on the ground.

All these plants and plant materials provide good soil cover.

242

maize

cotton

Soil cover cannot be provided in the fields at all times of the year. More particularly, it is hard to supply cover at the beginning of the rainy season, just when the rains are very intense and most dangerous.

243

anti-erosive strip

millet

However, steps can be taken to slow down runoff and force the water to deposit the soil particles it has picked up. **Anti-erosive strips** are planted to this effect.

Anti-erosive strips are permanent stands, pure or mixed, of herbaceous plants, trees and shrubs designed to slow down runoff and make the water penetrate the ground. The strips are planted along the **contour lines**. This means that a person walking along these strips always stays at the same height. He neither goes up nor down. It follows that anti-erosive strips planted on the contour lines cut across the runoff water which always flows down and away with the steepest incline unless it meets an obstacle on the way.

Figure 243 is an example of an anti-erosive strip with herbaceous plants. The strip is dense enough to slow down the sheet of water flowing over the fields of millet during the heaviest rains.

Once it has slowed down, the water penetrates the soil more easily and increases soil water reserves to the advantage of the millet.

Here is another field of millet pictured in the rain **(figure 244)**. The field is divided by herbaceous anti-erosive strips and the sheet of water flows away on successive levels. The levels are clearly visible because the anti-erosive strip in the middle is damaged. An anti-

244

sheet erosion

broken anti-erosive strip

millet

245

forestation

terraces

river

erosive strip must be **sufficiently wide** if it is to be really effective.

The sharper the incline, the closer the strips should be planted to one another. Close spacing is needed because the runoff must be stopped before it begins to gather speed and to form little channels. Water in channels flows much more quickly than water in sheets and, with increased speed, the water has greater force to pick up and transport soil particles.

The land in **figure 245** is well protected against runoff. There are trees on the hilltop and the lower slopes are divided by many anti-erosive strips planted very close to each other.

After a few years, this kind of land management leads to permanent horizontal terraces on which runoff is negligible.

246

The terraces **(figure 246)** were formed by planting anti-erosive strips of *Setaria* grass. Also, tillage helped to flatten the land on two different levels.

Mechanical ways of fighting runoff and erosion

Mechanical ways of fighting erosion exclude the use of plants. The aim is to reshape the soil surface by moving earth and stones to slow down or stop runoff. The following figures illustrate some mechanical ways of fighting erosion.

The way ploughing is done is important. In **figure 247**, the furrows have been carefully ploughed along the contour lines with the result that the water flowing on the ridges is slowed down in the furrows and has time to penetrate the ground. The ploughing pattern in this case is the exact opposite of the one practised in **figures 222 and 224** (beginning of Lesson 21).

247 ridges ploughed along the contour lines

Tie-ridging wil ensure even better water penetration. The water is trapped between the ridges by the partitions or mounds of earth that separate the furrows. These are called tie-ridges.

Extended **earth banks** can be built in cases of slow sheet erosion like the kind seen in **figure 217**, Lesson 21. The banks are low, varying in height between 30 and 60 centimetres, but the crest or summit must always be level, exactly the same height all along the bank, not going up or down. If the crest is uneven, water will go over the top and burst the bank.

248 tie-ridges

249 | ploughed furrows perpendicular to runoff

This type of earth bank is illustrated in **figure 249**. In heavy rain, the water flows from left to right. The presence of the bank means that the water is trapped in the fields on the left and runoff is prevented. The furrows run with the dike, so the ploughing pattern also slows down the runoff.

250

The purpose of earth banks is clearly demonstrated in **figure 250**. A circular bank, 30 cm high, was built on a gentle slope. The rain which has just fallen is trapped in the centre of the circle while outside the circle the same rainwater has already flowed away to nearby streamlets.

In Burkina Faso in Mossi country, two water barriers are used in the fields photographed in **figure 251**. In the centre, stones in straight lines are set at right angles to water runoff. In the foreground, branches and plants slow down the runoff.

Arranging stones in straight lines is an entirely mechanical way of fighting runoff. In fact, a good look at the picture shows that the water barrier is rather weak and can be easily broken by the force of the water running off in sheets.

On Santiago Island in Cape Verde, rainfall is very low and irregular.

The country is mountainous and, when it rains, all possible steps must be taken to ensure water penetration. Some of the anti-erosive measures in force are pictured here.

Take for example the landscape in **figure 252**. Low stone walls have been built along the slopes to check erosion. They are called **stone banks**. In this arid Cape Verde region, they are the best way to start fighting erosion because it would be very hard to grow anti-erosive plants on such drought-ridden land.

A big problem is how to maintain anti-erosive works. Goats, for instance, have broken through the stone bank in **figure 253**. When it rains, water will sweep through the hole and be wasted for farming, despite the enormous amount of work that went into constructing the embankment.

stones in line

251

stone banks

252

253

broken stone bank

Figure 254 illustrates another way of controlling runoff. A series of small check dams were built, one below the other down a gully. Realistically, these anti-erosive works will not achieve much because they have been sited too far down where runoff water is already flowing quite fast. It is better to contain erosion at the very start of runoff, i.e. in the fields themselves. If runoff had been stopped higher up, the gully would be empty.

When mechanical means alone are used to fight erosion, the results are not a hundred percent satisfactory and risks are involved including the danger of seeing earth banks, walls, embankments and ridging give way. So it is always worthwhile to back up mechanical measures with biological resources. For instance, the rough-and-ready wall in **figure 256** has been planted with aloe. A two-fold effect is achieved by this strategy. The stones give body to the wall, while the plants hold down the earth and mix in with the stones.

254

small-scale damming

256

aloe

255

terraces

gully

When more resources are available, large-scale anti-erosive works can be undertaken such as horizontal terraces delimited by stone walls. This type of management is used in Cape Verde Islands as shown in **figures 255**

106

and 257. The first photograph of the terraces was taken from the top of the hill. A large gully caused by erosion is seen in the background **(figure 255).**

The same terraces were then photographed from below **(figure 257).** Low stone walls bound the rectangular plots which can easily be used for growing vegetables. The plots are irrigated. In this dry region, there would obviously be no point in constructing elaborate field works if irrigation water was not available.

terraces

pawpaw

257

> **The fight against erosion, undertaken collectively for the whole rural settlement, is the one most likely to give the best results**

It is never a waste of time to combat runoff and soil erosion on one's own land, but individual measures are not enough to divert runoff away from a farm. Hence the need, generally speaking, to organize the fight on a collective basis. Erosion is a problem facing the entire community and, as such, must be discussed by all the concerned farmers. The collective approach also encourages good community water conservation.

---

258 **The layout of a hillside in anti-erosive strips**

Here the soil is well protected by a **cluster of trees and bananas** on the hill top.

The **stands of trees and shrubs** are composed of many plant varieties. Some are of direct use for human consumption such as pawpaw, pigeon pea and banana. Others are there to protect and improve the soil. Forage crops have been interplanted with the trees.

Forage crops and trees halt rain runoff on the ground. The earth running off the slope is stopped all along the line of herbaceous plants and trees that make up the anti-erosive strips. Terraces are formed in the long run.

These fields are for **seasonal crops**. The ground is cleared at the start of each growing season.

On these fields, the rotation of cultural associations and **fallow land** is practised.

The presence of stands of **trees and permanent plants** benefits the springs situated lower down the hill.

Figure 258 shows the layout of some hills in Rwanda agreed after lengthy, detailed discussions between all the farmers. Because rainfall in the region is high, plants were the obvious means to choose in the fight against erosion.

The farmlands on these slopes are subdivided into many sections corresponding on the one hand to cropping fields and on the other to anti-erosive strips. These strips were planted along the contour lines to cut across the watercourse at right angles and bar the way to runoff.

In the cropping fields, the ground is cleared at the start of each growing season and the risk of erosion is therefore high. In contrast, the ground under the anti-erosive strips has complete cover.

The cropping fields are planted with a wide variety of plants, usually in association. They include beans, sorghum, maize, soya beans, sweet potatoes, taro, yams, groundnut.

In the sections set aside for seasonal crops, the farmers can grow what they like on their own plots.

The anti-erosive strips, one to three or four metres wide, are composed of perennial plants (trees, shrubs and herbs) almost all of which are of direct use to the farmer. We find growing here banana, fruit trees, felling trees (for timber or firewood), medicinal herbs, legumes as soil fertilizers, fodder grasses. (See section on hedges in Lesson 17.)

This layout of terrain proves that **the fight against erosion and the pursuit of agricultural production can be combined successfully**.

Of course this sort of arrangement implies that all the farmers concerned are willing to support a layout plan and then carry it out under the supervision of a land authority recognized by the community.

Some farming practices to limit evaporation

- **Hoeing** consists of breaking up the top soil crust by using a hoe. When the crust is broken up, the water in the deep soil layers does not evaporate so easily. This is because the narrow canals through which the water rises into the air have been smashed and are thus protected from sun and wind **(figure 259)**.

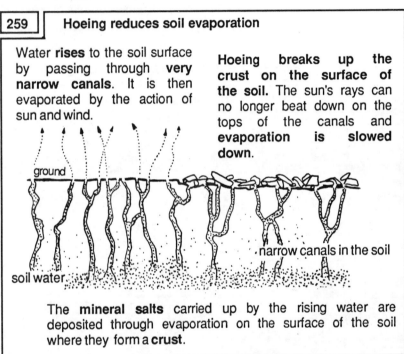

259 | **Hoeing reduces soil evaporation**

Water **rises** to the soil surface by passing through **very narrow canals**. It is then evaporated by the action of sun and wind.

Hoeing **breaks up the crust on the surface of the soil.** The sun's rays can no longer beat down on the tops of the canals and **evaporation is slowed down.**

ground

narrow canals in the soil

soil water

The **mineral salts** carried up by the rising water are deposited through evaporation on the surface of the soil where they form a **crust**.

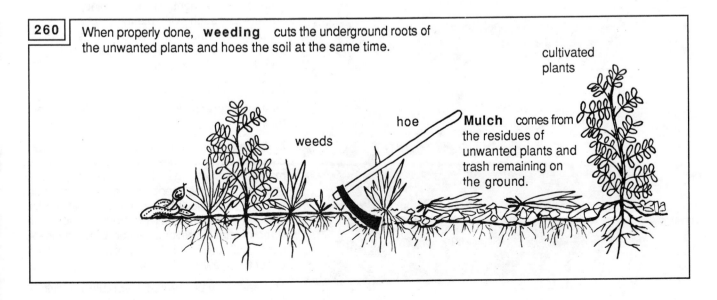

260 | When properly done, **weeding** cuts the underground roots of the unwanted plants and hoes the soil at the same time.

cultivated plants

hoe

weeds

Mulch comes from the residues of unwanted plants and trash remaining on the ground.

■ **Weeding** is an operation that involves cutting the unwanted plants that compete with cultivated plants for nutrients. Weeding is done with a short-handled hoe or a weeder **(figure 260)**.

There are many reasons why weeding reduces evaporation :

□ once cut, weeds stop transpiring and stop taking up soil water ;

□ when weeding takes place, it is really a combined weeding-hoeing operation. This is so because the best way of eliminating weeds is to cut their underground roots. The cutting action automatically loosens the top soil layer as in hoeing ;

□ lastly, the plants that have been dug up dry out on the ground and form a protective cover called **mulch**.

Some farming practices for promoting rainwater infiltration

Many farming practices promote rainwater infiltration.

Lesson 36 shows how humus can be produced to improve soil structure. All the practices that improve soil structure and increase the amount of available humus promote infiltration.

In this lesson some practices to prevent runoff have been studied. These methods are also helpful in improving infiltration. Hedgerows have already been discussed in this connection (Lesson 17).

Ploughing is another way of encouraging water penetration. By breaking up and loosening the soil, it makes cracks through which water infiltrates. **Figures 261 and 262** can be compared, to see the advantages of ploughing.

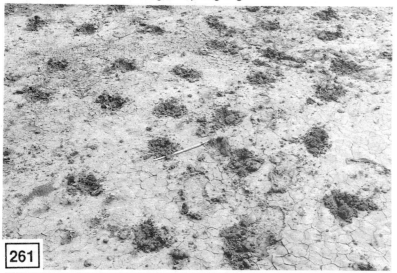

261

262

surface soil clogged by erosion

surface soil opened up by ploughing

The land in **figure 261** has been badly damaged by erosion. A thin layer of clay covers the surface soil as a result of raindrop splash and runoff.

The woman-farmer used a stick or a small implement to make little holes for sowing millet seeds. But for lack of strength or perhaps for lack of the right implement, she did not break up the clay crust properly.

When the rains come, the water will rapidly fill up the little holes and will find no infiltration channels in the ground.

In **figure 262**, the ground on the left is just as clogged up as the surface in the first photograph **(figure 261)**. In contrast, the ground on the right has been ploughed. The surface is rough with plenty of channels for water infiltration. When the rain beats down on this ploughed land, the soil will be sorted again and the cracks will gradually close up. So, although ploughing paves the way for water penetration, infiltration channels can block up again if no precautions are taken to keep them open.

Mulching is a soil cover made of straw, chopped-up branches or other protective materials such as jute sacking, plastic sheeting, gravel, pebbles and so on. The benefits of mulching can be seen in **figure 263** - it conserves soil moisture because the layer of straw prevents water evaporation. A good soil structure is therefore maintained under the straw.

This farmer has created favorable conditions

for the germination of vegetable seeds by spreading straw on the seedbed. Moisture and cool temperatures are maintained in this way. The same cannot be said of the uncovered soil to the right of the mulch.

The cultivators **(figure 264)** are preparing their fields for a millet crop by a weeding-hoeing operation while leaving some mulch on the ground.

Before sowing the millet, they can choose between burning, digging-in or spreading the mulch in rows between the lines of millet. Digging-in or leaving the mulch on thé ground will probably give better results than burning.

The soil profile **(figure 265)** is exposed to the sun. The mulch providing cover on the

mulching

264

straw

moist soil

dry soil

plantlets

263

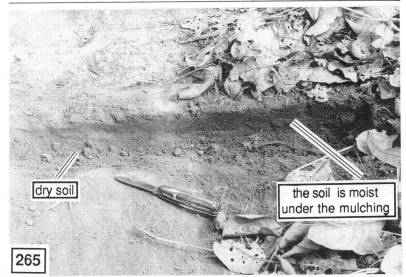

dry soil

the soil is moist under the mulching

265

right is formed of residue from trees. The soil on the left is bare. The difference in soil moisture is shown up by the colour of the soil.

Notes

Lessons 23 to 27

🔲🔲🔲🔲🔲🔲🔲🔲🔲🔲🔲🔲🔲🔲🔲🔲🔲🔲🔲🔲🔲🔲🔲🔲🔲🔲🔲🔲🔲🔲

Plant environment

The plant environment is composed of the following elements : the land where the plants grow, with its contours and composition (mountains, plains, lakes,...), the climate (light, heat, humidity, winds), other plants. Every region, every district, every plot of land has its own climate. Some plants adapt to the climate, but others do not because good growth conditions are missing.

Water and light are the most important elements for plant development. The amount of light and water

---

266   **What are the different parts of plants used for ?**

3 LEAVES

- Leaves are the **cooking pots of plants,** the places where the plant ingredients undergo the greatest modifications. In the leaves, **the plant transforms the unelaborated sap, composed of water and minerals salts, into elaborated sap composed of sugars, fats and proteins** that form the living matter of the plant.

- Plants use air and light to carry out this transformation (**figure 267**).

- The leaves of plants **breathe.** They take up air from their environment and breathe air back into the environment.

- Leaves **transpire.** They send water vapour into the air.

4 FLOWERS, FRUITS and SEEDS

- The **seed** is the part that **reproduces a plant.** Fruit is formed from the flowers. **The fruit nourishes the seeds.**

- Flowers, fruit and seeds are nourished by the elaborated sap sent out to them by the leaves, through the shoots, branches and trunk.

> **Start here , follow the arrows**

2 SHOOTS

- Stems, trunks and branches are **aerial or aboveground parts of the plant.**

- Stems (stalks) form the **framework** of the plant. They support the leaves, flowers and fruit. **Every plant has its own framework or structure.**

- Like the roots, **stems, branches, trunks and long shoots contain slender channels.** Some of these channels bring the mineral containing sap from the roots to the leaves. Others bring the sugar-enriched sap, prepared in the leaves, to the flowers, the fruit and the roots.

- Like roots, stems can **store up food.**

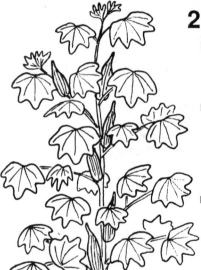

ground

1 ROOTS

- Roots form the **underground parts of plants**.
- Roots **fix the plant in the soil.**
- They **absorb soil water and nutrients** through the root hairs.
- They also **cast off waste into the soil.**
- They sometimes form a food store because root crops or tubers, like cassava, yam, etc. can be stored in the ground.
- **Slender channels run through the roots.** They carry the water and nutrients taken up from the soil to the stems.

that plants need varies with each stage in their life cycle. The climate may be favourable to plants at every stage of growth or it may be beneficial only at certain times.

The first lesson in this group (Lessons 23 to 27) deals with the main phases in the life of a plant. From there we move on to study the needs of plants and their development in time :

- the amount of light needed (Lesson 24), how to use light for agriculture (Lesson 25) ;
- the amount of water needed : this involves a study of rain calendars (Lesson 26) and how plants satisfy water requirements depending on rainfall intensity (Lesson 27) ;
- agricultural calendars, or the way the vegetative cycles of plants fit in with the rain calendar.

Lesson 23

Plants

Botany is the science dealing with plant and plant life. There are so many plants, and plant life is so complex, that thousands of books have been written on the subject. In this lesson, students will find a general introduction to botany to enable them to follow the other lessons in this section.

Plant parts and their uses

What are the different parts of plants used for ?

Figure 266 sets out the different plant parts and their uses.

The leaves are the cooking pots for plants

Figure 267 helps to explain the function or role of leaves. Imagine a woman getting a meal ready for her family. She needs foodstuffs and relishes to make the dishes, and a source of energy (heat) to cook them.

When the housewife is short of relishes, she cannot prepare tasty, nourishing food. If she is short of energy, the dishes will not be cooked through. Plants are like the woman cooking the family meal. If trees are short of light, they cannot elaborate the sap in the leaves. By referring to the figure above, the student will understand why such emphasis is laid in the following lessons on the need for cultivated plants to have light, water, air and mineral elements.

---

267 | **A leaf is like the housewife's cooking pot**

Here are the ingredients - the food and the relishes - used to make the dishes :

There is the food **absorbed from the air** :
- **oxygen ;**
- **carbon dioxide ;**
- **nitrogen** (used only by certain plants). This is obtainable from the soil air and is taken into roots and nodules directly.

There is the food **taken up from the soil** :
- **water ;**
- **mineral salts ;**
- **various organic substances** found in humus.

These foodstuffs reach the leaves in the form of **unelaborated sap.**

There is the **special relish** found in the leaves :
- **chlorophyll**, the substance that makes leaves green.

Energy is also needed to get the food ready :
- **Sunlight provides the energy**, or heat, used by the leaves.

Here are the dishes prepared from the ingredients on the left :
- sugars ;
- starches ;
- fats ;
- proteins ;
- vitamins ;
- substances for growth.

When it is ready, the food in the form of **elaborated sap** is sent through the leaves into all the plant parts, especially into the fruit, the seeds and the roots. In the stems or stalks, these dishes can also be changed into **cellulose** or **wood.**

Wood and bark

If we look at a tree, it is easy to distinguish wood from bark. Both wood and bark are streaked with vessels (narrow channels) to let the sap circulate.

The vessels in the wood carry the rising, unelaborated sap to the leaves whereas the vessels in the bark carry the descending, elaborated sap especially to the roots.

sawed branch

wood

bark

ringed tree

bark

wood

268

When trees are ringed, the roots are deprived of the elaborated sap they need. This is why **ringing** kills trees **(figure 268)**. They die for lack of food elaborated (manufactured) in the leaves.

It should be added, however, that in some trees, palms for instance, the vessels are not divided between the wood and the bark. The vessels carrying the elaborated sap run alongside those transporting the unelaborated sap inside the very stem of the tree. Hence a palm cannot be killed by ringing.

Following the sap and water in plants

269 **Water movement and its usefulness in plants**

4

The water transported by the ascending unelaborated sap fulfils many functions in plant life :

- **it transports the mineral salts;**

- **it must be there to transform the relishes, or unelaborated sap, into cooked dishes, or elaborated sap ;**

- it **keeps** the leaves and stems **erect**. Plants without water wilt ;

- it is needed for the **growth of buds and long shoots ;**

- **it transports the cooked dishes** from the leaves to the fruit, stems and roots ;

- water **cools** the leaves when it evaporates on their surfaces.

ground

3

The leaves use unelaborated sap and the relishes it contains. But **some water is evaporated into the air through the stomata,** the small openings on the surface of the leaves. The leaves are said to transpire.

2

The roots **force the unelaborated sap towards the aerial parts of the plant** through channels called **vessels.**

1

The roots take up the water and mineral salts in the soil. The **root hairs** do this job. Water and mineral salts combined form the unelaborated sap.

Start here ; follow the arrows.

Heat and water for plants

When the weather is hot, plants begin to **transpire** and the soil dries up. This happens because the water contained in the plant or in the soil is transformed into **vapour** and combines with the air.

Transpiration is the system that allows plants to keep cool. It can be compared to the way water cools off in an earthen jar. The water that evaporates through the earthen sides of the jar cools the water remaining inside.

In order to transpire, plants must take up water from the soil. Transpiration and evaporation therefore make the soil dry. If rainfall is generous, it offsets the dryness caused by evapotranspiration in soil and plants. On the other hand, if rainfall is low, it cannot satisfactorily replace the amount of water lost through evapotranspiration. At this stage, the **wilting** of plants is observed. They wilt and go limp because there is not enough water to keep the leaves and stems erect and turned to the light. The young okra plant **(figure 270)** is at the wilting stage. If it is watered quickly, it will recover **(figure 271)**. If not, it will die.

270

Permanent wilting takes place at the start of the planting season when crops are already seeded and have begun to germinate, but then get no rain for a couple of weeks. When this happens, the plantlets die from lack of water. If, by any chance, they survive the short period of drought, they will be puny and have poor yields.

When cultivated plants wilt because of excessive evapotranspiration brought on by heat, it is worthwhile intercropping them with other plants in order to lessen the effects of sun and heat near the ground.

Trees can play an important role in this respect, particularly if they have a deep rooting system and can reach down to new supplies of water.

271

Stages in the life of a plant

The life of a plant unfolds in successive stages in what is called the **life cycle**. **It is important to know the life cycle of cultivated**

272

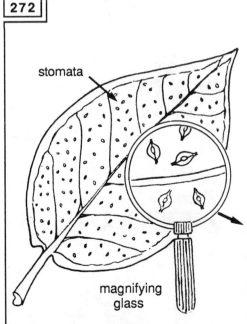

stomata

magnifying glass

How do plants breathe and transpire ?

Stomata are openings on the surface of leaves ; they can be examined under a magnifying glass or a microscope. The mouth-shaped openings, of which there are many on the leaf surface, open and close to allow the leaves to inhale the air.

Stomatal movement, the opening and closing of stomata, depends on the needs of the plant and on weather conditions. Generally speaking :

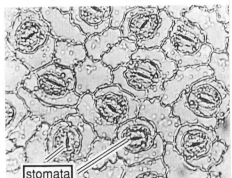

stomata

- stomata are open when the sun shines during the day ;
- they close when the plant is short of water ;
- when the plant is overheated and enough water is available in the soil, stomata open and the plant transpires. Transpiration in plants is like perspiration in people. It is a way of cooling off.

plants because a plant's need of water, food and heat varies with every stage in the cycle. In order to study life cycles, plants must be divided into two plant types - seasonal plants and perennial plants.

- **Seasonal plants** only live for a season or for some months. They disappear, abandoning seeds in the ground. Cereals fall into this category, such as maize, millet, rice ; also legumes such as bean, groundnut, pea ; and other plants like sesame, cotton, cucumber, eggplant, potato. Onion is also a seasonal plant that leaves bulbs in the ground after the harvest.

- **Perennial plants** remain in the ground for many seasons. This is so for all trees and many grasses, shrubs and herbaceous plants like sugar cane, sisal, *Citronella*, pepper, chillies and castor oil.

Some varieties of perennial plants can be cultivated like seasonal plants, for instance, varieties of yam, cotton, cassava, taro. When cultivated, these plants are seeded again every year, whereas left wild, they are perennial.

This sorghum plant has tillered.

The life cycle of seasonal plants

What are the main stages in the life of seasonal plants as illustrated by sorghum **(figures 273 to 276)** ?

- During the **germination** and **establishment** of the plant, the plantlet forms the first roots and stems. **Emergence** comes next, when the plant emerges from the ground, followed by **tillering** or splitting of the young stem into many parts at ground level. Only some *Gramineae* send out tillers, sorghum **(figure 273)**, rice and wheat for instance ; other plants, including maize, do not.

- **Vegetative growth** occurs when the plant develops its roots and foliage. In cereals, this is called **shooting** or **stem elongation**.

- **Flowering** or **ear emergence** is the time when flowers or ears appear **(figures 274 and 275)**. Immediately afterwards, the young fruit begins to form. If fruit setting does not take place or is unsatisfactory, fruit and cereal production is in danger.

273

274

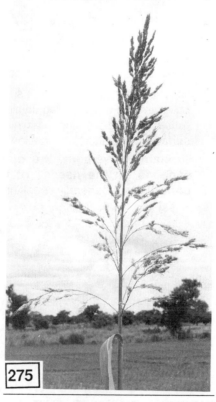

275

276

- What is important for tubers is not so much the flowering as **tuber formation**, the period when tubers form and swell.

- The next stage is **maturation** or **ripening** when the fruit, cereals and tubers ripen. The young fruit and grains formed after flowering build up reserves. The grains accumulate starches, sugars, fats and proteins, while the fruit manufactures pulp and juice. An ear of sorghum during maturation is seen in **figure 276**.

  As regards tubers, they thicken fully during maturation and finish storing up reserves.

- The last stage is **senescence** (ageing) when plants wilt.

The vegetative cycle of perennial plants : the multiannual cycle and seasonal cycle

Perennial species are those with living parts lasting from one year to the next without resowing. Their vegetative cycle is different from those of seasonal plants. A distinction must be made between the **multiannual cycle**, or what happens over a period of many years, and the **seasonal cycle**, or what happens every year.

The stages in the multiannual cycle are as follows :

- **germination** and **establishment** of the plantlet (first months of its life) ;

- **growth** and **formation** of the tree. This phase can take several years, - from four to seven depending on the species, and sometimes more ;

- the period of **leaf production** or **foliation** that can also last many years. The seasonal cycle occurs during this period (see below) ;

- **ageing** and **death**.

The seasonal cycle unfolds as described below and is illustrated by the orange tree photographed at different moments in its life cycle.

- Some trees lose all their leaves in a given season. Consequently, the period of **foliation** is the one when the leaves grow. However, with other trees, for example palms, foliage is permanent. Leaves fall, others grow at the same time. No particular time of foliation can be observed ;

- The flowers open during the **flowering** phase **(figure 277)**. Immediately afterwards, as with seasonal plants **(figure 278)**, the flowers **come to bearing.**

- The time of **fructification and maturation** (ripening) follows **(figure 279)** when the fruit forms and ripens.

Notes

277

278

279

Lesson 24

🔲🔲🔲

Plants need light

All plants need light. Some like full light, others prefer half light or deep shade. Still others are unaffected by light intensity. The shape of plants and their growth depend to a great extent on the amount of light they receive.

Two plant stems of the same species will develop differently if one is in the shade and one in full light. Supposing the plant likes light but grows in shade, the overshaded stalks will stretch out vigorously to reach the light. On the other hand, if the plant is in full light, the stalks will be short and the leaves tight and close along the stalk.

The shaded stems will probably have many leaves along the extended stalks and little fruit, while the stem in full light will have smaller, closer leafage and more fruit. **Figure 280** illustrates this point.

Maize and gourds are interplanted in **figure 281**. Maize likes light. Gourds do not mind one way or the other - they thrive in full light and in half-shade. Deep shade might prevent fructification.

280	Two groundnut stems : one in shade, one in the sun

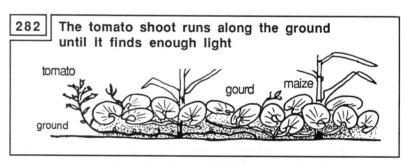

stem sheltered by leaves

In shade, the stalks stretch towards the light. Leaves are large, thinly distributed, light in colour.

In the sun, stalks are short. There is small, abundant leafage, very green in colour.

The maize gives light shade to the adaptable gourd, but the gourd with its large spreading leaves creates deep shade and stops other plants from growing underneath. A long tomato shoot is seen in the foreground. It has crept under the gourd leaves towards the light in order to spread its own leaves **(figure 282)**.

maize

gourd

tomato

281

282	The tomato shoot runs along the ground until it finds enough light

tomato

gourd

maize

ground

The gourd and the potato are competing for light in **figure 283**. The potato stalks are forced to stretch out beyond the canopy of spreading gourd leaves **(figure 284)**. In the struggle to reach the light, the potato has wasted time and energy which could have been spent on producing tubers.

Cocoa likes light but, because it cannot stand high temperatures, it is often planted in the shade. However, when heavily shaded, it grows badly and bears little fruit.

Not all trees give the right kind of shade, as can be seen by comparing **figures 285 and 286**. Some plants filter the light, like the tree above the cocoa **(figure 285)**. It lets the light through while lessening direct sunlight that might overheat the cocoa trees. This is good shading.

In contrast, the umbrella tree **(figure 286)** absorbs most of the light, leaving hardly any for the plants growing underneath.

potato

gourd

283

285

286

284 **The potato plant stretches its stalks in order to come out from under the gourd leaves into the light.**

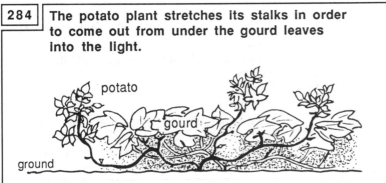

potato

gourd

ground

The light requirements (needs) of a plant vary with each phase of the life cycle

Table 287, column one, gives a list of the cultivated plants whose light needs will be studied. The next four columns give the phases of the life cycle as discussed in Lesson 23. A distinctive symbol is used for each phase. Comments are in the last column.

Germination and establishment	Growth of stems and leaves	Flowering, emergence of ears, formation of fruit, tuberization	Maturation (ripening)

The amount of light each plant needs at every phase of the life cycle is shown by a circle representing the sun as seen by the plant. The sun is sometimes veiled by the leaves of the plants in the upper canopy. The circles representing the sun are the following :

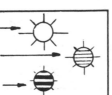

the plant demands exposure to full sunlight ;

the plant tolerates moderate shade ;

the plant tolerates shade ; it can remain in rather deep shade during the phase indicated.

With these explanations in mind, the table can now be examined.

Take the groundnut, for instance, in column one under the heading Legumes. By moving to the right along the same line, we learn that

287 | **Light needs of some seasonal plants at different times in their life cycle**

Cultivated plants	Life cycle	Comments

Cereals

maize, millet,
sorghum, rice,
finger millet

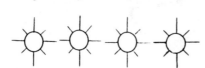

Cereals are affected by lack of light. Consequently, even a little shade results in lower yields due to retarded growth of stems and leaves.
However, rice tolerates some shade during the growing phase when temperatures are very high.

Tubers
cassava,
sweet poptatoe, taro

yam

cocoyam

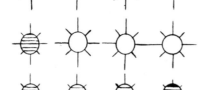

These plants demand light, but moderate shade barely affects the growth and production of taro, sweet potato and cassava.
Shaded yam cuttings ensure better emergence.
Cocoyam tolerates shade because it likes cool conditions. Shade must be less during growth than during the period of tuber formation (tuberization) and ripening.

Legumes
groundnut, earthpea,
chick pea, soya, cowpea,
bean, pigeon pea

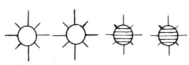
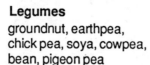

Legumes demand exposure to full sunlight, but moderate shade is beneficial, for instance, shade from cereals, in later phases of their life cycle.

Vegetables and condiments
chilies, cucumber, melon,
okra

tomato, onion, ginger

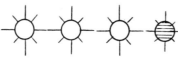
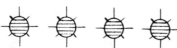

Tomato and onion demand light but respond badly to extremes of heat. For that reason, they tolerate some shade as protection from heat. Ginger adapts to both light and heat.

pepper

Pepper demands shade and filtered light for the whole duration of the vegetative cycle.

Other herbaceous plants

cotton, sesame, sunflower,
sisal, pineapple, sugar cane

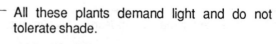

All these plants demand light and do not tolerate shade.

banana

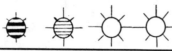

Banana demands light but benefits from moderate shade from overhead trees.

during germination and growth, the groundnut demands much light. This is shown by the circles representing full sunlight during both phases. However, in the columns corresponding to flowering and maturation, the sun is lightly veiled. This means that, after these two phases, the groundnut will tolerate moderate shade, the shade from an intercropped cereal, for example.

Table 288 deals with trees and their light needs during the four phases of their life cycle :

- germination in a nursery ;

- plantation, the period when roots and young stems are formed ;

- growth lasting many years and of varying duration, depending on the species involved ;

- full production.

The four phases are represented by these symbols :

During the period of full production, the light needs of trees like cocoa and particularly coffee are closely related to soil fertility. If the soil is fertile, these trees thrive in full light. But in poorer soils, they call for shade. The poorer the soil, the greater the need to filter the sun's rays.

Germination in a nursery	Plantation
Growth	Period of full production

288	The light needs of some perennial plants during their life cycle

	Germination	Plantation	Growth	Full production
cocoa				
coffee				
palm and coconut				
tea				
avocado mango lemon, orange, grapefruit				

A word of warning is necessary. The sun's rays give both light and heat. Though some plants demand light, they prefer cool conditions and look for shade. Cocoyam falls into this category. It demands light but preferably on shaded ground with cooler conditions. Consequently, cocoyam can be found under shade on dry soils, and fully exposed in a wet valley bottom.

Like everything else connected with agriculture, there are no hard and fast rules. These tables show that needs vary during the life cycle of a plant. But it must also be remembered that every plant species and variety lives in its own particular way. Even varieties of the same plant may have quite different needs.

It follows that general rules, applicable to all the plants mentioned, cannot be inferred from these tables. Careful observation of the behaviour of every variety in fields and farms is the best way of reaching sound conclusions valid for local conditions.

Lesson 25

The proper use of light

Plant life is impossible in darkness because light is essential to transform the unelaborated sap into elaborated sap and living matter. Some plants only thrive in full light. They are called **heliophilous** or **light-demanding plants**. Other plants need filtered light. They prefer shade and are called **skiophilous** or **shade plants**. They are shade-tolerant.

In fields and plantations, there are various ways of controlling the light falling on the leaves of cultivated plants. Some practices aim at giving more light to plants by reducing seed density and plant population, by pruning trees and by training. Other practices aim at reducing light intensity. This involves providing shade by the careful association of light-demanding and shade-tolerant plants.

How to give light to cultivated plants

Controlling density : wide or close spacing

Two fields of maize in pure stand can be compared for spacing.

In the first field, sowing was dense (**figure 289**). The stems are crowded and the leaves entwined. The upper foliage shades the lower leaves which are deprived of light, turn yellow and are useless. They no longer fulfil their role of nourishing the ears.

In the second field, sowing was not so dense (**figure 290**). Light penetrates deep between the leaves. All the leaves are active and therefore contribute to production.

A visit to an unshaded coffee plantation is also instructive. In the first plot, the trees were planted three metres apart (**figure 291**). There is no root competition because the root systems do not extend more than I.5 m round the stems and the branches are not intertwined. In this plot, branches and foliage are abundant everywhere along the stems, especially if the farmer is careful to keep the trees pruned.

There is enough light even in the understoreys to ensure a plentiful yield of berries. The soil has vegetative cover and intercropping with seasonal plants is possible.

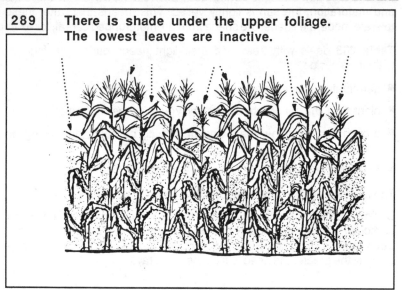

289 There is shade under the upper foliage. The lowest leaves are inactive.

290 Light penetrates deep between the maize stems. The lower leaves are active. The maize can be interplanted with other crops.

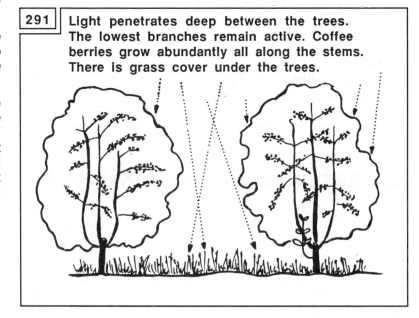

291 Light penetrates deep between the trees. The lowest branches remain active. Coffee berries grow abundantly all along the stems. There is grass cover under the trees.

The trees in the second plot were planted closer, two metres apart **(figure 292)**. At first when the trees were young, there was enough light and the stems were well-developed. But after four or five years, the branches began to intertwine. The shoots at the bottom of the stem were badly nourished because their leaves did not get enough light, and they died off.

In this case, the mature trees are bell-shaped. Leaves, flowers and fruit only appear on the crowns exposed to the light. The ground is bare because it gets little light. There is root competition through overcrowding.

Some coffee planters overcrowd the trees to eliminate the labour of weeding. Perhaps they are right to cut down on work in this way. But it might be better from the point of view of production to space the trees further apart and interplant with seasonal crops.

An idea for modernizing agriculture is to space tree plantations to leave room for interplanting with annual and seasonal crops and to choose the seed density that will make the best use of light.

To obtain good results, the farmer must ensure that roots are not competing for food (Lesson 32) nor leaves for light. When a plantation or a field is established with interplanting of crops and trees, the farmer must have an overall plan with answers to questions such as : What will the mature trees look like ? What spacing will leave room to grow lower layers of cultivated plants ? How can the soil be fertilized to benefit all the associated plants ?

Pruning trees

Overshading the leaves of light-demanding plants makes them inactive. They live at the expense of other, well-exposed leaves, and use up for themselves part of the elaborated sap designed to feed flowers and fruit. These leaves, growing in poor light conditions, are unproductive.

By pruning his trees, the planter manages, among other things, to expose all the remaining leaves to the light. He may also want to shape the trees so that the associated plants are not overshaded.

It is worth comparing the citrus trees (oranges, lemons, grapefruit) in **figures 293 and 294**.

The trees in **figure 293** have not been pruned. The crowns touch one another. The lower branches get little light and bear no fruit. Cereals cannot grow between the rows of trees because no light reaches the ground.

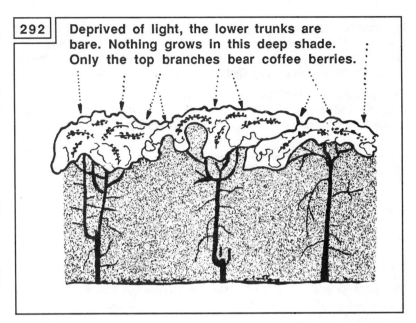

292 Deprived of light, the lower trunks are bare. Nothing grows in this deep shade. Only the top branches bear coffee berries.

293 These fruits trees, overcrowded and unpruned, create too much shade. Nothing grows at their base.

294 The same fruit trees have been pruned. Light reaches the ground so that cultivation between the trees is now possible.

In contrast, the trees in **figure 294** have been pruned. The planter has stopped the crowns from spreading. He has taken care to make the lower branches productive by giving them light. At the same time he has given the cereals enough light to ensure good growth.

The **pruning of leafage** above ground can be supplemented by **pruning the roots**. This means pruning any tree roots that might compete with the roots of seasonal plants by invading their root zone.

Staking

Staking consists of training plants up a stake or upright support, such as a pole, branch, stalk, tree.

Staking is suitable for climbing plants and plants that can be made to climb. Climbers include some varieties of yam, beans, gourd, peas, also pepper, giant granadilla, passion fruit and many other plants. Some plants, tomato for instance, do not climb naturally but can be artificially trained by staking.

Two bean plants climbing up a stake are shown in **figure 295**. All the leaves are well exposed to light whereas unstaked they would be entangled on the ground.

Figure 296 shows a common way of staking beans in Central Africa. The stakes or **props** are made from *Setaria* stalks tied together three

295

296

297

by three to form pyramids. *Setaria* is a fodder grass widely used in Rwanda to form antierosive strips.

Another kind of staking is illustrated in **figure 297**. Tree branches are used as supports for yam growing on ridges. This system has two advantages. Thanks to better exposure, the yam leaves will be more productive. Secondly, room is made for other crops. In this case, the farmer will sow a creeping variety of sweet potato between the yam stalks growing up the props. The sweet potato will provide cover on the ridges and in the furrows and manage with the light left by the climbing yam.

Shea butter trees are a common sight in the fields in the south of Burkina Faso. Farmers often sow a climbing yam at the foot of this tree that acts as a living pole **(figure 298)**. The yam leaves spread round the trunk to get the light on the sides of the tree.

Nourishing the soil

Manuring the soil with organic manures and nitrogen fertilizers helps to make better use of light. Nitrogen fosters leaf growth which, in turn, encourages increased plant activity when there is enough light and the leaves are well-exposed. The plentiful supply of sap elaborated in the leaves then goes on to nourish stems, flowers and fruit. In conclusion, nourishing plants correctly leads to better use of light. Leaves receiving unelaborated sap in

shea butter tree

yam

298

Mango trees grow best in full light but young grafted stems, not yet fully rooted, prefer shade. It must be remembered that active leaves use up much ascending unelaborated sap and by doing so, force the roots to take up a lot of water and mineral salts from the soil. The leaves must not be too demanding while the roots are still developing. By reducing the light, the leaves become less active and therefore less demanding on the roots.

Different shade for different soils

Plants that would normally demand shade can sometimes stand strong sun and high temperatures if water is available. Cocoyam and rice, for instance, thrive unshaded in damp valley bottom soils. This is because the plants can keep cool the whole time by evaporating large quantities of water through the stomata. On the other hand, in dry soils, short of water, shade is essential. Without shade, the leaves exposed to the sun will quickly wilt.

Other plants, growing on soils rich in mineral salts, develop well in full light. But on less fertile soils, the same plants demand shade. The amount of water and mineral salts available has considerable influence on the control and management of shade. In fact, where plants are concerned, shading is all the more beneficial when soils are poor.

abundance absorb a great deal of light in order to produce elaborated sap and send it back to the other plant parts.

Ways of shading

As was explained in Lesson 24, some plants like shade or filtered light. Other examples of plants growing in shady conditions were mentioned at the beginning of this book - cocoyam under banana, coffee in the shade of *Leucaena*, cocoa growing under tall forest trees that had been spared during clearing operations, rice under maize, pepper in woodland, etc.

Young plants that have just been sown or transplanted are nearly all sensitive to high temperatures and intense light. Hence, the need to look at common ways of shading plants. This can be done by creating **shade zones**.

Shade zones

The small nursery in **figure 299** consists of a roof of palm leaves as protection for vegetable seedbeds. The roof also stops raindrop splash from beating the soil of the seedbed.

A tree can also provide a shade zone. A nursery of young grafted mangoes has been placed under an African locust bean with a wide crown (**figure 300**).

299

300

124

Shade and plant growth

In Lesson 7, the importance of shade was already stressed in connection with Plot B. In that plantation, the farmer had interplanted cocoa plantlets with cocoyam and banana.

When first set in the ground, the cocoa plantlet objects to light. Deep shade is needed and is supplied by banana and cocoyam that form a many-layered canopy of shade vegetation.

When the plantlets begin to thrive, shade need not be so intense and bananas alone will decrease sunlight sufficiently.

Later on, the upper foliage of the cocoa trees shade the lower branches, the flowers growing on the tree trunks and the pods. This is called **self-shading**. Self-shading however is only effective when the leaves of the trees touch one another to form a closed canopy. If necessary, other shade trees may be planted. The development of this shade pattern is examined in **figure 301**.

| 301 | Development of shade in a cocoa plantation |

Shading of a cocoa tree when first planted (1st year)

Shading of a cocoa tree during growth (2nd, 3rd and 4th year)

Self-shading of the cocoa tree after 5 to 7 years
If self-shading is inadequate, **shade trees** can be planted between the cocoa.

Shade trees

All trees give shade but not necessarily the right kind of shade. **Figures 285 and 286** in Lesson 24 illustrate this point. The tree in **figure 285** gives light shade while the umbrella tree (**figure 286**) gives heavy shade, ill-suited to cultivated plants.

Figure 302 gives the characteristics most sought after in shade trees. However, as few trees present all the qualities listed, shade trees best adapted to the plants in question must be chosen. Soil, climate, slope of land, exposure to wind and other factors have all to be taken into account. Moreover, shade trees must not compete too vigorously with the cultivated plants they protect.

| 302 | Characteristics sought for in shade trees |

■ **It stands up well to pruning** and thinning.

■ **It yields produce immediately useful** to the farmer (wood for carpentry, fire wood, fruit,...).

■ **The root system does not spread too much** ; it avoids invasion of the root systems of cultivated plants and subsequent competition.

■ A good shade tree is **tall enough** to avoid inhibiting the growth of the plants it shades.

■ **Its roots are deeply penetrating** and take up water and mineral salts at great depth in the soil.

■ **Leafage is not too dense.** It filters the light. Trees with small or compound leaves are the best.

■ **It does not encourage diseases and pests.**

■ **It has one tall trunk** and consequently **no shoots** at the base.

■ **It produces rich, abundant litter.**

Lesson 26

🔲🔲🔲🔲🔲🔲🔲🔲🔲🔲🔲🔲🔲🔲🔲🔲🔲🔲🔲🔲🔲🔲🔲🔲🔲🔲🔲🔲🔲🔲🔲🔲🔲🔲🔲

Agricultural calendars ; rain calendars

Everyone knows that activity on the farm starts up again at about the same time every year, shortly before the rains set in. The **agricultural cycle** is the term applied to this recurring commencement.

Farm work depends on plant life and more specifically on the vegetative cycle (Lesson 23) whose course is linked to the **rain cycle** or **rain calendar**, that is to say, to the increase or decrease in the amount of rainwater falling on the rural settlement at different times of the year.

The agricultural calendar

The **calendar of time** comes, of course, from the division of the year into days and months starting each year in January and ending in December. All the activities connected with farming - clearing, tilling, sowing, weeding, harvesting, etc. - take place during the calendar year at times determined by the pattern of rainfall and the progression of the life cycle of every cultivated plant.

An **agricultural calendar** is drawn up by inserting on the calendar of time all the agricultural work accomplished on a farm in every field and for every crop. Here is an example of an agricultural calendar in a region of Sudan. It was drawn up after careful observation of four types of cropping fields and the exact times when farming operations are carried out. The names given to the fields are those used by the cultivators themselves.

Farmers often divide the year into lunar rather than solar months. The lunar month is the period that elapses between two new moons. Cultivators frequently know the lunar calendar much better than the civil calendar (from January to December). Also the lunar months are often called by special names in the vernacular (the native language of the country or place).

It is not hard to **draw up an agricultural calendar** provided one knows how to observe exactly what is happening in the fields. But it is a job that calls for accuracy and takes time. It should be tackled in two stages.

- The first step is to **identify the different field types** corresponding to their location on the settlement (plateau, slope, valley bottom, etc.), the kind of soil and degree of moisture, the associated crops, and the farming practices (flat tillage, ridging, mounding, strip cultivation). Usually the fields have **special names** known to the farmers.

- The second step is to **note the farmer's working hours** and the exact moment in the season when the various tasks are carried out. First, the major tasks of clearing, tilling, sowing, field maintenance, weeding, harvesting, mounding where practised should all be monitored. Then the tasks related to each crop such as pruning, manuring, field hygiene, staking, etc. should be recorded.

Observing an agricultural calendar may be easy ; explaining it is quite a different matter. The cultivator has to take into account a whole series of factors which he may feel intuitively without having consciously reasoned out his position or being able to put it into words. These factors include plant life, rain cycles, aims pursued, labour availability, the characteristics of the plant varieties on the farm, customs and so on.

Rainfall and rain measurement

Rainfall refers to the amount of rain associated with a given climate. Remarks like, 'It rains a lot or a little', 'It rains regularly or irregularly', 'Rains are light or heavy' are all connected with rainfall. However, if no attempt is made to measure rain, the concept or notion of rainfall remains vague. **Rain measurement** means measuring the amount of rain falling on a given place over a given period of time. There are many aspects to rain measurement.

- **The total amount of water falling on a given place during the year** can be measured to obtain the **annual rainfall**. This measure indicates whether it rains a lot or a little each year but says nothing about how the annual rainfall is distributed during the year.

1st field : Baawande

The main plants are groundnut and finger millet with secondary mixtures of maize, cassava and various herbaceous plants. Here is the work calendar in this field :

opening and hoeing burning and hoeing felling of trees	mid-March to end of May
sowing of maize	May to mid-June
planting of cassava sowing of groundnut	mid-May to mid-June
weeding of field where groundnut is the main crop	mid-June to end of July
groundnut harvested with immediate sowing of finger millet	mid-October to end of December
cleaning of finger millet	mid-October to mid-November
maize harvested	October to December
finger millet harvested	December and January

2nd field : Oti-moru

In this field, tilled on the flat, the main crops are maize, cassava and finger millet with secondary mixtures of sesame and other local plants. The work calendar in this field is as follows :

opening of the clearing	mid-May to end of July
hoe and burn	end of May to mid-August
felling of trees	mid-June to end of July
sowing of maize	beginning of July to mid-August
planting of cassava	August
sowing of finger millet, sesame and other plants	mid-June to end of July
weeding and cleaning	September to November
maize harvested	October to mid-December
finger millet harvested	mid-October to end of January
sesame and other plants harvested	December and January

3rd field : Nduka

This field is cultivated in ridges. Associated plants are maize, gourds and sweet potato. Here is the work calendar :

making of ridges sowing of gourds and maize	March, April, May
sowing of sweet potato	mid-May to mid-September
weeding and cleaning	mid-May to end of October
maize harvested	mid-October to mid-December
gourds harvested	October
sweet potato harvested	mid-December to mid-January

4th field : Cotton

This field has no local name because it was devised by Europeans in the first place. Cotton is grown alone (monocropping) except for some vegetables. It is established either in a forest clearing or in a field where cotton was grown the previous year. If it is first year cotton, trees are felled and burned. If it is second year cotton, a sanitary clean-up is carried out, i.e. the old cotton stalks are hoed out and the trash burnt.

This work calendar is as follows :

opening of field burning and tillage	May to June
sowing of cotton	June
weeding and cleaning at various intervals	mid-June to mid-October
cotton picking	mid-October to mid-December

In the same way, the amount of water **falling every month** can be measured to obtain the **monthly rainfall**. There are twelve monthly measures in a year. The total of these monthly measures gives the measure of annual rainfall. Thanks to these monthly measures, the periods of heavy and light rainfall can be determined and a distinction made between **rainy** and **dry seasons**.

In some parts of tropical Africa, there are two rainy seasons with two intervening dry seasons (bimodal rainfall). They are called **long** or **short** dry and wet seasons depending on their duration. In other parts there is only one wet and one dry season in the year.

■ Most cultivated plants depend essentially on **having a regular supply of water** during the period of growth when the soil must be permanently moist. Drought lasting eight to ten days may have disastrous effects on the crops at any time but especially if it occurs during germination and early establishment (Lesson 27).

■ Apart therefore from knowing annual and monthly rainfall, it is very useful to measure rains over periods of five, seven (**weekly rainfall**) and ten days.

■ The amount of rain falling on the ground over these short periods of five, seven and ten days is, indeed, the decisive factor in providing or not providing enough water to satisfy plant needs.

Measuring rainwater in periods pinpoints the sowing season. No farmer would dream of sowing when his knowledge of the climate tells him that the rains may well be unreliable at the very time when his crops are utterly dependent on water.

Another look at the fields in Sudan makes the situation clear. Even if it rains a little in February and March, it would be most unwise to sow at once, because the rains are not properly **established** until April or May when young plants no longer run the risk of wilting. The farmer needs more precise information. His need leads to the question : How is rain measured?

Measuring rain

The amount of rain is measured in millimetres. This can be done by placing a small empty drum flat on the ground in a place fully exposed to the rain and far from trees and buildings. After the fall of rain, the depth of water in the bottom of the container is measured with a millimetre ruler and gives the **amount of the rain (figure 303)**.

The scientific instrument used for measuring rain is called a **rain gauge** or **pluviometer (figure 304)**. It is a funnel standing in a deep measuring glass graduated in millimetres

305	Rain records over a forty-day period			
Month	**Daily rainfall in millimetres**	**Rainfall (5-day periods)**	**Rainfall (10-day periods)**	**Rainfall in July**
1 July	32 mm			
2 July	10 mm			
3 July	0 mm			
4 July	29 mm			
5 July	9 mm	80 mm		
6 July	0 mm			
7 July	0 mm			
8 July	0 mm			
9 July	0 mm			
10 July	0 mm	0 mm	80 mm	
11 July	0 mm			
12 July	4 mm			
13 July	18 mm			
14 July	9 mm			
15 July	14 mm	45 mm		
16 July	0 mm			
17 July	3 mm			
18 July	4 mm			
19 July	0 mm			
20 July	8 mm	15 mm	60 mm	
21 July	0 mm			
22 July	2 mm			
23 July	0 mm			
24 July	0 mm			
25 July	3 mm	5 mm		
26 July	15 mm			
27 July	0 mm			
28 July	4 mm			
29 July	4 mm			
30 July	11 mm	34 mm	39 mm	
31 July	60 mm			239 mm
1 August	14 mm			
2 August	46 mm			
3 August	0 mm			
4 August	5 mm	125 mm		
5 August	4 mm			1 mm
6 August	1 mm			
7 August	12 mm			
8 August	0 mm			
9 August	1 mm	18 mm	143 mm	

303	Measuring the amount of a fall of rain

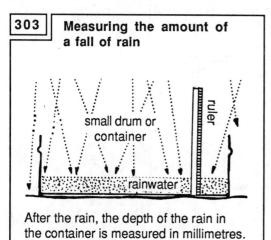

After the rain, the depth of the rain in the container is measured in millimetres.

(marked with lines and numbers). The funnnel collects the rain and empties it into the glass where the amount of the rain in millimetres can be read at a glance. The readings can be made every day or at the end of a given period and are marked on graph paper used for **recording rainfall** or **precipitation** (which includes hail).

Here is a July/early August rain graph for the region in Sudan mentioned earlier in this lesson **(table 305)**.

What conclusions can be drawn from these rain records ?

239 mm of rain (monthly amount of rainfall or precipitation) fell in July. This a satisfactory overall total. Note the very heavy rains (60 mm) on 31 July and six days with no rain from 6 to 11 July.

Whereas some five-day periods got plenty of rain (the first and the last but one), others got practically none. The second five-day period got no rain at all, and the fifth recorded two light rains totalling 5 mm.

If the monthly reading of 239 mm only was retained, one might reach the conclusion that it had rained enough in July. However, the day-to-day and period data reveal that rain distribution was such as to give drought conditions for many days at the beginning of July. If the soil was still moist from previous rains, that dry period may not have seriously affected well-established plants. On the other hand, it may have been disastrous for plantlets in the phases of germination and establishment when all plants need a lot of water.

By observing the rain calendar at regular intervals year after year, the cultivator can determine the periods with the greatest risk of drought. If he knows that every year in his part of the country, there are likely to be

304

many dry days at the beginning of July, he will take care to fix the sowing season so that plants suffer as little as possible from drought. He can decide to sow early to have plants well-established before the drought, or he can decide to postpone sowing to ensure that germination takes place when there is no risk of permanent wilting.

Making a graph of a rain calendar

Rainfall is measured in millimetres. Rather than giving the rain calendar and rain pattern in a table with figures as in **table 305**, it can be represented in **graph** form. The graph **(figure 306)** is a diagram in which two scales of measurement are used. The first scale is horizontal and is the **scale of time**. The second is vertical and gives the **scale of rainfall or precipitation**.

The precipitation readings consist of forty figures, one for each day from 1 July to 9 August. The level

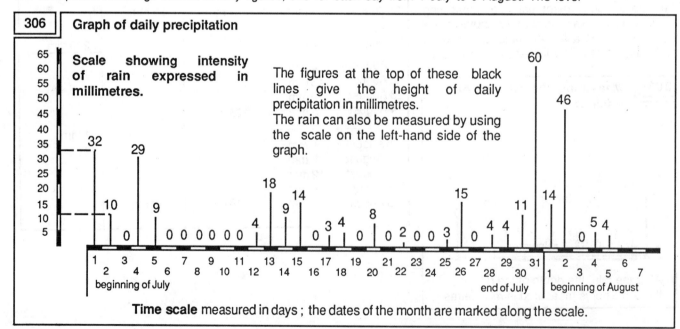

306 **Graph of daily precipitation**

Scale showing intensity of rain expressed in millimetres.

The figures at the top of these black lines give the height of daily precipitation in millimetres.
The rain can also be measured by using the scale on the left-hand side of the graph.

Time scale measured in days ; the dates of the month are marked along the scale.

of precipitation is indicated by a series of upright sticks, each stick or black line cut to scale at the height of precipitation for the day in question and placed on the corresponding date along the time scale.

The rain pattern for periods of any given length of time, five or ten days, for example, can also be represented in the form of a graph. This is done by adding the height marked on each stick for each day in the period chosen and by placing the sticks (giving the new total precipitation) on a time scale no longer divided into days but periods of days. This kind of graph is illustrated in **figures 307 and 308**.

To make the rain pattern stand out better, a dotted line links the tops of the sticks. By following the curve of the dotted line from the beginning of the time scale, we can

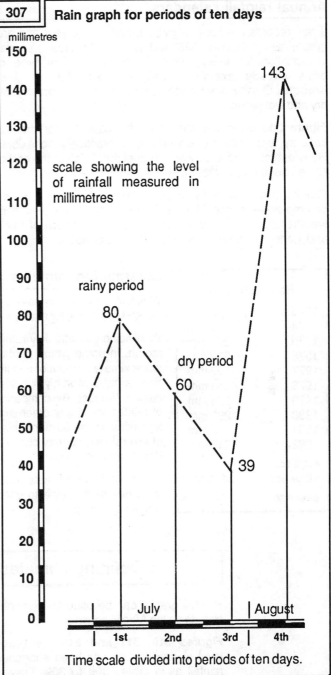

307 **Rain graph for periods of ten days**

scale showing the level of rainfall measured in millimetres

Time scale divided into periods of ten days.

see at once whether rains have increased or decreased in that time.

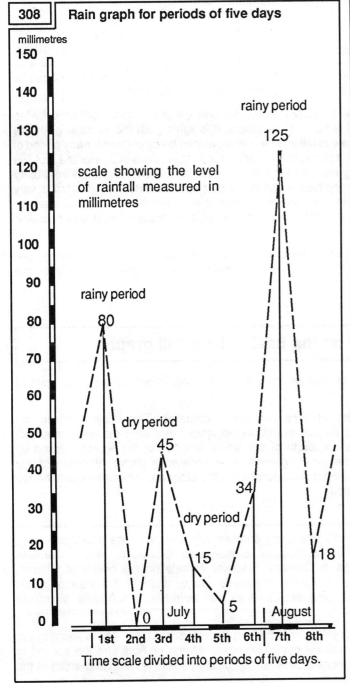

308 **Rain graph for periods of five days**

scale showing the level of rainfall measured in millimetres

Time scale divided into periods of five days.

Rainfall can be measured in many ways.

Rainfall records made for :

- □ *every day are called daily records ;*
- □ *every week are called weekly records ;*
- □ *every month are called monthly records ;*
- □ *every year are called yearly or annual records.*

Annual rainfall calendars

If rain records are kept for every month of the year and are then represented in the form of a table or graph, as in **figures 305 and 306**, the annual rain calendar can be drawn up giving the total precipitation for every month in the year. **This type of calendar is not as accurate as the ones already examined and, in particular, it gives no indication of short periods of drought.** On the other hand, it reveals the **seasonality** of a region, that is to say, the succession of dry and wet seasons.

Figure 309 gives the annual rainfall calendar for the same region in Sudan. Rains in December, January and February are very light, practically nonexistent. The rainy season is in March onwards. Heaviest rains are in July and August and finish at the end of October. The dry season proper lasts from November through December, January and February.

Graphs of annual rainfall records are needed to determine the climate of a region. **Weather stations**, or employees of the Ministry for Agriculture, are responsible for providing these data. The task of a weather station is to plot, day by day, the pattern of rainfall, wind, temperature, sunshine, humidity, and, generally speaking, all aspects of the climate.

Rainfall in July	
1973	235 mm
1974	266 mm
1975	201 mm
1976	289 mm
1977	199 mm
1978	253 mm
1979	226 mm
1980	290 mm
1981	212 mm
1982	239 mm
total for 10 years	**2410 mm**
average	**241 mm**

Determining rainfall averages

When rainfall records have been kept for many successive years, rainfall **averages** can be worked out. Take, for example, the July rainfall recorded from 1973 to 1982.

By dividing the total July rainfall by 10 (2410÷10), an average of 241 mm is obtained for the month. In some years, rainfall is heavier ; in others, it is lighter, but the average gives the approximate amount of rain likely to fall in July. Averages can be calculated for any period of time, annual, monthly, weekly, etc., or, for example, the first ten days in July or the first five days in August. By comparing annual or monthly figures for several years, the **reliability** of rainfall (or of any other climatic factor) can be calculated. The extent of variation is very significant to the farmer in his choice of crops, for example, where there is a high probability of low rainfall, he may choose a drought-tolerant species even though this is lower yielding than more risky crops.

Climatologists and meteorologists, who specialize in climate and weather sciences, use averages of this type to characterize climates all over the world, and particularly on the African continent.

Defining climates on the basis of rainfall graphs

Rainfall graphs can be used to examine climate on the basis of rainfall. There are six main types of climate on the African continent.

Figures 310, 311 and 312 are graphs of three of these climates. The scale showing the level of precipitation has been shortened so that the sticks no longer represent the real amount of rainfall as in **tables 306 to 309**. They are a fourth of the actual length, but the figures marked up the rain scale and the corresponding figures on the graphs show the level of precipitation without any problem. (The scale has been shortened for practical purposes ; the diagrams otherwise would be too big for the page. This is called a relative scale.)

Guinean climate

Guinea has **two rainy seasons** (one short, one long) and two dry seasons (one short, one long) during the year **(figure 310)**. This is the **wet tropical** climate also referred to as the **Guinean climate**. North of the equator, it is found in Conakry, Bouake, Tamale (in the centre of Ghana), Lagos, Yaounde, Bangui and Kampala. The same climate is found south of the equator round Libreville, Brazzaville, Kinshasa, Kananga (southern Zaire) as well as in Northern Angola, Mombasa and Dar-es-Salaam.

There are usually two agricultural seasons per year in the wet tropical zone. North of the equator, the first agricultural season (also called the **cultural cycle**) begins in March or April and the second in July or August. There is a time lag of six months between agricultural seasons north and south of the equator.

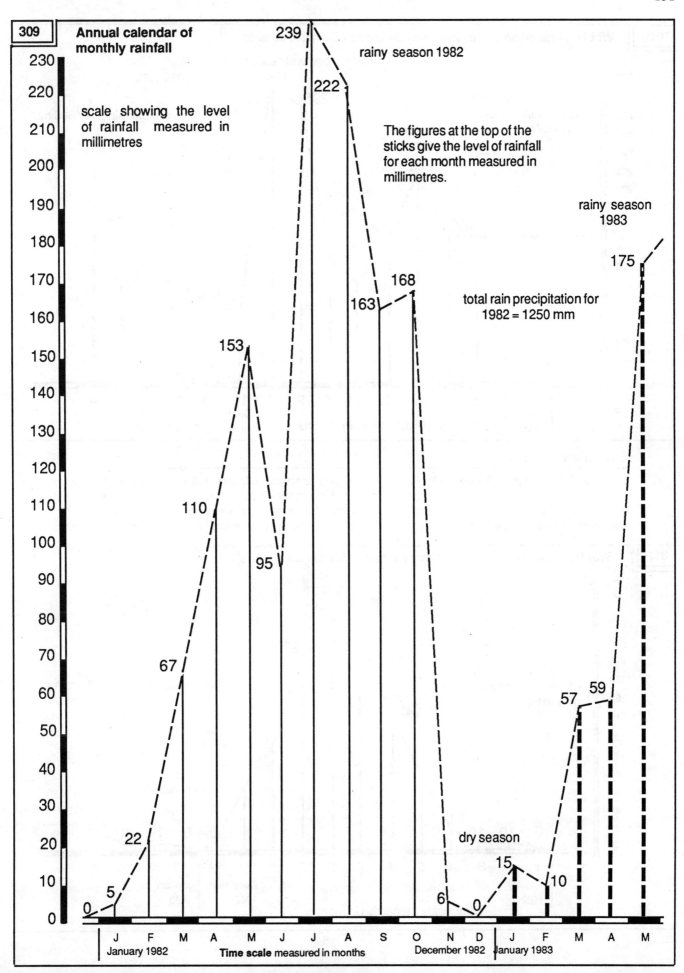

Annual calendar of monthly rainfall

309

scale showing the level of rainfall measured in millimetres

rainy season 1982

The figures at the top of the sticks give the level of rainfall for each month measured in millimetres.

rainy season 1983

total rain precipitation for 1982 = 1250 mm

dry season

scale showing the level of rainfall measured in millimetres

230
220
210
200
190
180
170
160
150
140
130
120
110
100
90
80
70
60
50
40
30
20
10
0

239
222
168
175
163
153
110
95
67
59
57
22
15
10
6
5
0
0

J F M A M J J A S O N D J F M A M

January 1982 **Time scale** measured in months December 1982 January 1983

132

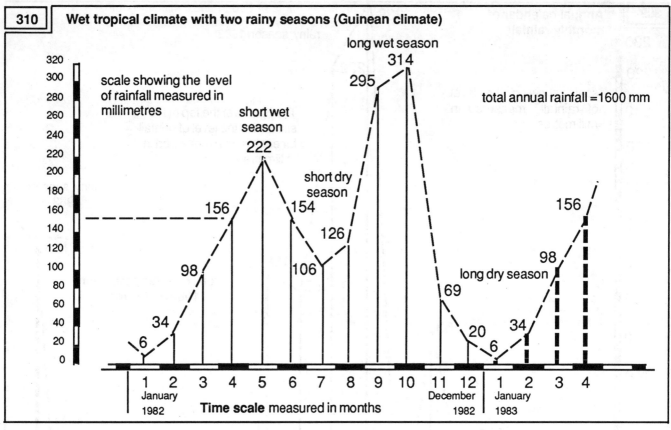

310 | Wet tropical climate with two rainy seasons (Guinean climate)

scale showing the level of rainfall measured in millimetres

long wet season
314
295
short wet season
222
short dry season
154
156
126
106
98
69
34
20
6
6
34
98
156

total annual rainfall =1600 mm

long dry season

1 2 3 4 5 6 7 8 9 10 11 12 | 1 2 3 4
January 1982 — December 1982 | January 1983
Time scale measured in months

Sudanian climate

The second example **(figure 311)** is that of a **wet tropical climate with one rainy season**. The only season of annual rains extends from May to November. In plains and on plateaus there is only one agricultural season unless irrigation is provided for during the long dry season lasting from December to April. This climate, also referred to as **Sudanian**, is found in the regions of Banjul,

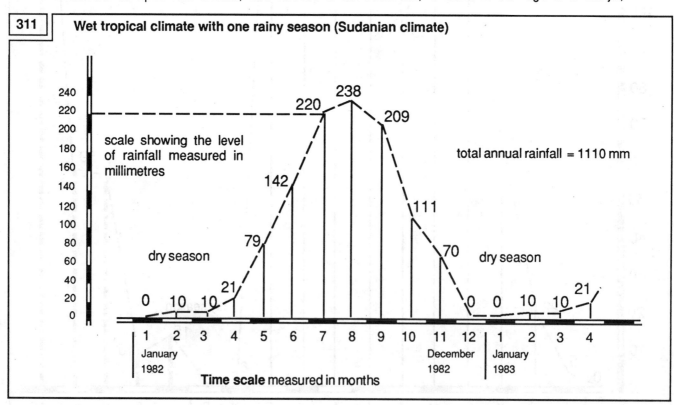

311 | Wet tropical climate with one rainy season (Sudanian climate)

scale showing the level of rainfall measured in millimetres

238
220
209
142
111
79
70
dry season
dry season
0 10 10 21
21
0 0 10 10

total annual rainfall = 1110 mm

1 2 3 4 5 6 7 8 9 10 11 12 | 1 2 3 4
January 1982 — December 1982 | January 1983
Time scale measured in months

Ziginchor, Bissao, Bamako, Bobo-Dioulasso, Kaduna, Sahr, and in parts of Sudan and Ethiopia. South of the equator, this climate prevails in the region of Nairobi, the greater part of Tanzania, around Lubumbashi, Harare, Beira, Maputo and Southern Angola.

Sahelian climate

The **Sahelian** climate is illustrated in **figure 312**. The rainy season lasts from June or July to September or October, but rains are often unreliable and sudden. Given rainfall duration and intensity, one short agricultural season is just about possible ; the crops preferred are millet and other cereals with short vegetative cycles. Vast zones of pastoral stockbreeding are found in Sahelian climates.

The regions with a Sahelian-type climate are Saint Louis, Nouakchott, Timbuktu, Niamey, Agades, Abeche, Khartoum, Mogadiscio and East Kenya.

Three other types of climate are found in Africa but their annual rainfall is not presented in graph form.

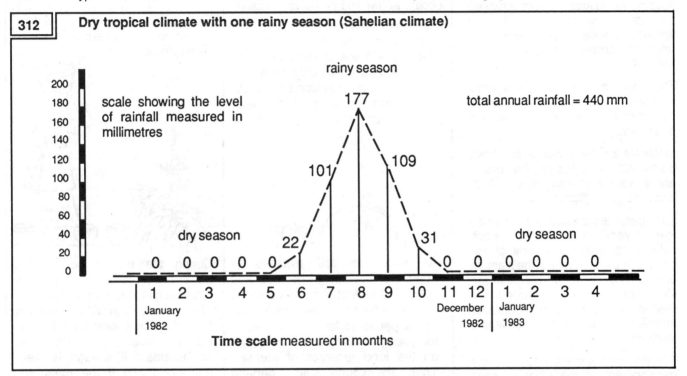

312 | **Dry tropical climate with one rainy season (Sahelian climate)**

scale showing the level of rainfall measured in millimetres

rainy season

177

total annual rainfall = 440 mm

101   109

dry season   22   31   dry season

0 0 0 0 0   0 0 0 0 0

1 2 3 4 5 6 7 8 9 10 11 12 | 1 2 3 4
January   December January
1982   1982   1983

Time scale measured in months

Equatorial climate

The **wet equatorial** climate is found near the equator, for instance, in regions such as Abidjan, Douala, Franceville, Ouesso, Bandundu, Kisangani. Rainfall is abundant and well-distributed throughout the year. Crops can be sown at different times in the year, and there are no well-defined seasons or agricultural cycles. Crops of the same species, for example, paddy rice, exist at many different stages of growth simultaneously in neighbouring fields.

Arid climate

The whole of the Sahara has an **arid** climate. Rains are very infrequent and very localized. This climate covers Egypt, Northern Sudan, Djibouti, Northern Somalia, North Niger and Chad, for example. It is also found in Southern Namibia and in part of South Africa. Agriculture is only feasible in the few places where water is available for irrigation.

Mediterranean climate

Both extremities of the continent enjoy a **Mediterranean** climate, North Africa, particularly round Rabat, Algiers, Tripoli and Bengazi ; South Africa in the area near the Cape of Good Hope.

This climate is influenced by the proximity of the sea. Conditions for farming are satisfactory, but the plants cultivated are different from those of tropical regions.

Finally, there are some local climates on which mountain altitude has a marked effect - in the Fouta Djalon region in Guinea, on the hills of Rwanda and Burundi, and on the mountains of East Africa, Ethiopia and on the island of Madagascar.

Lesson 27

🔲🔲🔲🔲🔲🔲🔲🔲🔲🔲🔲🔲🔲🔲🔲🔲🔲🔲🔲🔲🔲🔲🔲🔲🔲🔲🔲🔲🔲🔲🔲🔲🔲🔲🔲

When do plants need water ?

Plants and the water they need

Lesson 23 discussed **life cycles** and how plant life unfolds in successive stages. It was stressed that plant needs, especially the amount of water they require, vary from one phase of development to the next.

Then two important concepts were examined - the idea of rainfall calendars and the idea of agricultural calendars.

In this lesson, the seasonal life of two plants will be unfolded. The plants are a variety of yam **(figure 313)** and of sorghum **(figure 314)**.

The figures are divided into columns - five for yam, six for sorghum. Each column corresponds to one of the phases in the life cycle of the plant. In each column, various aspects of the life of the plant are mentioned and related to rainfall at that particular moment and the husbandry involved.

The drawings show the approximate development of the plant at each step in the cycle and the amount of water required in the form of water jars.

| 313 | The seasonal life of a variety of yam and the water it needs |

The first stage in the life cycle is the planting and establishment of the plant. It occurs at the end of the dry season

The yam cutting, or set, can be sown during the dry season, in which case it lies dormant until the soil becomes sufficiently moist. It should be dipped in wood ash first, to provides nutrients and some protection against some soil pests.

As soon as soil moisture is adequate, the set germinates and sends out its first shoots and roots. This is the phase of emergence.

For a period of four to six weeks, the young climbers and leaves feed on the food reserves of the set. Thus, the young stems depend very little on the roots that are given time to develop if soil moisture is adequate.

If the set is of good quality, food reserves enable the plantlet to avoid serious damage in case of drought.

The farmer often puts a layer of leaves or straw over the mound to save the plant from direct exposure. This practice is called **capping**. It helps to shade the emerging shoots and keep the mound moist.

Farm work involves making the mounds, cutting the sets, planting them or the top of the mounds and capping them.

The second stage is that of growth. It takes place when the rainy season is well-established.

During this phase, growth is first observed in climbers and roots, then in foliage.

During the growth period, the plant needs a lot of water to produce climbers and leaves. The food reserves that will be stored in the tubers are produced by the leaves. Consequently, they will be abundant if leafage is well-developed and if the roots are deeply penetrating.

This stage of growth, lasting about ten weeks, must occur at the height of the rainy season because the plant needs large amounts of water.

If foliage is thinly developed, the food reserves in the tuber will be low and result in a poor harvest.

Farm work at this time involves staking the plants, weeding and hilling up (earthing up) the mounds. Stakes are cut and set into the top of the mounds.

The third stage is the formation of tubers and it also takes place in the heart of the rainy season.	**The fourth stage is the ripening of the tubers.** They mature at the end of the rainy season and the beginning of the dry season.

	The tubers first swell up with food reserves and water.	Maturation comes next.

When climbers and leaves are plentiful, the plant can start forming tubers. Tuber formation coincides with flowering and lasts roughly two weeks.

This is the time when a shortage of water can inflict the worst damage and even cause crop failure. Thus the formation of tubers must occur during the wettest weeks of the season.

Farm work only entails maintaining mounds and furrows.

When the tubers begin to grow on the roots, they thicken, a process that lasts seven to eight weeks.

By now, water needs have decreased because less water is required to form tuber matter than leaf matter.

If the previous rains have moistened the soil deeply, the risk of a water shortage is slight because a deep rooting system will have had time to develop.

Little husbandry is required at this stage.

Ripening takes about four weeks. The plant carries to the tubers the food and reserves produced in the leaves and stems.

Now the roots are no longer very active and die. The tubers do not depend any longer on rain or soil moisture.

When tubers are fully enlarged, too much moisture can cause rotting. This is one reason why planting yams on mounds makes good farming sense, because mounded earth is always drier than earth in furrows.

Farm work is heavy during harvest and transport operations.

314	The life of a sorghum plant and its water needs

The first phase is the germination and the establishment of the plant

Germination takes place with the onset of the rains.

When planted, the grain of sorghum is dry. It must be given water in order to germinate. If the grain finds water in the soil, germination takes two or three days. Consequently, sorghum should only be sown after the first rains, when soil moisture has been restored.

When the seed is well swollen, it sends out a young root (a radical) from which the young plant will develop. The radical and stem appear after about five days.

Imbibition (swelling) of the seeds and germination combined take about eight days.

While the seed is dry, the plant is not in danger but as soon as the radical and stem appear, the young plant can suffer badly from drought.

The plantlet feeds exclusively on the scanty reserves stored in the seed. It is not strong enough to take up water and soil nutrients since the root system is still not well-established.
If soil water is not available at this point, the plant may wilt permanently.

Farm work mainly involves opening the field (clearing, cleaning and tilling) and planting. The farmer must wait until the very first rains have softened the ground for seedbed preparation.

Emergence takes place at the beginning of the rainy season when the superficial soil layers are quite moist.

The first leaves can now be observed. The roots are already able to take up water and soil nutrients, but they are short and sparse.

During emergence, the soil is bare. It is directly exposed to the sun that dries up the top soil layers where the young roots are growing.

The plant is very sensitive to drought in the top soil layers. If it stops raining, the plantlets have not enough water reserves to withstand drought, and run the risk of permanent wilting or damping off of seedlings.

Farming practices that help reduce the evaporation of soil water at this time are useful.

Husbandry is nil.

Tillering occurs when rainfall is heaviest.

Tillering is the process whereby the plant sends forth shoots from the base of the stem, thus increasing the number of grain-bearing heads.

At the same time, deeply penetrating roots begin to develop. The deep soil layers must be well supplied with water.

Growth is continuous from the 15th to the 35th day after seed imbibition, that is, from the third to the sixth week.

Tillering may be inhibited if rains were not sufficiently abundant to moisten the soil in depth. Tillering is also endangered if rainwater runs off into furrows instead of penetrating the soil.

Any farming practice that promotes water infiltration is beneficial at this point.

Farm work involves field maintenance such as weeding.

The second stage in the vegetative cycle is that of vegetative growth, also called shooting or stem elongation in cereals. This growth takes place during the wettest months of the season.

Shooting is characterized by the growth of the stems to a height of 2.5 m to 3 m, and occurs between the 35th and the 85th day (from the fifth to the thirteenth week) after the germination of this variety of sorghum.

Sorghum needs huge quantities of water and mineral salts to produce its tall stems and leaves. If a sorghum plant is weighed when stem elongation (and therefore growth) has ended, it would seem that the plant takes up and transpires into the air considerably more than 100 times its weight in water.

At this phase in the vegetative cycle, the plant soaks up a lot of water. However if there is no rain for a few days, plant development slows down and growth resumes when conditions become favorable again.

Risk of wilting is slight because the plant has already built up reserves, and deep soil layers make water available at least to some extent.

However, diseases and pests can be a serious threat.

Farm work entails field maintenance and weeding.

The third stage consists of ear emergence, pollination of the flowers, and formation of fertilized spikelets.
The rains have eased off and the end of the rainy season is in sight.

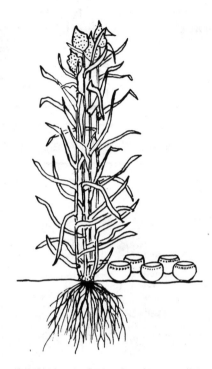

After stem elongation, ears begin to form from the 80th to the 105th day (from the 12th to the 16th week).
As in the previous phase, **water needs are high because the plant is producing a great deal of living matter.**

There are ample reserves of water in the soil supplied by generous rainfall over a period of many weeks resulting in deep soil infiltration.

Nevertheless, substantial losses in yield may occur when the flowers open for pollination. If the filaments or slender stalks bearing the pollen and the receiving stigmas are dried up, the grains may be empty.

The fourth stage is marked by maturation or ripening of the grains. It straddles the end of the rainy season and the beginning of the dry season.

At this time the grains enlarge with food reserves. To reach maturity, the grains take up the small amount of water needed from the stalks that dry up gradually. During the ripening process, the roots are inactive and die, while stalks are no longer rain-dependent.

In wet as in dry conditions, there is enough water in the long stalks and leaves to feed the grains properly.

Incidentally, sorghum grain, contains little water (unlike other crops, yam, for example). By the 140th day (20th to 21st week),

If everything has gone well in the previous phase of the cycle, sorghum is now drought-resistant. Intercropped plants, if any, can take up soil water without competing with the sorghum.

Farm work includes harvesting, transportation and storage.

138

Summary

Figures 315 and 316 summarize **figures 313 and 314** by representing the life cycles and water requirements of yam and sorghum measured in weeks.

The life cycles of these varieties of yam and sorghum last respectively 27 and 18 weeks from

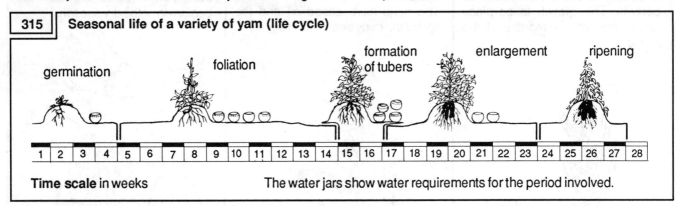

315 **Seasonal life of a variety of yam (life cycle)**

germination foliation formation of tubers enlargement ripening

| 1 | 2 | 3 | 4 | 5 | 6 | 7 | 8 | 9 | 10 | 11 | 12 | 13 | 14 | 15 | 16 | 17 | 18 | 19 | 20 | 21 | 22 | 23 | 24 | 25 | 26 | 27 | 28 |

Time scale in weeks   The water jars show water requirements for the period involved.

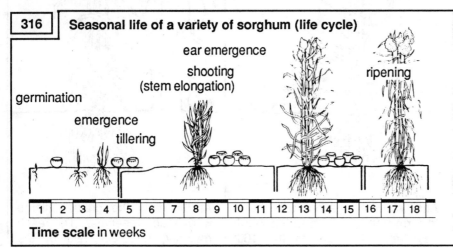

316 **Seasonal life of a variety of sorghum (life cycle)**

ear emergence
shooting (stem elongation)
germination
emergence
tillering
ripening

| 1 | 2 | 3 | 4 | 5 | 6 | 7 | 8 | 9 | 10 | 11 | 12 | 13 | 14 | 15 | 16 | 17 | 18 |

Time scale in weeks

germination to ripening. The total duration of the seasonal life of these varieties is therefore 27 weeks for yam and 18 weeks for sorghum.

Some varieties of yam have a vegetative cycle of 32, 36 or 40 weeks. They are **late** varieties compared to the one examined here. Others have a life cycle of less than 27 weeks and are referred to as **early** varieties.

Like all cultivated plants, the same is true of sorghum - there are early and late varieties with a corresponding long or short life cycle.

The life cycles of cultivated plants and rain calendars

Lessons 26 and 27 discussed rain calendars and their importance to the farmer. Following on from there, the life cycles of yam and sorghum are now plotted along their respective rain calendars (**figures 317 and 318**).

Figure 317 shows two yam cycles. The first crop was sown late, ten weeks after the onset of the rains. As a result, the plant cannot thrive at the end of the life cycle because it is short of water from the 17th week onwards, at the time of tuber enlargement.

In contrast, the second crop was sown early. The sets were already planted in April. The crop therefore got the full benefit of the rainy season and, as shown by the rain calendar, tuber enlargement took place in August and September when the rains were still in progress.

The life cycle of sorghum is illustrated at the bottom of **figure 318**, as though the crop had been sown in April with the first precursory showers of the season. It would then have suffered from drought during emergence. Early planting might have caused poor yields.

On the other hand, if the sowing season is postponed until May (7th week), when the rains are well-established, at no time is the crop threatened by drought, even during shooting (stem elongation) and ear emergence when water requirements are high.

However, if the sowing season had been too late, in July for instance, ear emergence would have coincided with abated rainfall, resulting in crop failure.

317 The life cycle of a variety of yam as plotted on a rain calendar

scale showing the level of rainfall measured in millimetres. The sticks give weekly rainfall.

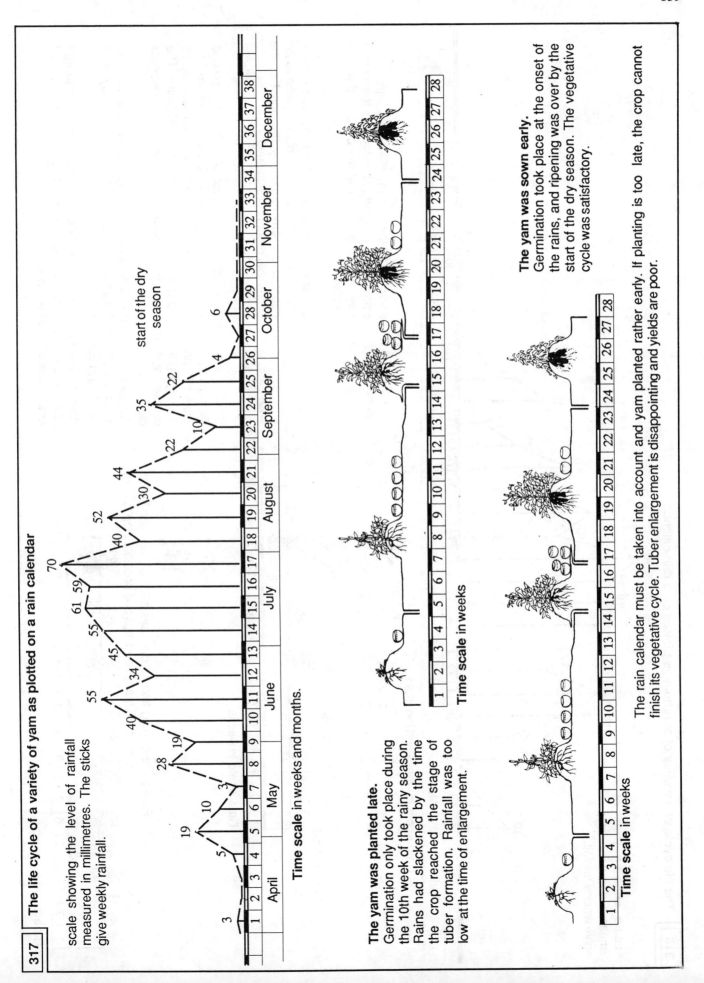

start of the dry season

Time scale in weeks and months.

The yam was planted late.
Germination only took place during the 10th week of the rainy season. Rains had slackened by the time the crop reached the stage of tuber formation. Rainfall was too low at the time of enlargement.

Time scale in weeks

The yam was sown early.
Germination took place at the onset of the rains, and ripening was over by the start of the dry season. The vegetative cycle was satisfactory.

Time scale in weeks

The rain calendar must be taken into account and yam planted rather early. If planting is too late, the crop cannot finish its vegetative cycle. Tuber enlargement is disappointing and yields are poor.

318 The life cycle of a variety of sorghum as plotted on a rain calendar

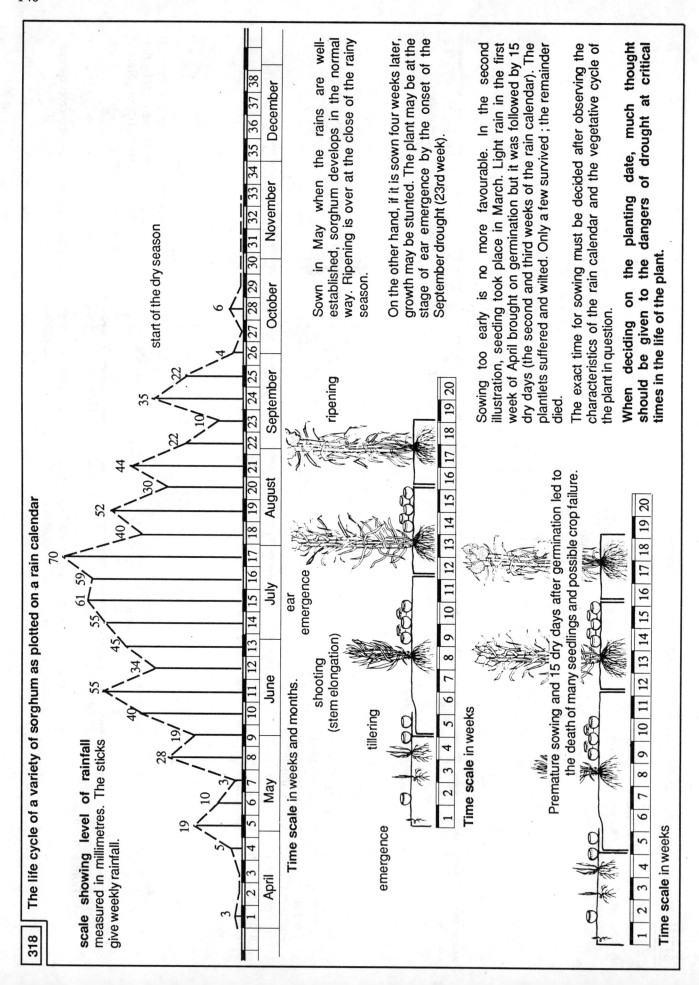

scale showing level of rainfall
measured in millimetres. The sticks give weekly rainfall.

start of the dry season

Time scale in weeks and months.

emergence

tillering

shooting (stem elongation)

ear emergence

ripening

Time scale in weeks

Premature sowing and 15 dry days after germination led to the death of many seedlings and possible crop failure.

Time scale in weeks

Sown in May when the rains are well-established, sorghum develops in the normal way. Ripening is over at the close of the rainy season.

On the other hand, if it is sown four weeks later, growth may be stunted. The plant may be at the stage of ear emergence by the onset of the September drought (23rd week).

Sowing too early is no more favourable. In the second illustration, seeding took place in March. Light rain in the first week of April brought on germination but it was followed by 15 dry days (the second and third weeks of the rain calendar). The plantlets suffered and wilted. Only a few survived; the remainder died.

The exact time for sowing must be decided after observing the characteristics of the rain calendar and the vegetative cycle of the plant in question.

When deciding on the planting date, much thought should be given to the dangers of drought at critical times in the life of the plant.

Labour availability

By looking at the agricultural calendars of the four fields in the Sudan and, indeed, by observing rural life in general, it becomes obvious that the amount of work to be done in the fields varies from one period of the agricultural year to the next. Farm work begins in March, shortly after the first rains that fall in February. Seedbed preparation and planting stretch from March to July. All the workers on the farm are very busy at this time.

The crops grow during the months of August, September and October. Rainfall is abundant. Field operations are limited to weeding and maintenance. Farm workers are underemployed.

However, when harvesting begins, mainly in November and December, activity is intense once again - the crops are harvested, transported from the fields to the village and then stored. The rains end in November ; it is the start of the dry season. It will not rain again until the end of February.

In January and February, farming is almost at a standstill. This is the time for a range of other activities such as fishing, repairing household dwellings, social gatherings, journeys, collective tasks and so on.

The shorter the rainy season, the more acute are the fluctuations in farm work. Farm work is entirely dependent on **rainfall intensity**, a point that will be discussed later on.

In the case of the Sudanese village, the lull in farm work is fairly short. It lasts two months and is taken up with non-agricultural tasks.

However, in more arid zones where rains last only a matter of weeks, work on the farm stops for a much longer period, even up to five or six months.

This happens in Sahel zones and explains why so many agricultural workers are forced to abandon their villages in the dry season to look for work in towns or elsewhere, because economic idleness is never a source of revenue.

This undesirable situation also underlines the value of irrigation in the dry season when it enables the farmer to extend the cropping season. Individual or collective operations that give work during the dry season, thus lengthening the duration of agricultural work on farms and particularly the cultivation of valley bottoms, are always a worthwhile contribution towards modernizing agriculture.

Remember

■ *Agriculture is a perpetual cycle of repetition. Every year at a given time in the civil calendar, the farmer has to begin the same operations all over again - clearance, tilling, sowing, field maintenance, harvest. This recurring activity is referred to as the agricultural or cultural cycle.*

■ *The agricultural cycle depends on plants. Every year, they too live once more through the successive phases of their vegetative cycle - germination, emergence, growth, fruit formation, maturation.*

■ *The progression of the life cycle of plants depends on the cycle of seasons that, in turn, depends on the rotation of the earth round the sun.*

■ *Every region, every locality, has a specific agricultural calendar into which the farm worker's labour is inserted. The agricultural calendar is itself shaped by what the cultivator wants to do, what he can do, what he cannot do (see Lesson 2). In particular, he cannot force plants to grow in an unsuitable climate.*

Notes

Lessons 28 to 30

Plant varieties and seeds for farming

Seeds are those plant parts which serve to reproduce and multiply plants. Obtaining good seeds, adapted to local environment, is one of the most important aspects of farming.

Good, well-adapted seeds are those with all the qualities looked for in agricultural plants. The qualities are those required by the consumer product and by the crops, the qualities which contribute to soil conservation, and all the qualities that help achieve agricultural goals as put forward in Lesson 2.

The first step is to classify cultivated plants into **species** and **varieties** (Lesson 28). The **seeds** of these plants, their origins and qualities will be discussed next (Lesson 29). Lastly, the economic problems connected with seed selection will be examined. This is a problem that farmers face all the time. Is it advisable to have one's own supply of seed on the farm or to look for it elsewhere ? The answer to this question depends on many factors (Lesson 30).

Lesson 28

Plant varieties

Plants are grouped into families, for example, *Gramineae*, with sub-groups called **genera** (singular **genus**), for example, *Sorghum*. In each genus there are distinct types called **species**, for example, *bicolour*, and each species has many varying forms called **varieties**, for example, Short Kaura.

Sorghum, banana, bean, fan palm, okra, shea butter, in fact , every known plant, bears the name of a **species**. The meaning of species is illustrated by comparing fan palm and bean. They are unmistakably different plants that cannot be interbred any more than cats and dogs. They belong to different species.

All the plant species mentioned in this book are listed on page 290 with the name in English or in the vernacular and the scientific name in Latin.

Some plants belonging to different species look alike although they cannot be interbred. They are said to belong to the same **family**. For instance, *Gramineae* **(Grasses)** form a family of plants and include sorghum, millet, wheat and maize, etc. **Palms** form another family with species such as oil, coconut, fan,

319

320

321

raffia and date palms. *Leguminosae* **(Legumes)** are another family with many species. The student is already familiar with bean, groundnut, soya beans, pea, earth pea, tropical kudzu *(Pueraria)*, *Acacia*, locust bean and other members of this family of plants.

Plants of the same species are not necessarily identical. Some features may differ, for example, shape of leaves and flowers, colour of seeds, shape of stems. This is so when the same species comprises many **varieties**. **Figures 319, 320 and 321** illustrate three varieties of yam, each with different leaves.

It is essential for the cultivator to know the **varietal characters** or distinctive features of every variety grown on the farm.

With rare exceptions, varieties of the same species can interbreed. When interbreeding takes place, the characteristics of the two varieties are combined.

322

Determining the characteristics of the plant varieties in use

Endless questions can be asked to determine varietal characteristics since every plant has its own characteristics and particular utility for man. **Here is a series of questions that will help the cultivator to determine at least some varietal characteristics of special interest to him.**

■ Is it an **early (early-maturing)** or a **late** variety ?

■ Does it demand **light** or **shade ?** The answer to this question will decide how the plant should be treated - exposed to the sun in the first case, kept in shady conditions in the second. The yam in **figure 322** is a shade variety ; the three varieties in **figures 319, 320 and 321** are light demanding.

■ Has the variety a **deep** or **shallow root system ?** Lessons 31 and 32 explain the importance of this characteristic. Mixing varieties with different rooting systems makes for optimal use of soil space.

■ Is the variety **social or not ?** Some varieties tolerate the presence of other plants ; some do not. Knowledge of this varietal characteristic influences the choice of farming practices. Some varieties are not merely social ; they are positively beneficial for associated plants **(figure 329, page 147)**.

■ Is the variety **pest-resistant or not ?** Resistance should be examined for every pest connected with the plant and particularly for pests active in the locality.

■ Are stems **erect, climbing** or **creeping ?** The answer is important when it comes to growing plants in mixture to avoid interplant disturbance. It must be remembered that climbing, creeping and erect varieties of the same species use light in different ways.

■ Varieties differ in **adaptability**. This means the quality of adapting to climatic conditions without undue harm. For example, a variety of maize is deemed non-adaptable if ten dry days at the

323

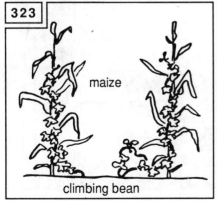

maize

climbing bean

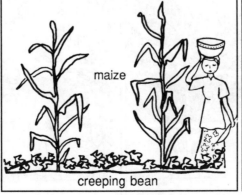

maize

creeping bean

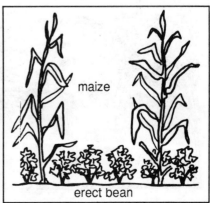

maize

erect bean

shooting stage (stem elongation) cause poor yields. Another variety is said to be adaptable if, after the same ten days drought, it resumes growth, makes up for lost time and gives satisfactory yields.

- Can the variety be **propagated by cuttings or not** ?

- Is the grain **soft** or **hard** ? This is an important point where storage, milling and cooking are concerned.

- Is the variety **sweet** or **bitter** ? For instance, cassava tubers of the bitter variety cannot be eaten straight away ; they contain a toxic acid that must first be destroyed by fermentation in water.

- Is the variety **fertile** or **sterile** ? It is fertile if it produces seeds that germinate. It is sterile when it does not produce any seeds or produces seeds that fail to germinate.

- Varieties of maize, sorghum, millet, rice, avocado and cocoa produce fertile seeds capable of germinating. The cultivated

tea plantation

drought-resistant variety

324

variety struck by drought

banana is always sterile. It can only be propagated from suckers, though some varieties of wild banana produce seeds. The same is true of most varieties of cocoyam and pineapple.

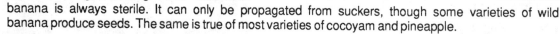

- Is the variety **drought-resistant or not** ? Two varieties of tea are illustrated in **figure 324**. The variety on the left cannot withstand drought. It has perished after four months drought. The variety on the right is resistant. It is still green after the same period of drought.

- How are the varieties **used** ? Varieties can be characterized by the ways in which the produce is used. For example, groundnut varieties are sorted into those good for eating and those better for oil extraction. The eating varieties have a higher protein content, the others are richer in oil (Lesson 47). Some varieties of banana are eaten raw, others are always cooked or used for beer-making. Sugar cane is produced for manufacturing sugar or cultivated to be eaten raw like a piece of fruit.

- As for trees, varieties can be characterized by **tall** or **short stems**. Tall varieties may grow to great heights, while short varieties remain dwarfed. Height must be taken into account for fruit picking. Tall trees must be climbed to pick fruit ; fruit on low trees can be picked from the ground. The height of trunks also influences agricultural practices as seen in **figure 325**.

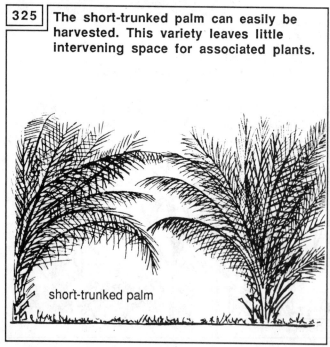

325 **The short-trunked palm can easily be harvested. This variety leaves little intervening space for associated plants.**

short-trunked palm

The tall-trunked palm is harder to harvest. There is more room for annual and perennial intercrops.

tall-trunked palm

■ Is the variety **stable or unstable** ? In cases where the qualities of the plants propagated from seed are identical to those of the parent plants that produced the seed, the variety is said to be stable. Seeds can be planted every season and the qualities of the variety maintained. However, where seeds produce plants with qualities different from those of the parent plants, the variety is said to be unstable.

If the cultivator has a wide choice of plant varieties at his disposal, he will be able to plan and manage cultivation better because he is in a position to decide on the crop combination best suited to each plot of land. Consequently, **if the future of farming is to be guaranteed, farmers must know exactly what plant varieties grow on the settlement, keep these varieties and not adopt new, outside varieties until their value and endurance have been tested.** It is not generally wise to grow only one variety on a large scale. There would be the danger of severe damage by a disease. This would be unlikely to damage different varieties to the same extent.

Selection or improvement of varieties

The **selection** of varieties is usually carried out by research stations or seed firms using a large number of varieties collected from many parts of the world. Researchers select, from among the varieties at their disposal, those with characteristics best adapted to the regions demanding improved varieties.

A research station, in West Africa, for example, will grow varieties of sorghum sent in from Central and Southern Africa, Asia, North and South America. The behaviour of the varieties is monitored at the station in order to determine varietal characteristics - are the varieties early or late, sensitive to drought, long- or short-stemmed, disease-resistant, large- or small-eared, and so forth.

| 326 | **Crossing two varieties of sorghum** |

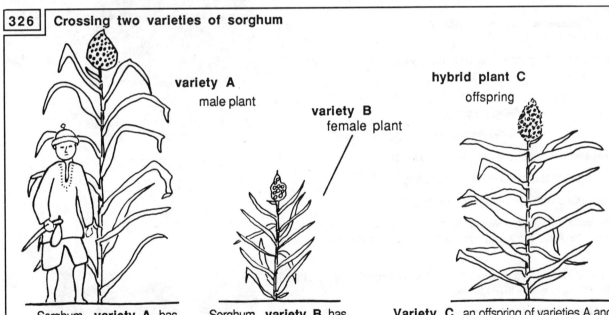

variety A
male plant

variety B
female plant

hybrid plant C
offspring

Sorghum **variety A** has red seeds, wide curving leaves, tall stems. Ears are large but sensitive to drought and disease.

Sorghum **variety B** has white seeds, erect slender leaves, short stems and small ears. This is a hardy variety.

Variety C , an offspring of varieties A and B, is **hybrid** . It has spreading leaves, medium height stems and pink seeds. Ears are rather large with small thorns. Drought-endurance is satisfactory but the variety is prone to disease.

The hybrid (crossed) variety may prove more advantageous than the male and female varieties and will then be referred to as an **improved variety** with **improved seed** because field performance is better than that of the parent plants.

Still, we must beware of hasty conclusions. If plants of varieties A and B are interbred, they always produce C. But the offspring of variety C may well lose the qualities of that variety. For that reason, the hybrid variety is said to be unstable. Hence, if the performance obtained from C is to be maintained, the cultivator **is forced at regular intervals to buy new hybrids seeds** interbred from A to B.

It should be noted that every valuable hybrid variety may be offset by many of no reproductive value.

When monitoring is complete, the most promising varieties are selected and distributed to local farmers to be tested. If results are satisfactory, the seeds are then multiplied in fields attached to the station or on special **seed farms** before being distributed to farmers for sowing in their fields.

By careful observation, it is also possible to select certain plants with a view to combining their characteristics because it is believed that the offspring (in the form of new varieties) will respond more fully to farming goals. Plants of different varieties are crossbred and the characteristics and behaviour of the new varieties are observed in experimental fields. This technique is called **intervarietal crossing, hybridization** or **crossbreeding**. A hybrid or interbred plant is one that combines the qualities of the parent plants.

Figure 326 gives an example of crossbreeding between two varieties of sorghum.

Smallholders would find the kind of seed multiplication described above too complex and too costly, in terms of production factors, to start carrying it out on their own farms. However, they should persistently demand the selection and multiplication of seeds with characteristics adapted to the farming conditions with which they contend.

Concrete examples of seed selection and improvement that should be feasible :

■ a cereal with vigorous head growth and well-held grain that will not shed during harvesting, and with wind-tolerant, disease-resistant stems ;

■ a variety of cassava, less bitter, with vigorous tuber development, resistant to mosaic virus ;

■ an early variety of upland rice with high yields and a rather short vegetative cycle ;

327

■ oil palms with ample bunches, fruit with high oil content, nuts easy to crack and short trunks to facilitate cutting ;

■ maize with soft grain that take less time to cook than hard varieties ;

■ forage grass with more abundant, juicier blades ;

■ a variety of bean that grows well in association ;

■ a variety of potato with good keeping qualities ;

■ a mango with fat, juicy, sweet fruit, small-seeded compared to the size of the fruit, trunk not too high in order to facilitate picking **(327)** ;

■ a high-yield variety of yam, climbing rather than creeping, so that it can be grown more easily in mixtures ;

■ a variety of legume that fertilizes the soil better ;

■ an erect variety of groundnut with high yields, resistant to rosette virus ;

■ a productive cocoa tree resistant to black pod disease ;

■ dwarf sorghum with heavy tillering, ears well protected from bird damage by hairs and barbs **(328)**.

Generally speaking, all the varietal characteristics mentioned at the beginning of this lesson can be obtained by the selection and improvement of plant varieties and seeds. However, in order to breed seeds really adapted to local environments, farmers and researchers together must **decide what improvements they are looking for.** Without preliminary consultation, research into selection and improvement could produce so-called improved seeds that are, in fact, less efficient on farms than local varieties.

tall-stemmed sorghum

short-stemmed sorghum

328

Selection and improvement programmes should be carried out **near the farms**, so that they can be geared to the farmers' needs, goals and working conditions.

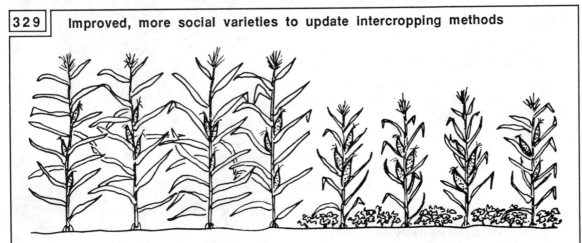

| 329 | **Improved, more social varieties to update intercropping methods** |

Tall maize stalks make interplanting with legumes difficult, because the legumes are deprived of light.

The shorter stalks of the improved variety give less shade and are better suited to the interplanted legumes. This line of maize can be planted more densely. Legume yields remained unchanged but maize production rose thanks to selection.

Lesson 29
Planting material

cotyledon

plumule

integument

radicle

331

There are **two main types of planting material :**

- **seeds** resulting from the fertilization of flowers ;

- **vegetative parts** that are pieces of stems, leaves or roots. People most often refer to cuttings, but there are other kinds of vegetative parts that will also be discussed in this lesson.

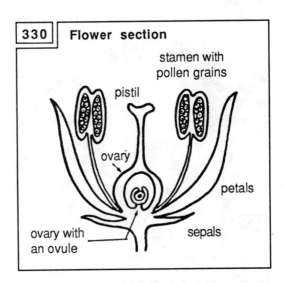

| 330 | **Flower section** |

stamen with pollen grains

pistil

ovary

petals

ovary with an ovule

sepals

Some plants cannot be reproduced by seed, cocoyam, banana and sugar cane, for example. Some cannot be propagated by cuttings, for instance, bean, maize, okra. Others can be multiplied easily by seed or by vegetative propagation as desired, coffee, guava, onion and so on.

When a plant is propagated by seed, one can never be sure that the characteristics of the offspring (progeny) will be exactly the same as those of the parent plant because interbreeding has taken place. On the other hand, vegetative reproduction ensures that the characteristics of the offspring are absolutely identical to those of the parent plant. (This is called cloning.)

Seeds

Where do seeds come from ?

Seeds are the offspring of plants, born in flowers. They always result from the **union of male and female reproductive cells**.

These cells can be observed in flowers around us, beginning with very simple specimens. A flower section is illustrated in **figure 330** as though it had been cut in two with a razor blade.

■ The male reproductive cell is the **pollen** grain, the yellow or pinkish dust that is shed when flowers open out. Pollen is formed in the **anthers** that burst, shedding the pollen shortly before or just when flowers open. Pollen is very light and is transported by wind, insects and water.

■ The female reproductive cell is the **ovule**. The ovule or ovules are contained in that part of the flower called the **ovary**. Ovules are usually very tiny and can only be examined under a magnifying glass or a microscope once the ovary has been sectioned in two. The ovary itself is generally visible to the naked eye at the base of the flower or when the **petals** (coloured parts of flowers) and the **sepals** (green parts of flowers) have been removed.

The ovary is always covered by the **pistil**, an organ of varying length and thickness that receives the pollen grains.

332

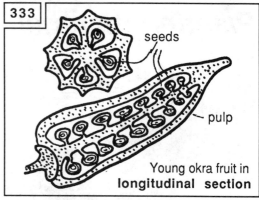

333

seeds

pulp

Young okra fruit in **longitudinal section**

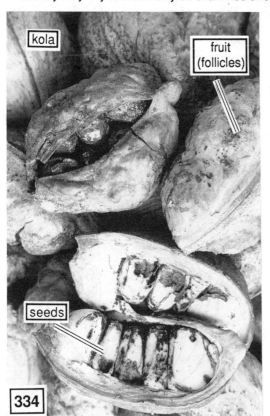

334

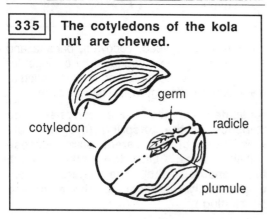

335 | **The cotyledons of the kola nut are chewed.**

germ

cotyledon

radicle

plumule

Sometimes, fertilization takes place between the ovules and the pollen of the same plant. The seed formed of this union will produce a plant very like the parent plant (self-fertilization).

The pollen fertilizing the ovule often comes from another plant, because pollen grain is carried some distance by wind, insects and rain. The characteristics of the plantlet born of this fertilization will be a combination of maternal and paternal characteristics following hybridization. This is a common occurrence when many varieties of the same plant grow in the same field.

The union of a pollen grain and an ovule inside the ovary forms an **egg**. This is the process of **fertilization**. As it grows, the egg becomes a seed that always contains a **germ and reserves**. The germ is made of a **radicle** that produces the primary plant root at a later stage, a **plumule** that develops into the first shoot, and one or two **cotyledons**, sometimes more than two **(figure 331)**. The cotyledons are sacs whose function is to nourish the newborn plantlet until such time as the first root hairs are able to take up soil nutrients. The ovule is the future seed and the ovary the future **fruit**.

336

338

The difference between seed and fruit

The okra fruit illustrated in **figure 332** has been cut in two. Note the seeds, about fifteen or so, still attached to the fruit that feeds them, just as the unborn child is connected to its mothers by the umbilical cord.

Figure 333 shows an okra fruit in longitudinal section.

Kola follicles are pictured in **figure 334**. **Follicles** are fruit that must be split open to remove the **nuts** or kola seeds. The seeds can be cracked quite easily into several parts or cotyledons between which the germ, with its radicle and plumule, lies.

Figure 335 represents an open kola seed and germ.

Cotyledons often look like tiny fleshy leaves that develop during germination and spread close to the ground. They contain the nutritional reserves that allow the plantlet to send out its first roots and make them active.

Some plants germinate but keep the cotyledons inside the seed coat buried in the ground.

The roots of the young groundnut plant **(figure 336)** have been shaken free of earth. The cotyledons are visible just between the roots and the first leaves.

Some flowers in detail

On the flamboyant flower in **figure 337**, a pistil and eight stamens hanging from long white filaments can be distinguished. Below the flower in bloom, is a withered flower with dried-up filaments and the enlarged ovary containing the ovules. The ovary is just under the withered bloom at the end of the short stem supporting the flower.

Figure 338 illustrates a plant from

337

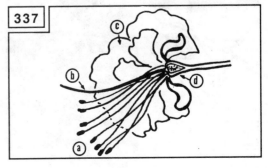

337

Key to the letters in the figures on this page

a stamens
b pistil
c petals
d ovary
e ovary enlarged at the stage of fruit formation
f developing fruit
g insects

the sweet potato family, showing the different organs and the fruit ir formation - flowers, stamen and pistil, ovaries.

The ovaries are well enlarged and already look like the fruit into which they will grow. Ovary enlargement is called the stage of fruit formation. As was stressed in Lesson 24, this phase is of great importance in the life cycle of plants cultivated for their fruit or seeds.

An okra flower and developing fruit are pictured in **figure 339**. Stamens surround the pistil. Many honey-gathering insects can be seen crawling over the flower. They will pick up pollen and carry it to other flowers in their search for nectar.

okra flower

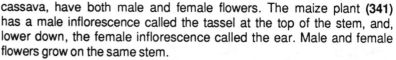

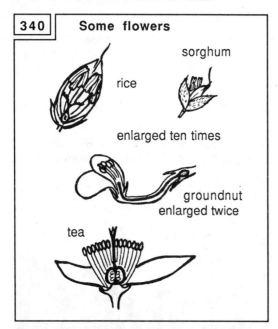

340 **Some flowers**

rice

sorghum

enlarged ten times

groundnut
enlarged twice

tea

In the same figure, fruit in formation can be distinguished. Fruit grows fast - in a matter of days or weeks - and therefore needs a great deal of water at this stage in its development (this point was discussed in Lesson 27).

In many plants, the flowers contain stamens and pollen, as well as an ovary and ovules. **Figure 340** gives examples of this structure.

Other plants such as maize, palm and cassava, have both male and female flowers. The maize plant **(341)** has a male inflorescence called the tassel at the top of the stem, and, lower down, the female inflorescence called the ear. Male and female flowers grow on the same stem.

This is not the case for all plants with male and female inflorescences. Male and female flowers of the pawpaw are illustrated in **figures 342 and 343**. They are growing on separate stems as is usual with this plant.

339

female flowers of the pawpaw

342

male flowers of the pawpaw

343

Seed structure

All seeds have the same basic structure. It consists of :

- an **integument**, or seed coat, of varying softness and hardness ;

- one or more **embryos** or germs. The embryo contains the radicle, or beginnings of a root, the plumule, or tiny shoot, and the first leaves of the future plant ;

- nutritional reserves to nourish the embryo during growth. The reserves are stored either in the seed coats or in the **albumen** found more particularly in cereals.

The avocado seed in **figure 344** was photographed during germination. Note the plumule between two large cotyledons and the radicles.

The reserves stored in the cotyledons nourish the germ during germination until the plantlet is able to fend for itself.

The food stored in the seeds must release enough energy to enable the plant to absorb the water needed during germination, and to form the young stems and roots.

Consequently, and as a general rule, small seeds are weaker and more vulnerable than large seeds. It is essential at all times to help seedlings by reducing the effort demanded of them in two ways - preparing well-loosened soil where radicles can develop easily, and burying the seedling at the right depth, not too far down, so that the plumule will not have spent itself by the time it comes above ground.

It follows that good seeds are those that store the most nutrients as they ripen in the fruit. Seeds harvested too early cannot lay in enough food reserves. A plant hit by drought as the fruit ripens deprives the seeds of nutrients. If weevils destroy some of the reserves, the germs will be badly nourished during the growth period.

The reserves stored in the seed are often the very nutrients of value to human beings. Farmers and scientists therefore share a common aim in their search for plants producing many seeds, well-stocked with reserves abounding in nutritive elements for man.

Two seeds are shown in section in **figure 345**. The maize seed has one cotyledon and a large store of flour. The bean seed has two cotyledons tightly packed into the seed coat. One cotyledon has been removed to reveal the germ.

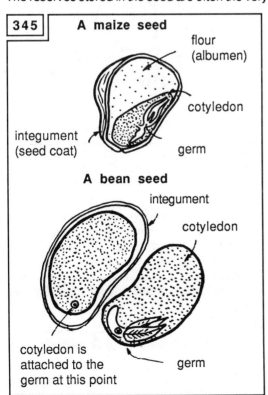

345 **A maize seed**

flour (albumen)

cotyledon

integument (seed coat)

germ

A bean seed

integument

cotyledon

cotyledon is attached to the germ at this point

germ

maize

tassel or male flower

ear or female flower

341

avocado seed during germination

cotyledons

plumule

radicles

344

Notes

Vegetative seeds

Vegetative seeds are those obtained from a plant part without fertilization. This type of propagation ensures that the living matter of the progeny (offspring) contains exactly the same qualities as those of the parent plant.

Reasons for using vegetative propagation

- Some plants produce few seeds or none at all. Yet they can be multiplied quickly by cuttings.

- When two plants of different varieties are interbred, the resulting combination of seed characteristics may prove unsatisfactory. Vegetative propagation is a way of avoiding such unreliable hybrids.

- When a plant variety has agronomically desirable characteristics, the use of vegetative propagation ensures fast multiplication of exact replicas, because many cuttings can be made from a single tree. On the other hand, propagation by seed sometimes requires lengthy research programmes. Another advantage of vegetative propagation is that it can be carried out on the farm, whereas seed selection can only be done at specialized stations.

Different kinds of vegetative seeds

bulbs

346

---

347 **Examples of cuttings**

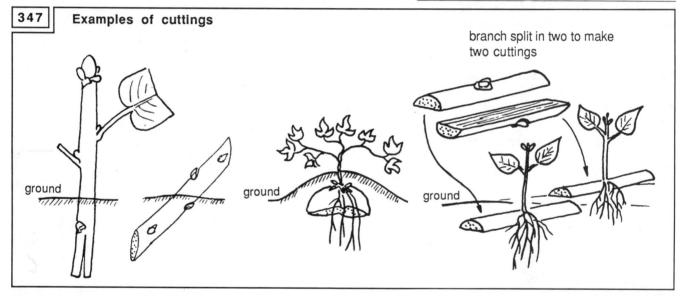

branch split in two to make two cuttings

ground

ground

ground

---

348   **Rich soil** is needed to plant cuttings. They must be shaded and watered regulary. Here are some suitable containers.

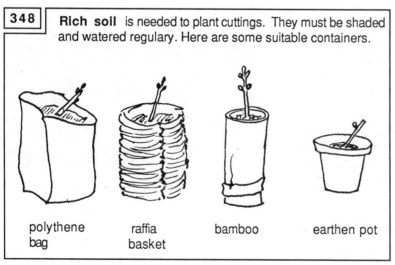

polythene bag   raffia basket   bamboo   earthen pot

- **Bulbs** that form at the base of some species, e.g. onion and garlic **(figure 346)** ;

- **Cuttings**, or parts of stems or roots, pieces of tubers, sometimes even leaves, from which buds and roots can grow. Some cultivated plants are almost exclusively propagated by cuttings, for instance, cassava, yam, potato, sugar cane **(figures 347 and 348)** ;

- **Bulbils (figure 349)** found at the base of stems or in the leaf axils of some species, for example, sisal ;

- **Shoots and suckers** that rise from the parent plant at different levels, e.g. banana and pineapple **(figure 350)** ;

bulbils

349

shoots

350

- **Layers** or cuttings from a branch that remains attached to the parent stem **(figures 351 and 352)** ;

- **Scions** or parts of plants (stems, buds, etc.) grafted onto a rooted plant called the

351 **Layering**

The bark of a branch is removed without wounding the wood.

The branch, earthed and covered in sacking, is watered regulary.

When roots have formed after a few mouths, the branch is severed and the layer planted.

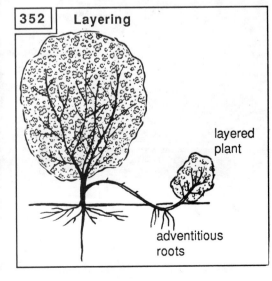

352 **Layering**

layered plant

adventitious roots

rootstock, the idea being to combine the qualities of the scion and the rootstock. This is commonly done with *Citrus* species ;

- **Stolons** or **runners** - kinds of sprouts and stems growing at a distance from the parent plant but still nourished by it (**figure 353**).

How to choose good cutting material

- Good cuttings are always taken from plants known for their **high yields and the quality of their produce**.

- They should come from **healthy stock** showing no signs of disease or pest damage. Compare the cassava plants in **figures 354 and 355**. The first plant is

stolons (runners)

353

diseased cassava

354

healthy cassava

355

diseased and should not be used for cuttings (**figure 354**). The second one is perfectly healthy and is good reproductive material (**figure 355**).

- Cuttings must be **in an active state of growth**, but not too young. For trees and shrubs, the wood and bark of the cuttings should be well developed, though the bark must not be too hard. A stem right for cutting is **mature**.

- They must have **one or more buds** able to form stems and roots.

- They must have **enough nutritional reserves** to last until the roots are well established.

- Finally, **cuttings must not be exposed to the sun after harvesting** or attacked by moulds or other adverse conditions affecting growth activity.

Generally speaking, cuttings must come from perfectly healthy stock of proven quality.

Grafting

Grafting is a special form of vegetative propagation where the qualities of two plants are fused without fertilization. Usually a **rootstock**, hardy and tolerant to local conditions, is united to a **scion** prized for the quality of its produce and high yields. A rootstock is a young well-rooted plant ; only the stem stripped of branches and leaves is used. The scion is a small piece of stem and bark with a bud that, as it grows, replaces the severed stem of the rootstock.

Notes

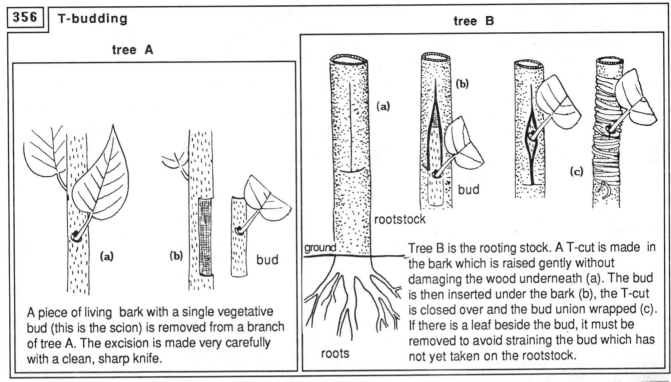

356 T-budding

tree A

(a) (b) bud

A piece of living bark with a single vegetative bud (this is the scion) is removed from a branch of tree A. The excision is made very carefully with a clean, sharp knife.

tree B

(a) (b) bud (c)

rootstock

ground

roots

Tree B is the rooting stock. A T-cut is made in the bark which is raised gently without damaging the wood underneath (a). The bud is then inserted under the bark (b), the T-cut is closed over and the bud union wrapped (c). If there is a leaf beside the bud, it must be removed to avoid straining the bud which has not yet taken on the rootstock.

The grafting technique illustrated in **figure 356** is called T-budding.

A grafted *Hevea* seedling (rubber tree) is illustrated in **figure 357**. The rootstock is planted in the ground and severed about 50 cm from the base.

The scion, a piece of high quality *Hevea*, has been grafted half-way up the stem of the stock plant. The scion has budded and grown into a stem from which the new tree trunk will develop. The rooted stem retains the qualities of the stock plant while the new budded stem has those of the scion.

Mango grafting is common practice in the Orodara region of Burkina Faso **(figure 358)**. The trees are budded on the farms and sold in home and foreign markets - in Ivory Coast, Mali and Ghana.

grafted *Hevea* seedling

scion

rootstock

357

grafted mango

scion

rootstock

358

Good seeds and how to obtain them

Good seeds have three basic qualities.

■ They produce the kind of plant the farmer is looking for, in other words, plant qualities and yields are satisfactory.

156

- They are tolerant to local farming conditions.

- They are healthy, their nutritional reserves ensure germination and they are not diseased.

In order to judge seeds and the plants they produce, one must have a clear idea of the characteristics and qualities desireable in cultivated plants. The following aspects should be taken into account :

- **size, volume and bulk of the parts for consumption** - size of cereal heads, quantity of grain on the heads, size of tubers and fruit, amount of fruit per tree, length of fibres, comparative bulk of the usable and unusable parts, e.g. the amount of pulp extracted from mangos compared to skins and stones, or relative fibre bulk from palm fruit and size of nuts. These are all qualities connected with **plant yields** from the point of view of the quantity of produce obtained. This is referred to as **quantity yields** (Lesson 46) ;

- quality characteristics - **taste, firmness** to touch (tubers, for instance), **flavour** in pineapple, **strength** of cotton yarn, **resistance** of rice grains to breakage when milled, **sweet or bitter quality** of cassava, **seed examined for content** - for **protein, mineral salts, fats and sugars** (Lesson 47) ;

- the characteristics mentioned so far relate to the quantity and quality of the harvested produce. However, other characteristics connected with plant life in the local environment must not be forgotten. **Hardiness** and **resistance** are particularly important.

A plant is **hardy** when it thrives in its environment and when growth is not adversely affected by competition from other plants. Hardiness is a quality associated with wild species.

Hardiness and resistance often go hand in hand. **Resistance** is the capacity of a plant to withstand disease and pest damage. In the same way, a hardy plant usually stands up fairly well to the diseases associated with it. It should be noted that some plant varieties resist specific diseases while being prone to others.

359

360

361

Plant resistance can be attributed to many factors. For instance, leaves, stems, flowers or fruit may produce substances inhibiting the activity of pest carriers. Resistance may also be due to plant structures - hairs that keep insects away from leaves, or thorns that repel birds and rodents. Many other characteristics contribute to plant resistance.

Ways of procuring good seeds on the farm

Selecting plants for seed

Diseased, sickly, low-yield plants are poor reproductive material. Their seeds are of no interest to the farmer. Consequently, the first objective is to pick seeds from healthy plants with all the agronomically desirable qualities, especially sturdiness, resistance to disease, abundant heading, plenty of fruit and seeds, good quality produce.

It is always worthwhile to mark the positions of seed plants in the fields and harvest them separately before the main harvest. An easy technique is to inspect the fields a few days before harvesting and to mark the seed plants with raffia fibre or paper.

Harvesting mature seeds

Immature (unripe) seeds have not laid in all the food reserves required. If used for seeding, they will be unable to nourish the plantlet well until it reaches self-sufficiency. These seeds must be ruled out.

Another disadvantage of immature, as compared to mature seeds, is that they are often moister and poorly protected by the thin texture of the seed coats. As a result, these seeds are hard to store and are a ready prey for fungi, moulds, bacteria and insects. Remember that ripe grain is easier to harvest and is less easily damaged, but if it is left too long, many species are liable to shed their seed before harvest.

About drying seeds (figures 359, 360 and 361).

Seeds must be dried thoroughly before storage. Drying is imperative for many reasons.

Dry grain is living but very inactive with greatly reduced rates of respiration. Consequently, dry grain is less subject to attacks from microorganisms (fungi, moulds and bacteria) that develop faster in a humid atmosphere.

Properly dried seeds will not germinate during storage. The integuments or seed coats are harder and less easily destroyed by **weevils**, insects that bore through grains and stems.

Grain should preferably be dried in the shade to avoid overheating because excessive heat is deadly. By drying grain, its moisture content can be lowered to establish a balance with the level of dryness of the ambient air. In humid climates, it is wise to finish drying artificially by circulating warm, dry air to ensure storage in optimal conditions. The temperature of seed grain must never exceed 35°C to 40°C, that is to say, the temperature of human blood.

About sorting seeds

Cultivators can sort seeds in two ways. One method is based on external seed characteristics and the other on signs of apparent health.

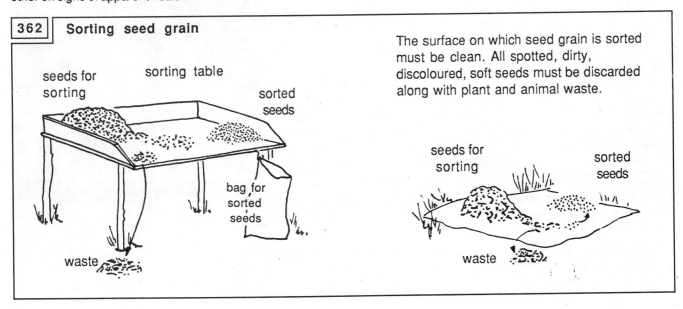

362 │ Sorting seed grain

seeds for sorting

sorting table

sorted seeds

bag for sorted seeds

waste

The surface on which seed grain is sorted must be clean. All spotted, dirty, discoloured, soft seeds must be discarded along with plant and animal waste.

seeds for sorting

sorted seeds

waste

With plants like bean, groundnut, maize, rice or sorghum, where the seed grain itself constitutes the agricultural produce, it is a simple matter to select a handful of seeds that look like the produce one would like to obtain in succeeding harvests. The seeds selected are a **seed sample**. Then, as the harvested produce comes in, seeds resembling the samples as closely as possible are selected. This is called **mass selection**.

However, when the desired agricultural products are leaves, flowers or fruit, mass selection is not so satisfactory because the qualities demanded from such produce cannot be compared with those of the seed samples.

It always pays to sort seeds. All damaged seeds - discoloured, scratched, soft, empty, broken, dirty and so on - are automatically discarded. Only whole, full, clean seeds are retained. Sorting can be done by hand on a flat surface, using various implements such as winnowers and sieves that eliminate a maximum number of light seeds, those not filled to capacity with food reserves.

Some seeds can even be plunged into a bucket of clean water and stirred for a few minutes. Seeds floating on the surface are removed, the rest are taken out and dried carefully. This is also a way of eliminating waste, such as leaves, bits of wood and insects, because they float on the surface too.

Hand-sorting is a long, tedious job but it is always rewarding. It reduces the amount of seed grain to be stored and to be sown. It also improves storage conditions because it gets rid of sources of seed infestation (**figure 362**). The seed waste can be fed to livestock such as poultry.

About cleaning seeds

Dirt of any kind left in stored seeds can cause infestation. Dirt includes vegetable and animal waste of all sorts, earth, humus and organic matter.

Great care should be taken not to dirty seeds during harvest operations and to keep them away from sources of infestation such as waste lying in the fields, old sacks, and used, dirty containers.

After harvesting, beating must be kept to a minimum so that waste in the form of stems, leaves and ears is not mixed with the seed grain for storage. Above all, beating must be limited to avoid damaging the seeds.

About storing seed grain in good conditions

Living seeds always withstand pest attacks better than dead grain. To build up resistance, the right storage facilities must be provided and certain rules applied.

- **The storage environment must be as dry as possible** for the reasons given above.

- **Temperatures must be kept low.** Low storage temperatures reduce seed activity whereas high temperatures may be fatal.

- With one or two exceptions, seeds must be stored in **as airtight conditions as possible**. Since insects need air to develop, depriving them of air helps to control them. A traditional method, practised, for example, in Togo and Benin, consists of an admixture of grain with dry, clean, white sand or ashes. The gaps between the seeds are filled up and the amount of air in the store is reduced.

- Drums, bags, demijohns, butts and granaries must be perfectly clean, disinfected and insect-proofed before the new grain arrives. All containers must be washed with soap and water, Javel water, cresol or some other disinfectant. The walls are scrubbed vigorously to eliminate all waste and dust from previous stores and are sprayed with fungicide and insecticide before storage takes place.

- Grain stores must be sealed as tightly as possible to avoid spoilage from rain, rising damp and humidity, and deterioration caused by insects, rats and other rodents.

In dry zones, seed storage in earthen jars and granaries inside dwelling houses is often effective because smoke from open hearths drives pests away.

363

When large quantities of seed are involved, it is better to make several lots and store them separately to spread the risk of total loss. If one lot is destroyed by disease or vermin, then hopefully the others will still be there.

Large earthen jars used for storing seeds and grain were seen in a house in the Dagari region of Burkina Faso (**figure 363**).

The granaries in **figure 365** are made of wood, matting and thatch. They are not as airtight as the earthen jars, and protection against insects, rodents and birds is not as effective as in earthen containers and granaries.

364

365

The granary in **figure 364** was built in a forest zone in Cameroun. It rests on supporting posts to escape rising damp but it is very exposed to grain pests.

Grain seed must be fresh when sown

Only seeds that have retained all their germination energy should be used, in order to guarantee good plant emergence. When seeds are stored too long, they gradually lose their energy. The term **germination capacity** or **energy** is used to denote their energy and value at any particular time. Germination capacity can be calculated quite easily for seeds that germinate quickly such as maize, rice, millet, sorghum, corn, bean and pea. It is not so easy to determine for large seeds with hard shells such as mango stones, cocoa and palm nuts.

Here is a way of calculating germination capacity. A batch of, say, a 100 seeds is taken from a bag of seed grain. They are then spread on a damp cloth on a plate lying in the shade. The cloth must be kept damp throughout the test. After a few days, the germinated seeds are counted. Results are satisfactory if almost all the seeds have germinated, but considered poor if many seeds are still dormant (inactive). The germination capacity is said to be 95%, if 95 out of 100 tested seeds germinate.

A cultivator who knows before seeding that 11 out of 100 seeds will not germinate, will obviously sow more than 100 seeds to be sure of having 100 plantlets.

The total seed requirement for a field is determined by the germination capacity of a test batch. If 60 kg of good seed are usually required to sow a particular field, then 120 kg are needed if germination capacity is 50%. If germination capacity is 89%, as in **figure 366**, then the amount of seed required is calculated as follows :

366 | **A hundred seeds are left to germinate on a damp cloth lying on a plate in the shade**

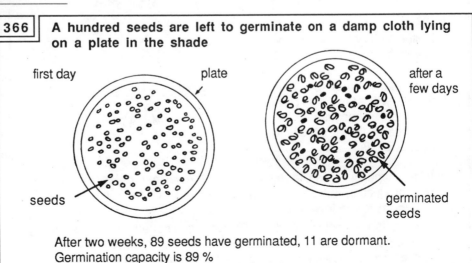

first day   plate   after a few days

seeds   germinated seeds

After two weeks, 89 seeds have germinated, 11 are dormant.
Germination capacity is 89 %

$$\frac{60 \text{ kg of seed}}{89\% \text{ germinating seeds}} = \frac{60 \times 100}{89} = 67.5 \text{ kg of grain seed}$$

Establishment percentage, the percentage of viable seed producing strong, young plants, is less than the germination percentage of the seed. This is because various field factors, such as poor drainage, over-deep sowing, pests and seedling diseases, cause early plant loss.

Lesson 30

Seeds and farm economy

Generally speaking, as plants become more domesticated and are more highly bred to improve yields in quantity and quality, their hardiness declines.

This is why varieties of high-yield cereals cultivated in countries with agricultural industries are not usually hardy or resistant in tropical Africa. In order to maximize yields from these selected lines, the fields must be treated with all kinds of **herbicides** to kill weeds, with **insecticides** to kill pests and with **fungicides** to kill fungi. These chemical products are used to get rid of all disease carriers and weeds competing with the crops.

As a result of these farming methods, the balance existing naturally between cultivated plants and their environment because of natural plant hardiness and resistance is replaced by the artificial balance created by means of chemicals on sale in commercial and industrial systems.

Figures 367 and 368 demonstrate why the qualities of a good seed can only be determined **against the background of local farming conditions in which the seed is planted.**

Two contrasting situations emerge from these figures. **Figure 367** presents the situation on most African farms where seeds from local sources are used. **Figure 368** describes the position in

---

| 367 | **Here is the background in which local seeds are used on African farms** |

All the production factors are found on the rural settlement :

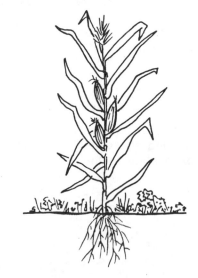

- Farm work is practically **all done by hand** because machines are a rarity.

- **The seeds are produced on the farm** from seed grain stored for that purpose after the harvest.

- **Plant nutrients** are limited to those found in the soil itself.

- **The plant must be able to withstand pest attacks** because no chemicals are used to destroy pests.

- **Weeds invade the fields.** Therefore, cultivated plants must be helped by weeding and, at the same time, must be hardy enough to tolerate some competition.

- **Plants rely exclusively on natural soil resources** because no, or only negligible quantities of chemical fertilizers are used. Proper use of compost can greatly improve this situation.

- **The plant must be hardy in order to thrive.**

In this type of closed farming system, **the farmer relies on market forces hardly at all** in order to buy production factors. Only small amounts of farm produce are for sale. **Monetary income and expenditure are low.**

The balance between the different components of the living environment - insects, birds, microorganisms, plant population - is barely affected. On the whole, **the living environment is respected** unless it is damaged by fire or by harmful human intervention.

industrialized countries where many farmers buy their seeds every year from seed merchants or from stations specializing in seed production.

The seeds adapted to the natural agricultural environment of African farms usually produce plants with low yields compared with those cultivated in other parts of the world on farms using large quantities of artificial products, e.g. chemical fertilizers, pesticides, insecticides and fungicides. However, the average African farmer avoids dependence on these inputs.

However, comparisons cannot be limited to crop yields only. The link between income and expenditure must also be taken into account. If many artificial products are used to increase crop yields, expenses and energy consumption go up in proportion (Lesson 48).

Seeds selected for farms in industrialized countries more often than not call for the application of artificial production factors not always available in rural Africa. Moreover, these seeds are often incapable of manifesting their qualities in the African environment. In contrast, although local seeds would not adapt to the climates and artificial conditions of farms in industrialized countries, they tolerate growth conditons in Africa.

Seeds in Africa deserve to be called improved only when they have been subjected to the conditions prevailing on the farms where they are planted.

Figures 367 and 368 lead to the conclusion that choosing good seeds calls for expert knowledge of local farming conditions and environment. It is pointless to buy highly selected seeds if the fertilizers they require are not available. On the other hand, it is also a waste of money to spread fertilizers on hardy plants that have not been selected for high yields and without knowing how they will react to these fertilizers.

Good seeds are those performing well in the given agricultural surroundings. Seed selection must always be carried out bearing those surroundings in mind and in accordance with the technical and economic choices of the cultivators.

| 368 | Here is the background in which seeds are used on farms in industrialized countries |

Many of the production factors employed are bought off-farm :

- **Farm work is practically all mechanized.** Manual work is reduced to a minimum.

- **Weeds are controlled by using herbicides.**

- **Each year, seeds are bought from stations specialized in seed production.**

Yields and income are high on these farms, but monetary outlay is considerable. In fact, **this type of farming requires huge sums of money.**

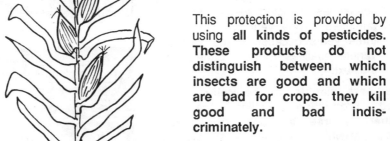

- The plants cultivated have been selected for their yields. However, they have most probably lost the quality of hardiness and therefore are more sensitive to pests. They need to be well protected.

  This protection is provided by using **all kinds of pesticides. These products do not distinguish between which insects are good and which are bad for crops. they kill good and bad indiscriminately.**

- High-yield plants take up large quantities of mineral salts from the soil and export them. To compensate for nutrient uptake, **many fertilizers are needed to prevent soil exhaustion.**

In farming systems on an industrial scale, **farmers are entirely dependent on trade circuits** both to buy production factors and to sell harvested produce.

Natural environmental balance has given way to an **artificial balance** made possible by the use of chemicals. Plants no longer need to be so closely adapted to their environment in order to thrive and produce.

Should seeds be produced on the farm ?

In the past, cultivators only sowed seeds carefully stored from the previous harvest.

Today, for a variety of reasons, cultivators no longer stock the seed they will require later on.

- Sometimes there is such a shortage of food that, in order to bridge the gap until the next harvest, farmers and their families are obliged to eat seeds set aside for the next planting season. To avoid this risk and forego the temptation of eating seed grain, it is advisable to deposit seeds in good storage conditions in a **seed bank** and only withdraw them for sowing. This is one way of overcoming the temptation of eating seeds.

- Often too, cultivators sell their provision of seed grain. Here again, a seed bank would help them to avoid this danger. As a general rule, farmers who sell seed lose a lot of money because they have to buy a fresh supply for planting.

- Some farmers rely too heavily on seed merchants or on State seed stations for seed supplies. These seeds may be excellent, especially when an official seed station does a conscientious job. Unfortunately, difficulties frequently arise :

  - seeds do not always meet the farmer's requirements and are not adapted to conditions on his farm ;

  - quality is not guaranteed ;

  - seeds may arrive late and in insufficient quantities ;

  - sometimes, selected seeds have to be purchased every year or every two or three years because seeds lose their varietal qualities from one generation to the next.

The best aim is to be self-reliant in seed production. Failing that, the farmer is free to look for whatever good quality supplies are available from outside sources. Seed selection is therefore an important farm activity. Its main aspects and possible improvements can be summed up as follows :

- Select and store seeds on the farm (Lesson 29).

- Develop budding and grafting.

- Establish **seed fields** reserved for seed production, with the help of agriculturalists and researchers.

369

370

371

cocoa plantlet

cotyledons

polythene bag

When trees are budded, a **tree stock** can be established, i.e. an orchard where trees are exploited for budding purposes rather than for their fruit. This is a common horticultural practice for trees of proven quality.

- Establish **seedbeds** and **nurseries** to make the best possible use of seeds **(figures 369, 370 and 371)**.

- Establish and improve special granaries for stocking seed. Use natural or artificial products to protect seeds (in particular, insecticides and fungicides).

- Take part in **multilocal tests** organized by official agencies. Multilocal tests refer to the testing of new seeds in many localities to see if they are adapted to the environment.

All the activities mentioned so far can be undertaken on farms with the backing of government services. Other steps connected with seed storage can be undertaken on a collective basis to protect all the local farmers from the risk of seed shortages in the planting season. Such steps could include :

- setting up **collective reserves**. There is a very old tradition in Africa whereby village elders are responsible for **communal granaries**. Collective reserves can also be started by cooperative movements. Members leave their seeds at the cooperative where they can be sure of sound storage conditions and retain personal ownership of the bags deposited ;

- **seed banks**, run on different lines. They buy seeds from farmers and sell them back after sorting and checking the grain.

To complete this picture of seed selection within the scope of farmers and collectives, the work of seed stations and firms specialized in seed production should be mentioned. They operate on a worldwide basis in the search for varieties with qualities adapted to local needs. This is a hard, slow job that only succeeds when undertaken in close collaboration with the farmers concerned (Lesson 28).

Farm production of seeds is not worthwhile in every case

This lesson has been concerned with propagating material for staple crops, such as cereals, tubers, bananas, etc., where farmers cannot take any risks because food for their families depends on these crops.

However, there is far less risk attached to condiment plants and certain fruits. The quantity of seed needed for these plants is insignificant and it is usually worth buying them every year from seed producers. Seeds in this category include radish, cabbage, onion, tomato and all the vegetables known as European.

Unfortunately, there are still too few stations producing selected seeds for native African vegetables. These seeds are mostly found locally.

Reminder

- *Seeds which give rise to doubts about their quality should be avoided.*

- *Good seeds should be produced on the farm.*

- *Producing seeds on the farm is a task as important as producing foodstuffs, fodder, manure, etc.*

- *Cooperation with technical bodies specialized in seed improvement is also important. Cooperation does not mean sacrificing self-reliance.*

- *Seeds are considered to be good or improved when they have been tested satisfactorily on local farms.*

- *Seeds bought off-farm should be reasonably priced, of tested performance and in supply when needed. Germination capacity should be checked.*

- *In the absence of these criteria, seeds produced on the farm or in the village are often more reliable.*

Lessons 31 and 32

🔲🔲🔲🔲🔲🔲🔲🔲🔲🔲🔲🔲🔲🔲🔲🔲🔲🔲🔲🔲🔲🔲🔲🔲🔲🔲🔲🔲🔲🔲🔲🔲

How plants use the soil

Lessons 31 and 32 examine the ways in which agricultural plants exploit the soil. This study involves observing roots, the volume of the area they occupy (root occupancy) and the transformations they bring about in the soil as they develop.

Lesson 31   🔲🔲🔲🔲🔲🔲🔲🔲🔲🔲🔲🔲🔲🔲🔲🔲🔲

Roots

Root system or **rooting** is the term used to describe the roots of a plant taken as a whole. Just as the shape of stems, leaves and flowers is different for every plant species, so the shape of the roots and the nature of the root system vary from species to species. Root systems determine the volume of soil exploited by plants and for that reason it is important to study them.

Observation of root systems begins by digging the soil as described in Lesson 18, taking care to dig at places where roots have not met any obstruction due to soil composition. Otherwise, the roots examined would be deformed.

It is particularly useful to study the root systems of adult plants to see what lessons can be learned and applied to farming practices. These aspects will be covered in detail :

- the shape of roots ;
- the shape of root systems, i.e. their volume and the area they occupy ;
- the growing point of the roots, that is, the way they are attached to the stem ;
- the way roots are specialized.

maize   pea

372

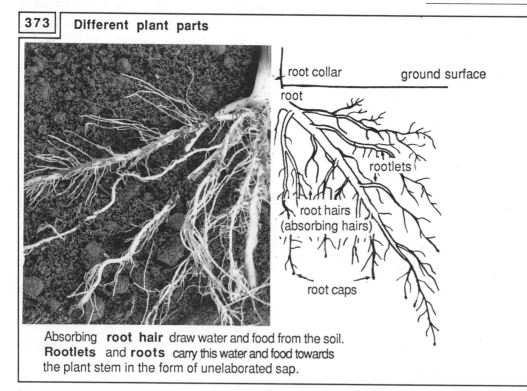

373 | Different plant parts

root collar   ground surface

root

rootlets

root hairs
(absorbing hairs)

root caps

Absorbing **root hair** draw water and food from the soil.
Rootlets and **roots** carry this water and food towards the plant stem in the form of unelaborated sap.

Roots and root systems

Two plants, one maize and one pea, were freed of earth and washed in water **(figure 372)**.

The roots are subdivided into finer rootlets that can be subdivided again **(figure 373)**.

A close look at the picture and particularly at the rootlet tips reveals many slender hairs. They are the **root hairs**, also called **absorbing hairs** because they absorb water and nutrients from the soil to meet plant requirements. Where cultivated plants are concerned, **it is advisable** to check and **see** where the formation of root hairs is thickest because it is there that plants take up nutrients most vigorously.

Note the tiny head called the **root cap**, at the end of every rootlet. The energy of the rootlets is concentrated in the root caps as the roots penetrate the soil. Cut off the root caps and the rootlets will stop growing.

The specialized roots of legume crops should also be mentioned. Their root nodules enable plants to take in nitrogen from the air. The root system of a pea plantlet with many nodules can be seen in **figure 374**.

The root system of plants occupies a certain soil area, depending on the nature of the rooting, taproot, fibrous, deep, shallow, branched or not branched. This area is called the **root extent** or **spread** because it is within this area that roots absorb food. Many cultural practices, particularly plant arrangements in fields, depend on the external form of root extent.

Primary and secondary roots

The germination and development of a bean plantlet unfold in **figures 375 and 376**. After imbibition of the seed, the first or **primary root** appears covered with absorbing hairs.

The **secondary roots** appear after a few days. Later, as the plant develops, the primary root penetrates the soil, the secondary roots spread laterally and the nodules are formed **(figure 375)**.

The growing point of all the bean roots, both primary and secondary, is below the **root collar (figure 376)** where plumule and radicle meet in the seed (Lesson 29).

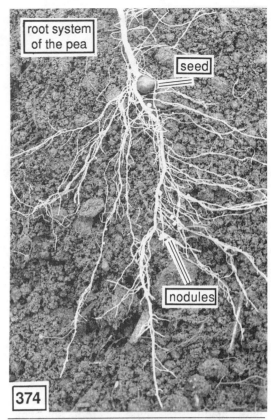

root system of the pea
seed
nodules

374

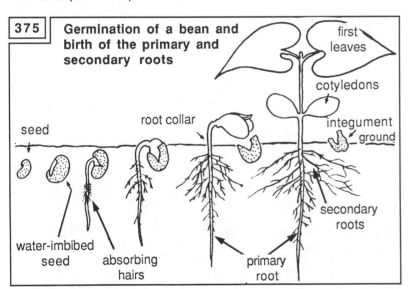

375 Germination of a bean and birth of the primary and secondary roots

first leaves
cotyledons
root collar
integument
ground
seed
secondary roots
water-imbibed seed
absorbing hairs
primary root

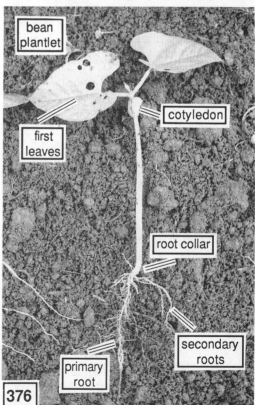

bean plantlet
first leaves
cotyledon
root collar
primary root
secondary roots

376

Adventitious roots

Roots growing from stems or parts of stems are adventitious and so, all plants propagated by cuttings live thanks to this type of root.

Young adventitious roots can also be seen on the rooted cassava cutting **(figure 378)**. They have not yet accumulated any food reserves because they are too busy contributing to stem growth.

Figure 377 shows a sugar cane cutting with young adventitious roots.

Notes

rooted cassava cutting

cutting

adventitious roots

378

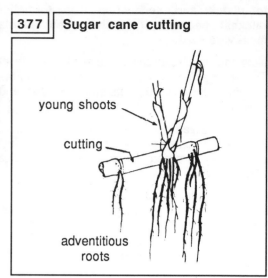

377 **Sugar cane cutting**

young shoots

cutting

adventitious roots

The first rootlet emerging from the maize seed is a primary root but its contribution to plant growth is negligible **(figures 379, 380, 381)**. Other roots grow on the plumule above the place where the seed is attached. These are the **adventitious roots** that will be entirely responsible for feeding the plant at a later stage. This type of root sometimes appears above ground on the first nodes of the stem. The maize plantlet has no root collar, or rather, the collar disappears at the same time as the primary root.

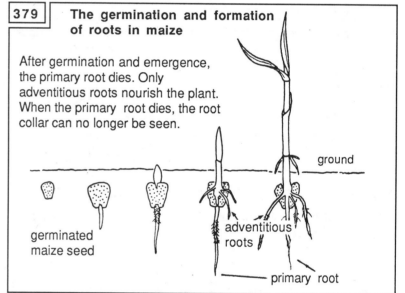

379 **The germination and formation of roots in maize**

After germination and emergence, the primary root dies. Only adventitious roots nourish the plant. When the primary root dies, the root collar can no longer be seen.

ground

germinated maize seed

adventitious roots

primary root

maize plantlet

380

roots hairs

381 germination of maize

Suitable cultural methods, such as earthing up plants, promote the growth of adventitious roots and reinforce the base of cultivated plants **(figure 382)**.

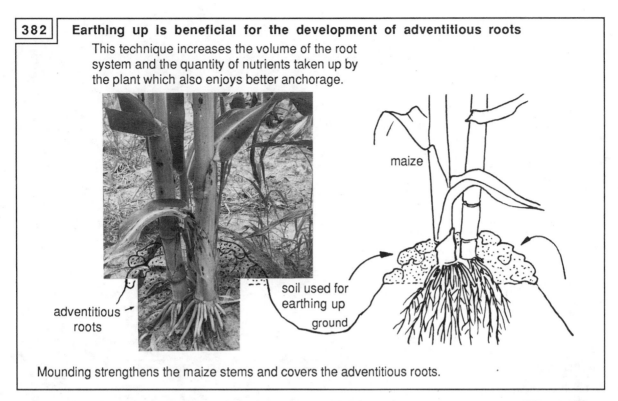

382 **Earthing up is beneficial for the development of adventitious roots**

This technique increases the volume of the root system and the quantity of nutrients taken up by the plant which also enjoys better anchorage.

maize

soil used for earthing up
ground

adventitious roots

Mounding strengthens the maize stems and covers the adventitious roots.

The growth of bean and maize is followed to maturity. The bean has a **taproot system (figure 383)** ; the maize has a **fibrous root system (figure 384)**.

The **taproot** anchors the plant deeply in the ground while **secondary roots** spread out laterally from the taproot.

The term **fibrous root system** is used when there is no taproot and when the roots penetrate the soil like a broom.

bean

ground

ground

maize

383 taproot system

384 fibrous root system

168

In both cases, rooting can be **deep** or **shallow**. When rooting is deep, plants absorb water and nutrients from deep soil layers, but when it is shallow, nutrients are taken up from the top layers. Plants with shallow root systems suffer more from dry soil than plants with deep root systems, but they are not so adversely affected by the presence of a shallow water table.

Some roots have a specialized role

Among the best known specialized roots are **tuberous roots** capable of **storing up food**. Common examples are yam, cassava, sweet potato.

Other roots are specialized in **reaching down deeply for water** in deep soil layers and are able to penetrate to the water tables (water reserves).

Still other roots are specialized in **breathing on the soil surface** like the breathing roots of the red mangrove (**figure 385**).

Rhizomes are underground stems that look very much like roots. The rhizomes of Spear grass (lalang, *Imperata cylindrica*) are a familiar example. Potato, ginger and sisal are other examples of rhizomatous plants. Potato rhizomes are tuberous, i.e. they can store up food. **Figure 386** shows a rhizome of Spear grass dug up and left lying on the ground.

385   red mango   breathing roots

386   rhizome

387   pawpaw

388   tomato

The root systems of some familiar plants

The taproot system of a pawpaw plantlet is exposed in **figure 387**. The position of the slightly swollen root collar and the secondary roots that have formed underneath it can be identified. Note the difference between this and the bean plantlet displayed in **figure 376** at the beginning of the lesson. It too has a taproot, but its secondary roots develop close to the soil surface whereas those of the pawpaw are more deeply penetrating.

The primary root of the tomato is a taproot, but secondary roots develop into a fibrous system that does not stand out clearly in **figure 388** because the plants are still young.

Carrot, radish and turnip each have a tuberous taproot **(figure 389)**.

cocoyam

390

389 — carrot

Maize, millet, sorghum, rice, finger millet and, generally speaking, all plants belonging to the *Gramineae* family (Grasses) have a fibrous root system. The same is true of cocoyam and banana illustrated in **figures 390 and 391**, and also of onion.

Cassava and yam have combined fibrous and tuberous root systems.

All the plants propagated by cuttings and named in this lesson (banana, cocoyam, yam and cassava) and others like taro and sweet potato only find nutrients through **adventi-**

391   banana

392   potato

tous roots. Sugar cane and bamboo fall into this category too.

The root system of the potato is composed of adventitious roots growing from rhizomes (underground stems). Here and there **(figure**

392), the rhizomes are enlarged by food reserves and form the potato itself.

The root system of a shea butter tree that had been cut down is displayed with its penetrating taproot, and its secondary roots spreading horizontally through the soil.

It must be stressed that changes in rooting structures and distribution take place as the plant develops. The root systems of plantlets occupy a smaller area than those of mature plants. Hence the need to observe root systems and habits over the whole life cycle of plants. Nonetheless, when roots are developing, root systems play a vital role in determing future plant yields.

shea butter tree

393

Common plants with taproots : groundnut, okra, bean, lettuce, carrot, soya bean, cowpea, tobacco, tomato, cotton, sunflower, castor.

Plants with fibrous root systems : rice, maize, sorghum, millets, finger millet, wheat, sesame, taro, cocoyam, onion, elephant grass, banana, cucumber, etc.

Plants with tuberous root systems : cassava, yam, sweet potato, taro, cocoyam, Goa bean, etc.

Plants with rhizome systems : spear grass, ginger, sisal, mint, pineapple, potato, etc.

Root systems respond to soil structure

Correct root penetration and spreading are essential for proper plant production. However, soil structure can obstruct root growth and modify normal root systems.

taproot deformed by large stone

395

The taproot of the pawpaw plantlet (**figure 395**) is quite misshapen because it struck a large stone.

The taproot of the bean illustrated in **figure 394a** has spread extensively in rich well-structured soil. There are many enlarged nodules. This is the kind of rooting one would expect to find in a well-fed plant.

In poor, sandy soil (**figure 394b**), roots find few nutrients. Because leaf development is retarded, the roots suffer from a shortage of elaborated sap. They remain small and thin although they could easily penetrate the soil. Nodules are sparse and tiny. In this case, the only solution is to nourish the soil by using organic matter and mineral salts (fertilizers).

Suppose, on the contrary, that the same bean is planted in heavy, clayey soil rich in mineral salts (**figure 394c**). Its roots will thicken without spreading, root extent (area) is restricted, the rootlets are exhausted as they try to develop and penetrate the compact soil. Because the plant is growing in rich, poorly aerated soil, the plant does not need to gain nitrogen from the air and consequently there are few nodules. This type of soil must be used for cultivated plants tolerating a clayey structure, like cocoyam or rice.

It may be that roots are obstructed by a water table near the soil surface (**figure 394d**). The water table marks the limit of root penetration because the roots would be asphyxiated if they entered the water. They are forced to spread horizontally above the water level. **Subsoil tilling** or **subsoil drainage** would give roots more room for spreading and ensure better development. Subsoil tilling is a

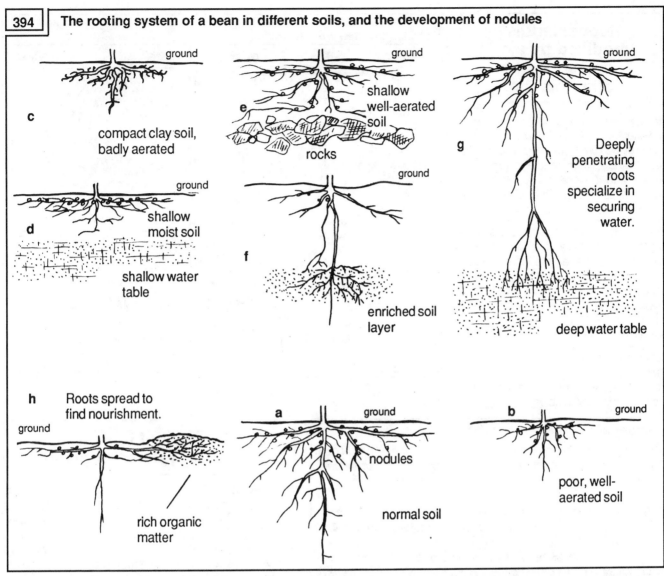

394 The rooting system of a bean in different soils, and the development of nodules

c — compact clay soil, badly aerated

e — shallow well-aerated soil / rocks

g — Deeply penetrating roots specialize in securing water.

d — shallow moist soil / shallow water table

f — enriched soil layer

deep water table

h — Roots spread to find nourishment. / rich organic matter

a — nodules / normal soil

b — poor, well-aerated soil

technique involving the break-up of hard or impermeable soil layers with the help of powerful machinery.

A layer of rock near the soil surface may prevent the roots from penetrating the soil deeply **(figure 394e)**. The taproot is distorted and grows sideways. This process is typical of growth in mountainous country where the soil layer on rocky ground is thin (shallow soil). On this type of land, it is often good farming practice to mound or ridge the soil in order to give roots a greater volume of earth in which to develop.

When the water table is at a considerable depth and soil structure is not too compacted, the plant can send down its specialized roots until they secure water **(figure 394g)**.

It has been shown that when a plant enters a richer soil layer, its roots develop more abundantly there than in less fertile layers **(figure 394f)**. The ability of roots to look for food on the surface of fields has been observed too. Root production in the form of rootlets and absorbing hairs is much greater in places fertilized by organic waste and/or ashes compared to root activity on unenriched surfaces **(figure 394h)**.

396 Plants adapt root distribution in response to associated plants

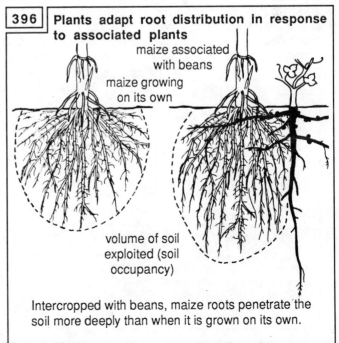

maize associated with beans

maize growing on its own

volume of soil exploited (soil occupancy)

Intercropped with beans, maize roots penetrate the soil more deeply than when it is grown on its own.

Root extent is modified by the presence of other plants

Figure 396 shows how the area exploited by maize roots intercropped with bean is wider and deeper than when maize is grown in pure stand (on its own). Moreover, the maize crop influences the development of the bean roots.

Root systems frequently react to one another, either by growing further apart as in this example where maize roots grow to a greater depth in association than on their own, or by drawing closer to each other, one plant thus enriching the soil for the benefit of the other.

Differences can be observed in the response of root systems in a community at different stages in the life cycle of every species.

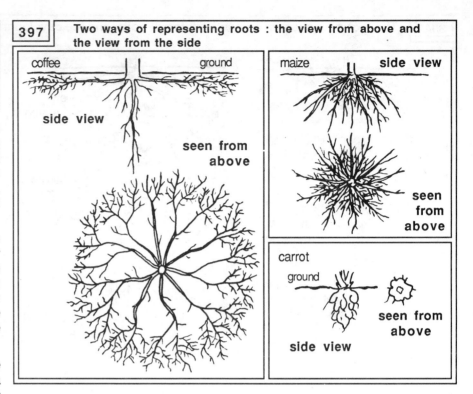

397 Two ways of representing roots : the view from above and the view from the side

coffee — ground — **side view** — **seen from above**

maize — **side view** — **seen from above**

carrot — ground — **seen from above** — **side view**

Important

When we speak of the area exploited by the roots of a given plant, it must be remembered that roots can draw up water from a certain distance, be it millimetres or centimetres.

*Cassava roots, for instance, attract water and food several centimetres away. In **figure 399**, the dotted line marks the limits of the underground area affected by the presence of cassava roots at the time of enlargement. It is obvious from this drawing that, if other plants draw up water and food from the same area, they will suffer from drought and undernourishment even if their roots do not actually touch the cassava roots. This is why cassava is said to be unsociable at the time of tuber enlargement.*

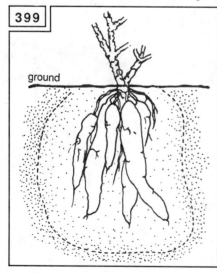

399

ground

Another way of representing the root systems of plants

Root systems have been shown, so far in this lesson, by drawing sections in the soil where stems emerge. They could be represented in another way, as though they had been observed from above by placing an imaginary eye to the top of the stem. This is a way of showing root extent or occupancy, that is, the soil area occupied by the root system.

Here are three types of root extent, each one represented in the two ways described above - a view from the side and a view from above.

Representing root systems in both these ways gives a good idea of the underground area occupied by plants.

The variable areas occupied by coffee, maize and carrot (**figure 397**) show up clearly when the root systems are drawn from above and placed side by side in the same field (**figure 398**).

Cultural practices

Root distribution is sometimes modified by man when seedlings are transplanted or pricked out. As a general rule, roots grow downwards. Indeed, when they curl upwards, they die. Consequently, the way the transplanting operations are carried out is extremely important.

Attention should therefore be paid to the art of

173

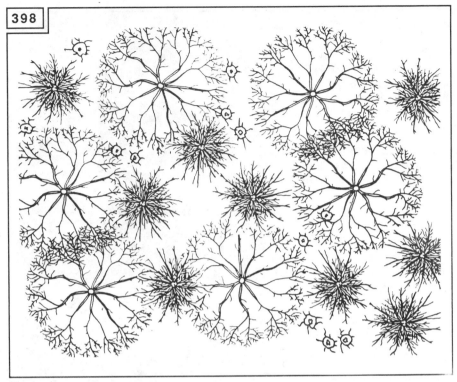

398

Points to remember

- *There are many kinds of root system, the main groups being the taproot system and the fibrous root system.*

- *Roots can be deep or shallow.*

- *Tuberous roots are storage roots that enlarge with food reserves. Certain roots are specialized in finding water, others in breathing.*

- *Rhizomes are creeping stems at or under the surface of the soil ; they look like roots.*

- *Roots can be primary, or secondary when they grow under the root collar of the plant. Roots born on the stem are adventitious.*

Figure 401 illustrates some rooting systems typical of herbaceous plants as well as trees and shrubs. Only root extent varies from species to species.

- *The root systems of plants respond to soil characteristics.*

- *They also respond to associated plants.*

A knowledge of the root systems of cultivated plants, their development and distribution, are of great importance in indicating how fields should be planted, which crops should be cultivated and which should be sown in association.

transplanting **(figure 400)**. If there is not enough room to let all the roots settle in a downward position, partial clipping is advisable because then other roots will grow and inject new life into the root system. Shoots and leaves can be trimmed at the same time. In this way, the roots are able to recover before the leaves begin to transpire abundantly and exhaustion of the reborn rootlets is avoided.

401 **There are numerous root systems.** The dots outline the shape of root extent.

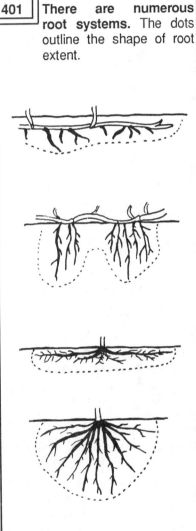

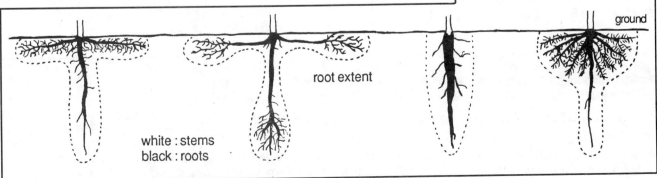

root extent

white : stems
black : roots

ground

| 400 | **The corect way to transplant** |

Sometimes plants have to be transplanted for cultural reasons. Millet plants (figure **400**) are being pricked out in a field with many **misses** (plants that failed to germinate or emerge).

Great care should be taken to **set the roots downwards (a) and make sure that none are bent back (b).**

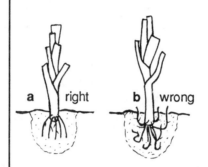

400

♦♦

Lesson 32   ♦♦♦

Knowing root zones

How best to sow and plant

The importance of proper spacing is illustrated in **figure 402**. The roots of the three maize plants on the left are entangled while, on the right, two seedlings were spaced 60 cm apart, so that their roots do not overlap.

The plants on the left were sown by **dibbling** (two or more seeds were sown in the same hole) whereas the plants on the right were sown one seed per hole 60 cm apart.

The seedlings with entangled roots all need the same amounts of water and food at the same time of the

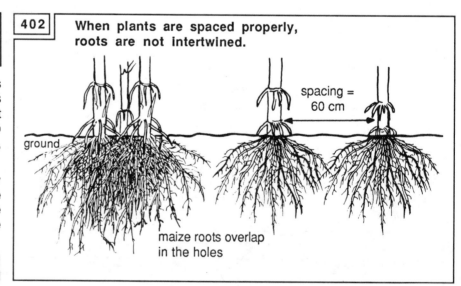

| 402 | **When plants are spaced properly, roots are not intertwined.** |

spacing = 60 cm

ground

maize roots overlap in the holes

agricultural season because their vegetative cycles evolve simultaneously (Lesson 23). Since food in the area exploited by the roots is limited, the three plants have to share the resources available.

The opposite is true of seedlings correctly spaced. Their roots are not intertwined and do not compete for food.

It is true to say that, on the whole, **two plants of the same species tend to compete for food when their roots are entangled.** Admittedly, dibbling is sometimes advisable, provided only a few seeds are sown per hole and the holes are far enough apart to allow the plant room for spreading.

403

Dibbling and thinning

At times, dibbling is recommended as a precautionary measure. A few seedlings may fail to germinate or plantlets be attacked, but at least the remainder will develop. Remember, however, that if the surviving seedlings are overcrowded, they will compete with one another and growth will be retarded.

Thinning is practised when there are too many plants per hole. The woman **(figure 403)** is thinning her millet field. She sowed five to ten seeds per hole to ensure a good rate of establishment. Now she is removing the weaker seedlings in order to leave three or four stems in each hole. If she left any more, the roots would compete for soil nutrients and the plants would remain undersized, spending more energy on fighting for food than on growing vigorously. If the roots are not damaged, the seedlings she digs up can be transplanted to fill in gaps where seeds have failed to germinate. The leaves and roots of transplanted plants are clipped.

Good intercropping practices

Cocoyam and sweet potato are intercropped and exploit the soil together **(figure 404)**. These plants have fibrous root systems that penetrate in depth rather than spread through the soil. Both manage to thrive while making mutual concessions.

As both these plants have abundant leafing, the main task in this association is to prevent the sweet potato leaves from choking the cocoyam foliage, and the cocoyam leaves from robbing its companion of light. The farmer can intervene with appropriate steps such as pushing back the creeping potato stems that might choke the cocoyam, or conversely cutting back the cocoyam leaves.

sweet potato

cocoyam

404

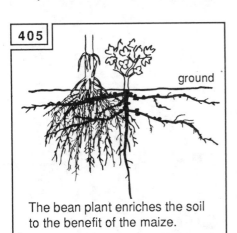

405

ground

The bean plant enriches the soil to the benefit of the maize.

When maize and bean are interplanted, their roots overlap **(figure 405)**. However, the two species differ widely in many respects - in their growth patterns and nutrient requirements, in the quality of the nutrients looked for, in the way their roots take up these nutrients. The most active periods of nutrient uptake do not coincide. All these differences ensure that maize and bean can be intercropped successfully just like other cereal and legume associations discussed earlier on.

How best to occupy the soil surface

The root systems of coffee trees planted in two different ways are illustrated in **figure 406**. The two rows on the right were planted so close together that all the soil surface is occupied, but the rootlets and root hairs form a tangle. Consequently, the trees are in competition for the nutrients in the soil where the roots overlap. Furthermore, trees of the same species have similar needs at any given time.

Hence, despite good occupancy of the soil, the plants suffer because of the competition for food. They are also likely to get in each other's way in the aerial environment, as was described in Lesson 25, **figure 292**.

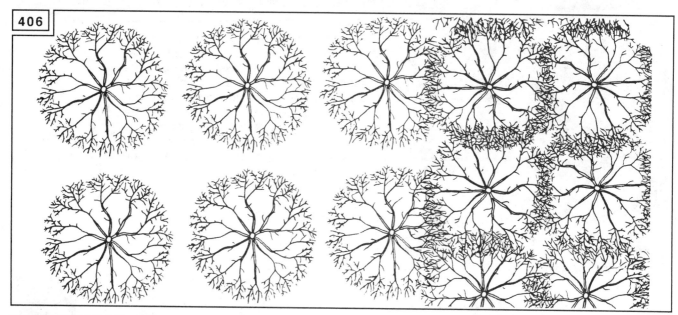

406

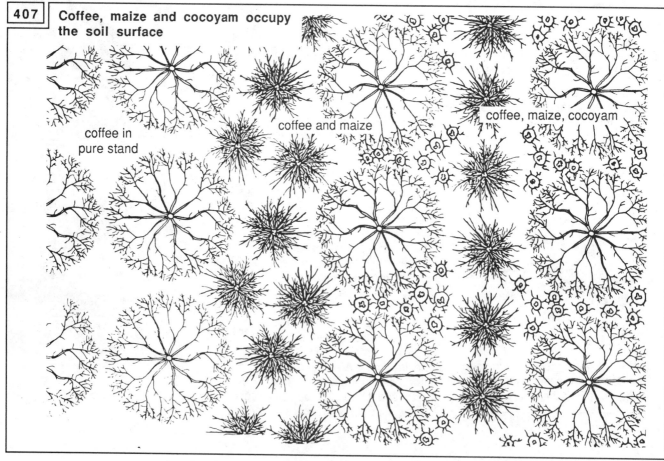

407 Coffee, maize and cocoyam occupy the soil surface

coffee in pure stand

coffee and maize

coffee, maize, cocoyam

The trees in the three bottom rows were spaced so that their roots are kept apart. The risk of competition between the plantlets has been eliminated because the root zones are separate. However, large areas of the soil surface are empty, so, although each individual plantlet will give of its best, the soil is not necessarily exploited to maximum capacity.

In order to exploit the soil surface fully, plants can be grown in mixtures so that the space left empty by one plant can be filled by another. In **figure 407**, for example, the coffee trees on the left, planted as a pure stand, were properly spaced and the root zones do not mingle. On the right, however, better occupancy of the surface was achieved by interplanting the trees with maize in the middle, and with a combination of maize and cocoyam on the right.

## How best to occupy the volume of available soil	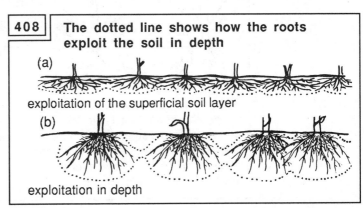 **408** The dotted line shows how the roots exploit the soil in depth (a) exploitation of the superficial soil layer (b) exploitation in depth

Good soil occupancy can be observed by comparing **figs 408 and 409**.

Figure 408 represents two root systems. The first plant **(a)** has short, shallow roots with soil occupancy restricted to a few centimetres below the surface. The second plant **(b)** has longer, deeper roots that exploit a greater volume of earth although there are empty soil areas between the roots.

In figure **409**, the situation is quite different. Here several plant species have been associated and, as a result, the volume of soil occupied is much greater. Increased occupancy has many advantages. Soil nutrients (mineral salts) that will be transformed into living matter are present in larger quantities when there are many plant species as opposed to only one species. Leaching of the upper soil layers is curbed. Finally, the total amount of living matter produced on soil occupied in this way is greater than in cases of single-plant occupancy

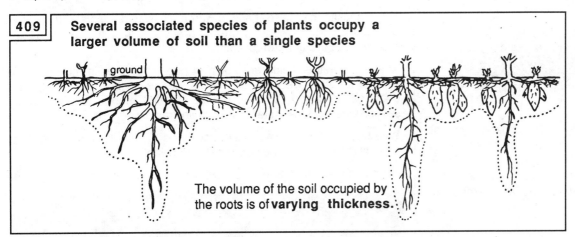

409 Several associated species of plants occupy a larger volume of soil than a single species

ground

The volume of the soil occupied by the roots is of **varying thickness.**

The age factor in crop mixtures

Even when plants grow well in mixtures, they interfere with each other at certain times in their vegetative cycles.

This kind of interference shows up in **figure 410**. Groundnut seedlings still poorly rooted **(a)** have to contend with the extensive roots of maize plants. Or, on the other hand, young maize shoots must snatch food from groundnut with well-established roots **(b)**. It is best to plan sowing so that groundnut and bean can develop without plant dominance on either side **(c)**.

Good intercropping practices cut root competition to a minimum particularly during the most important phases of the life cycles, that is to say, during rapid growth and fruit formation.

410 | **The effects of plant age on crop mixtures**

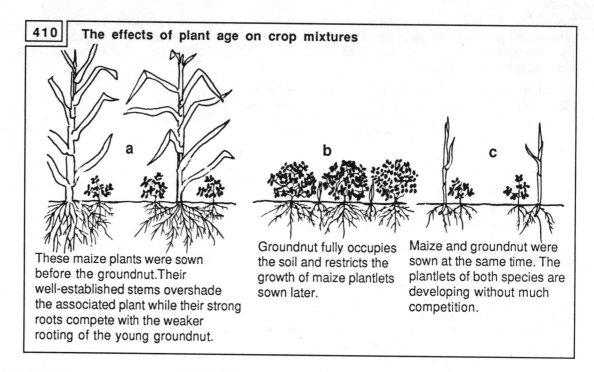

a These maize plants were sown before the groundnut. Their well-established stems overshade the associated plant while their strong roots compete with the weaker rooting of the young groundnut.

b Groundnut fully occupies the soil and restricts the growth of maize plantlets sown later.

c Maize and groundnut were sown at the same time. The plantlets of both species are developing without much competition.

411 | **Weeds must be eliminated to ensure good growth conditions for cultivated plants**

Here, groundnut is choked by tall weeds. It will not be easy to get rid of them because rooting is so well-established. Groundnut growth is at risk because weeding was not carried out on time.

Note the well-developed groundnut plantlets. They are dominant because an **early weeding** operation has eliminated unwanted plants before the latter became firmly established.

Cultivated plants confronted with weeds

The presence of weeds (unwanted plants) in cropping fields is inevitable and may impede the development of cultivated plants, more especially during the phase of maximum growth **(figure 411)**.

The opposite can also happen when cultivated plants are thriving. Competition from weeds is greatly reduced ; in fact, the weeds are impeded by the cultivated species.

Anyhow, whether fields are sown in pure stands or in mixtures, it is important that weeding is done before cultivated plants are endangered, that is to say, during the first months of the cultural season when roots and stems are growing rapidly.

Weeding means cutting back undesired plants to ground level to prevent them from overpowering cultivated plants.

If weeding is careless or too late, plant roots willl compete with one

cotton plantlets

mulch

weeds

413

another. In **figures 411 and 412**, the roots of groundnut and maize seedlings are struggling against the roots of uncultivated plants. The situation after weeding is illustrated on the right of both figures.

Weeding is in operation in a cotton field **(figure 413)**. Unwanted plants have been removed from the ground on the right, but on the left, they have completely overrun the cotton crop. Weeding was done too late.

| **412** | **The effects of weeding** |

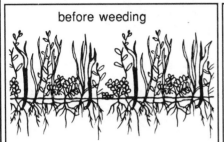

before weeding

Weeds compete actively with maize plantlets.

after weeding

After early weeding, competition from weeds is greatly reduced.

Mulch (plant material such as straw) is spread on the ground and prevents recovery of young weeds.

Remember

A knowledge of root systems and habits is invaluable to farmers for many reasons.

- *The shape of the root system at various times in the cultural season determines the method of planting and crop associations.*

- *The shape and development of the root zones (root extent) of cultivated plants and weeds determine how and when weeding should be carried out.*

- *The shape of the root system determines the pattern of tilling, ridging and mounding.*

To exploit land fully, it must be examined in all its dimensions - length, width, height/depth. Remember that, in order to obtain satisfactory yields, cultivators must :

- *arrange plants so that the stems grow freely in the aerial environment (Lesson 25) ;*

- *space plants correctly ;*

- *allow the roots of useful plants to exploit the soil more deeply ;*

- *associate plants so that their roots exploit different soil areas, and leaching of the upper soil layers is limited.*

The right arrangement of cultivated plants depends on these factors :

- *where plants find nutrients, i.e. root extent or zone ;*

- *how root extent is modified in the course of time (the root zone of a plantlet differs from that of the mature plant),*

- *what nutrients are taken up and when exactly the uptake takes place during the vegetative cycles of cultivated plants.*

The following rules must be observed to obtain maximum yields when crops are sown in mixtures :

- *make sure that plants of the same species are not competing for soil nutrients and respect spacing ;*

- *associate plants with different root systems - taproot, fibrous, deep, shallow, etc. ;*

- *associate plants whose life cycles do not follow the same pattern. Steps can be taken to ensure that the progression of the vegetative cycle of one plant does not disturb the life cycle of the other plant or plants grown in association ;*

- *associate perennial plants (trees, shrubs, herbaceous varieties) and seasonal herbaceous plants ;*

- *associate plants whose water and nutrient requirements are not all concentrated into the same periods of the cultural season.*

All these points presuppose sound knowledge of cultivated plants, of plant varieties, and of their behaviour. Observation and experimentation on farms will indicate the best cultural methods for crop associations.

Lessons 33 to 36

Soil life and humus

Roots are not the only living things in the soil. A host of living beings occupy and exploit the superficial soil layers - small animals, insects, earthworms, microorganisms, termites and the like. Their presence is good for plant life.

Everything living in and on the ground dies eventually and produces humus. And, because humus is of enormous value to agriculture, its production must be fostered.

Factors that help increase humus production are studied under the following headings :

- Small animal, fungi and microoganisms in the soil (Lesson 33) ;
- Plants transform the soil (Lesson 34).

In addition, many cultural methods contribute to soil life, and especially to root life, by helping the formation of humus. They are discussed in two lessons :

- fallow land (Lesson 35) ;
- other cultural methods (Lesson 36).

Lesson 33

Small animals, fungi and microorganisms in the soil

A great many living beings inhabit the soil, feeding in it and leaving decaying matter there. They burrow, transport and transform the soil.

Life in the soil is of great importance for fertility. Bad cultural methods can kill that life and this puts one's farming future at risk because living things create humus. Humus is of vital importance for soil structure and for providing uninterrupted plant nourishment. In fact, **humus** is matter produced by all living beings in the soil through the break-down of plant and animal residues.

What beings live in the soil ? What is humus ? How does it affect soil structure (organization of mineral particles) and plant nourishment ? After answering these questions, good agricultural practices and also practices which destroy soil life will be examined.

Everything living in the soil

414

Roots

Roots run through the soil and break up the clods of earth. As they edge their way through the ground, they allow water to infiltrate. Other living organisms use the cracks made by the roots to penetrate the ground (Lesson 31).

Small mammals

Small mammals include rats, mice, hares, moles and other warm-blooded animals. They burrow underground to live there or to look for food. They eat roots, seeds, larvae and insects. As they burrow, they increase water penetration and aerate the soil. They mix and open the soil, also combining it with faeces that contribute to the formation of humus **(figures 414 and 415)**.

Termites and ants

There are innumerable species of termites and ants, some native to forests (**figure 416**), others to savanna (**figure 417**). These insects are specialized in transporting soil.

415

416

Termites are noted eaters of wood. They cut it up into small, thin pieces, hastening its transformation into humus. Termites burrow underground passages, transform earth and mix it up with wood residues and other living matter.

Ants do a similar job, but they prefer fresh material like leaves, seeds, moist trash, larvae, remains of dead animals, fungi and moulds.

Termite waste as well as dead termites and ants are an excellent source of food for the living beings mentioned later in this lesson. If plant residues were not broken down by ants and termites, other living organisms would have difficulty finding food.

Termites and ants are therefore vital links in the long food chain that manufactures humus and makes soil fit for agriculture.

The fields in **figure 418** were subjected to untimely bush fires and deforestation. There are several termite mounds, some of them are near the few low shrubs that survived the slash-and-burn operation.

termite mound

417

termite mound

termite mound

418

forest degraded by fire and erosion, and invaded by termites

182

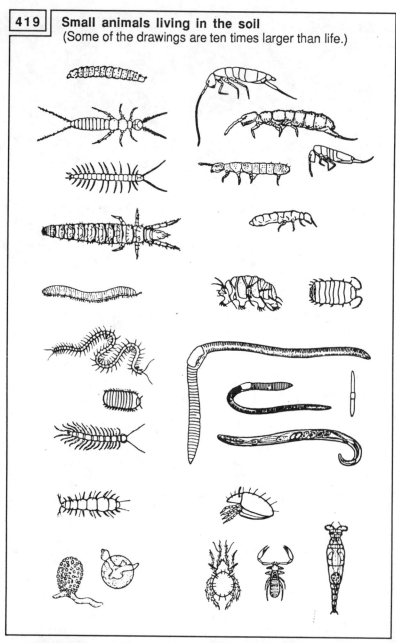

419 | **Small animals living in the soil**
(Some of the drawings are ten times larger than life.)

Insects, slugs and snails

Hundreds of insect species inhabit the soil, often as larva. While some are visible to the naked eye, others are so tiny they can only be seen under a magnifying glass. Most of these animals live on waste from plants, from other animals, from fungi and moulds. Their faeces turn quickly into humus.

Many insects and larvae burrow the soil, aerate it, and break down living matter. **Figures 419 to 426** show them at work in the ground.

420

421

422

423

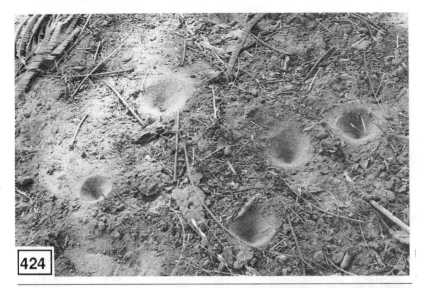

424

426

425

Earthworms

Earthworms transport soil the whole time **(figures 427 and 428)**. They swallow earth and eject it after mixing it with faeces. They feed on all the microscopic organisms living in the soil (see below).

earthworm

gallery

427

earthworm

earth and faeces ejected by earthworm

428

A single earthworm is able to swallow and mix up with its faeces about one kilogram of earth per year. In loose, fertile soil, 25 to 30 earthworms occupy one square metre (a square with sides one metre long). Together they transform 25 kg to 30 kg of earth per year.

Earthworms are also useful because they never stop burrowing, thus allowing water and air to circulate freely.

Other worms also work their way through the soil but their activity is not so easily observed. **Figure 428** shows how some worms drag back the soil they swallowed to the field surface.

Fungi

Fungi living in the soil feed mostly on residues from wood and leaves. Everyone knows what fungi look like **(figures 429 and 430)**. On the outside they are like white or coloured threads. Sometimes, only these threads can be seen because the greater part of the fungus is invisible, hidden inside the decaying matter.

Larvae of insects, ants and termites eat these fungi which are valuable because they hasten the decay of living matter and the production of humus.

Microorganisms

Microorganisms are living things so tiny they cannot be seen by the human eye, even with the help of a strong magnifying glass. In fact, they can only be examined under a microscope **(figures 431 to 434)**. Yet, these living organisms are a crucial factor in soil fertility. All waste, whatever its nature, contains millions of these organisms. They also invade water, whether stagnant or flowing.

Microorganisms feed on decaying matter and gradually convert it back into mineral salts from

fungi

429

seed

fungus

430

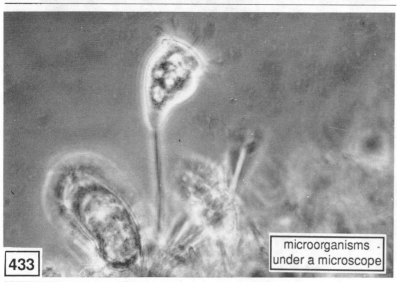

microorganisms under a microscope

433

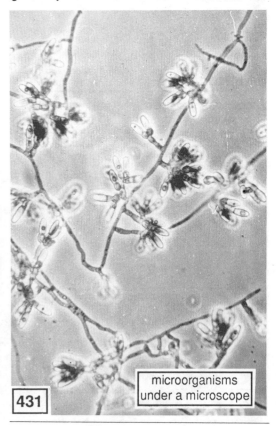

microorganisms under a microscope

431

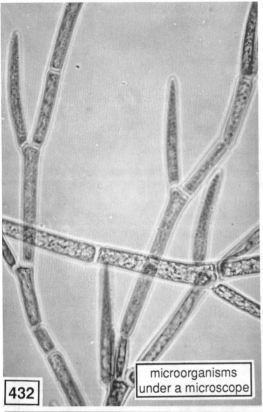

microorganisms
under a microscope

432

leaf

435

fungi

Ants attack the branch a few days later, cut the leaves into small pieces and drag them off to ant hills where they are eaten. The ants also carry away fungi growing on the decaying leaves to cultivate them in the ant colony.

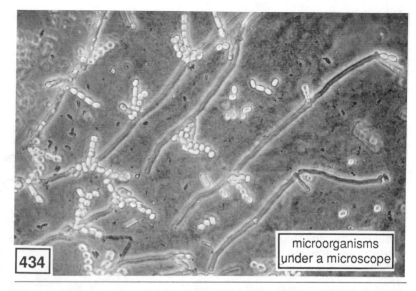

microorganisms
under a microscope

434

which plants can derive nutrients once more. This rotting process or decomposition includes the **mineralization** of organic matter, which is the release from the decaying matter of the minerals it contains.

It should be noted that roots and bacteria thrive in proximity. On the one hand, roots extend to areas where bacteria have broken down organic waste transformed into humus. On the other hand, bacteria are active near roots which exude large quantities of waste into the surrounding soil.

All soil inhabitants work together to transform the soil and benefit plant growth. Cultivators should therefore be concerned to promote this organic life by all the means at their disposal. Destroying even part of this community impoverishes the soil.

Humus

Humus comes from the decay of organic (once living) debris brought about by all the inhabitants of the soil.

Here is a particular example of this process. A withered branch falls to the ground. If it rains or the ground is damp, white or brown spots appear on the leaves **(figure 435)**. These are fungi starting to decompose the leaves.

straw being broken down
by termites

436

Next, the ants come back to attack the branch and destroy the wood itself. Just like the leaves, the branch is broken down into tiny pieces that are carried off to the ant hills and stored there **(figure 436)**.

The decaying remains and faeces of ants, termites, other insects and worms are gradually mixed with clay and sand. Many microorganisms feed on these soil mixtures, while earthworms swallow the earth to eat the microorganisms combined with it.

As with men and animals, living things in the soil prefer certain foods and avoid others. They are therefore specialized.

Humus is what remains after their activity. Humus is a black or brown substance containing water and mineral salts. The process whereby humus is produced by the whole line of living things in the soil is called **humification** or **formation of humus**.

By a continual process of formation and transformation, part of the humus is broken down into mineral salts, another part remains stable in the soil for some time, while a third part can be removed by percolating water or by runoff.

Humus always benefits soil structure (Lesson 19) and is an important source of nutrients for plants.

Two factors go to make good humus :

- **a variety of residues**, i.e. waste coming from different plants and different animals. Compare the diversified litter on a forest floor **(figure 437)** and the litter found in a plantation with only one species of trees **(figure 438)**. Humus from the first source is much richer than humus from the second ;

- **different consumers** as described above.

One leaf species broken down by one kind of ant or one kind of fungus never produces good humus. On the contrary, good humus is produced by the combined work of all the soil inhabitants that consume a wide range of nutrients. In a similar way, growing crops in association and practising rotation provide the variety of foodstuffs that allow soil organisms to produce humus.

Humus, an essential part of agriculture, springs from soil life. That is why cultural practices destroying that life endanger agriculture.

437

diversified litter
(with many components)

438

undiversified litter
(with few components)

Soil life is fragile

Some agricultural practices destroy soil life.

Untimely bush fires

Fire is a terrible **waster**. It destroys the organic matter providing food for all living elements in the soil, kills or drives away a large number of animals, insects, microorganisms and useful plants.

Fire sweeping over the same land year after year leads to the permanent loss of some plant and animal species. Only a handful of fire-resistant plants and organisms remain in the fields.

Every year, farmers set fire to the land shown in **figure 439**. Only a few plants are still growing there. The grass grows in tufts instead of being spread all over the ground surface.

Termites are the sole occupiers of the land. Other insects, including their eggs and larvae, were burnt to death. Some mammals (mice, rats, and so forth) moved away to other habitats. Erosion has triumphed.

Methods limiting the variety of cultivated plants in any particular field from one season to the next

Growing a **variety** of plants promotes a variety of living organisms in the soil. Similarly, decreasing the number of plant species reduces the range of other organisms in the soil. The difference, for instance, between forest and savanna soils is noticeable.

439

440 exhausted soil

Any cultural method causing erosion is harmful to soil life

Erosion brings destruction to soil structure because soil crumbs are broken up and air can no longer circulate in the soil (Lesson 21). Life in the soil is always handicapped when confronted with poor soil structure.

Another harmful effect of erosion is observed when water runs off slopes. It carries away waste and matter on which microorganisms feed. Decaying material is lost to the farmer for good. Eroded soil is more or less sterile **(figure 441)**.

Soil life declines when the same crop is grown continuously on the same land that is never fertilized with trash or manure coming from other land. Hence the need for the **rotation** of cultivated plants from season to season, or alternatively, the association of many species in the same season.

Plants have to be weeded to ensure good growth conditions but when weeding is done in a rough-and-ready way, it may be detrimental to soil life because it reduces the variety of food designed to feed soil organisms and be changed into humus.

Weed control by using pesticides is particularly destructive of soil life because these products kill indiscriminately and poison plants whose residues should serve to feed soil inhabitants.

Here is land cultivated without crop rotation for many years. Only a few tall herbs remain, a sign of soil exhaustion **(figure 440)**.

Chemical fertilizers sometimes destroy soil life

Chemical fertilizers are known to poison microorganisms but in some cases they overactivate the same microorganisms that then starve to death for lack of food in the form of organic waste. Like all remedies, chemical fertilizers have good and bad effects on the soil depending on whether they are used properly or not (Lesson 43).

441 exhausted soil

442

Direct exposure to sunlight is bad for soil because soil life is killed by high temperatures and drought.

Tilling without due regard for soil life can be harmful

Ways of tilling will be studied in Lesson 37. Here we shall simply point out that tilling strongly activates organisms living in the soil because it increases air penetration. However, too great an increase in soil life is a threat to cultivators who do not bother to provide food for soil organisms. What happens in this case ?

- Microorganisms become more active and use up more organic matter and plant residues.

- They produce more food for the roots.

- Cultivated plants take up more nutrients from the soil and crop yields are higher.

- Agricultural products leave the field in greater quantities as a result of tilling.

Greater amounts of mineral salts contained in agricultural products also leave the soil. If these salts are not replaced by fresh mineral salts, the soil will soon be exhausted.

Tilling without nourishing the inhabitants of the soil is a threat to the life of the farm. The danger is all the greater when tilling is carried out by heavy machinery ploughing in depth.

Bulldozers were used to clear virgin forest **(figure 443)**. Then a disc plough was used to till the land. Upland rice was cultivated in the first

organic matter mixed with earth

444 well-nourished soil

The destruction of trees

Trees and hedges contribute to soil life in many ways. Their roots are deeply penetrating and they produce abundant waste, thus nourishing the countless inhabitants of the soil. They facilitate water infiltration and shelter many birds, insects and other useful forms of animal life. Felling trees and hedges pointlessly reduces the food available to organisms living in the soil **(figure 442)**.

Exposing soil unnecessarily to direct sunlight

soil exhausted by too much tillage

443

cultural season and groundnut in the second. Chemical fertilizers were spread.

The sorry state of this field is obvious. Sandy soil deprived of organic matter has replaced forest land. Spear grass (lalang), a sure sign of soil impoverishment, is already running riot.

The man who cultivated this field made a bad mistake. He thought he could let machinery and chemical fertilizers do the work of the inhabitants of the soil.

Whatever machinery and chemical fertilizers are used, it must never be forgotten that, first and foremost, agriculture concerns life. Nourishing soil life by providing animal and vegetable waste takes priority over the use of other production factors in farming **(figure 444)**.

Lesson 34

Plants transform the soil

Everything living in the ground transforms the soil. Roots transform the soil and modify it to such an extent that without roots, there would be no soil.

Here is a profile in mountainous soil **(figure 445)**. There is a marked difference between the white rock and the black earth. This soil layer results largely from the patient work of roots. They eroded the rock, made cracks in it, slipped into the tiny gaps and, helped by infiltrating water, separated it into smaller and smaller particles **(figure 446)**.

cracks in the rocks through which roots will work their way

soil layer loosened by roots

445

herbs

soil with crumb structure

roots

rock

446

In contrast to the white rock, the soil is black because humus was gradually formed from the residues of leaves and stems, and from the roots themselves.

Root energy in transforming the soil is extraordinary. If roots are not interfered with, they are capable of restoring soil fertility even when it has been severely damaged by bad agricultural methods or overgrazing (grazed by too many cattle). Letting land return to fallow is a way of making many plants transform soil structure (Lesson 35).

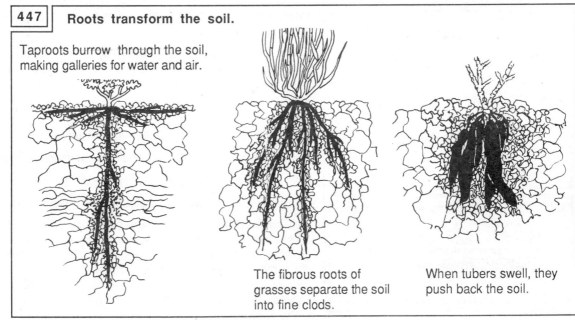

447 **Roots transform the soil.**

Taproots burrow through the soil, making galleries for water and air.

The fibrous roots of grasses separate the soil into fine clods.

When tubers swell, they push back the soil.

Plants transform the soil in several ways :

■ Penetration, water uptake, production of organic root exudates, death and decay of plant parts all modify soil structure ;

■ Leaf canopies protect surface soil structure, whilst roots bind the soil against erosion.

Plants separate the soil

■ Some plants **break up the hard soil layers**. Their taproots penetrate stony or clay layers and make channels for water penetration **(figure 447)**.

■ Plants with fibrous root systems **unite the soil into aggregates**, that is, they transform it into crumbs.

■ Some plants **loosen the soil**. Others need the land to have been cultivated to a depth if they are to be grown successfully. For instance, plants with tuberous roots press hard against the surrounding soil. When the tubers are pulled, the soil is loosened as though it had been ploughed.

When roots are strong and plentiful, they are better able to separate the soil and encourage soil life. **Figure 448** shows the difference between deep soil layers, paler in colour and hard in texture, and the darker superficial layers broken down by roots.

soil loosened by roots

hard soil

448

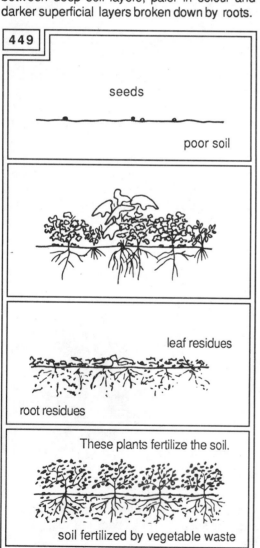

449

seeds

poor soil

leaf residues

root residues

These plants fertilize the soil.

soil fertilized by vegetable waste

Roots change soil composition and structure

Roots are like women preparing food for other people. Foodstuffs and spices are used for making dishes that turn out quite unlike the ingredients.

This transformation is illustrated in **figure 449**. This starts out with bare soil containing only sand and clay with a few scattered seeds.

Plants growing in this soil are nourished by a single dish containing a mixture of sand and clay.

The roots of these pioneer plants only find nutrients in the mineral salts taken up from the mineral fraction of the soil. These salts form part of the living matter of plants.

These pioneer plants die and rot on the ground where the decayed waste is mixed with sand and clay to form humus. Thanks to this organic waste, the soil is enriched and its structure improved.

New plants growing on the same soil are nourished on two main dishes :

■ the mineral salts contained in the clay ;

■ the mineral salts contained in the humus.

Two different dishes are always better than one to feed living beings and make them healthy. What is true of people is true of plants. Plants growing in soils enriched by organic matter, including humus, are not the same as those growing on impoverished land. As the range of plant species on this land becomes more varied, the number of inhabitants (insects, worms, fungi, other microorganisms, small mammals and so forth) increases.

Roots improve water circulation in the soil

By forcing their way through the soil, roots form many channels through which water can percolate. This process explains why plants improve water infiltration.

If we could see what happens in the ground after rainfall, we would observe the following pattern of water penetration after one, two, three and four hours (figure 450).

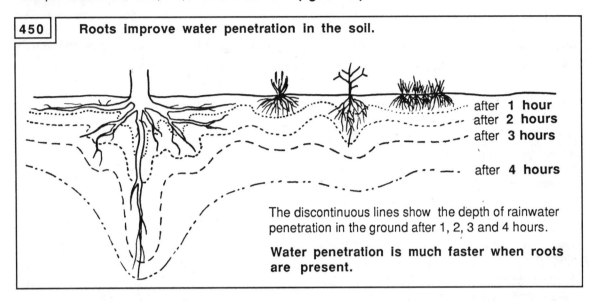

450   **Roots improve water penetration in the soil.**

— after **1 hour**
— after **2 hours**
— after **3 hours**
— after **4 hours**

The discontinuous lines show the depth of rainwater penetration in the ground after 1, 2, 3 and 4 hours.

Water penetration is much faster when roots are present.

When roots have many branches, the number of channels for water circulation is increased. In fact, water infiltration depends on the depth and profusion of branching. When roots die, they rot and leave many channels contributing still more to water infiltration and distribution in the soil.

Since roots are able to improve water infiltration in the soil, they are able to drain it. Sometimes they pump up water and transpire it through their leaves. Consequently, when land suffers from an excess of water, draining is often helped by planting trees with high water requirements (figure 451).

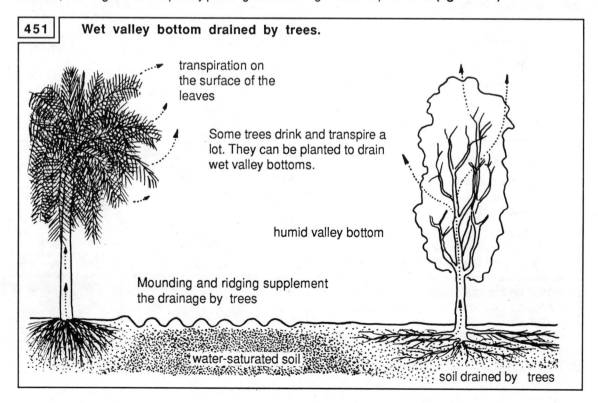

451   **Wet valley bottom drained by trees.**

transpiration on the surface of the leaves

Some trees drink and transpire a lot. They can be planted to drain wet valley bottoms.

humid valley bottom

Mounding and ridging supplement the drainage by trees

water-saturated soil

soil drained by trees

Associating useful species in relation to their role in the soil

During planting operations, soil conditions and the effects of plants on the soil must be taken into account.

As successive agricultural seasons come and go, care should be taken to diversify the plants associated in the field because this is the best way to preserve and maintain soil structure. If the same species is grown on a plot of land season after season, soil structure becomes poorer and poorer with fewer components. Soil structure becomes more and more basic because identical roots do identical jobs, and poorer and poorer because organic residues remain unchanged instead of being diversified.

What happens when various plants are intercropped with coffee **(figure 452)** ? The annual plants are sown between the rows of coffee trees in areas extensively occupied by coffee rootlets.

Before the planting season, hoeing is carried out and the coffee rootlets are clipped. This is called **root clipping**. It activates the trees and causes the renewal of much of the root system. When old roots are trimmed, new, more active rootlets appear.

All the plants intercropped with the coffee have a particular effect on the soil and on the trees themselves. Bean from the legume family enriches the soil thanks to its active nodules. Cocoyam loosens the soil because, when it is harvested, the soil is broken up and coffee roots benefit in the process. Maize roots penetrate deeper than coffee with its shallow rooting system and help produce a crumb soil structure.

452

Banana competes with coffee but, on the other hand, it produces large quantities of living matter that are restored to the soil. Its leaves make excellent mulch **(figure 453)**.

Every plant grown with the coffee therefore plays a specific role in the association. Even though yields are not necessarily the best that might be attained, they come in addition to the coffee and improve the soil.

Important points

Plant roots are active in the soil and transform it in many ways :

- *they modify soil structure and composition ;*

- *they modify soil nutrients ;*

- *they modify water stores in the soil by increasing water infiltration and decreasing runoff.*

By adapting agricultural methods to farming conditions, cultivators can use plants to drain land and can grow them in association, with consideration to their effects on the soil - opening up, aeration, nitrogen fixation, enriched organic matter, etc.

453

Lesson 35

⬚⬚

Fallow land

The use of fallow land is one of the many agricultural methods that increase the amount of humus on fields for the benefit of cultivated plants.

Leaving land fallow is a cheap way of restoring the natural fertility of the soil lost through cultivation. Fallow land is a piece of land left to rest after a period of cropping (Lessons 7 and 8). All kinds of vegetation grow there - herbaceous plants, shrubs, trees, creepers, and so forth. Fallow seeds come from surrounding land or are preserved in the soil during the cultivation period.

Leaving land fallow affects soil fertility and increases its ability to provide cultivated plants with a favourable environment (good soil structure) and sufficient food. However, every plant has characteristics and requirements peculiar to itself. That is why **land may be fertile for one plant species and infertile for another.** Or, land fertile for a plant species at a given phase in the vegetative cycle may be sterile for the same plant later on when it has taken up all the available nutrients in the soil layer exploited by its roots. The same soil may prove fertile for another plant species with other nutrient requirements. This point was mentioned in Lessons 7 and 14 in connection with pioneer plants while methods of crop rotation and leaving land fallow were explained in Lesson 8.

Why leave cultivated land fallow ? (figure 454)

454	Why leave cultivated land fallow ?

This is what happens when the land is cultivated :

■ Every cultivated species takes up specific nutrients from a given soil layer. The species exhausts the soil in a selective way, i.e. a way peculiar to itself.

■ The production of living matter is minimal because farmers control the species that will not be harvested.

■ Many agricultural products are obtained from the soil. They leave the land, carrying away the nutrients coming exclusively from the area exploited by the cultivated plants.

■ The pests associated with the cultivated plants increase rapidly.

■ While under crop, the soil is exposed to sun, rain, and farm implements. Organic residues are quickly destroyed by many microorganisms. The crops use up humus, and soil structure deteriorates.

This is what happens during the fallow interval :

■ During the fallow interval, plants exploit many soil layers and take up a wide variety of nutrients.

■ Biomass (living matter) is produced in large quantities. Residues are abundant and varied. This intense activity modifies the qualities of the soil.

■ All the living matter and residue from it remain on the land and rot there. Nutrients taken up by deep rooting plants are restored to the layers near the surface.

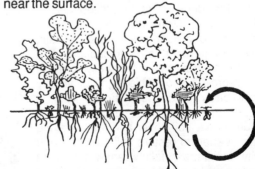

■ The pests associated with cultivated plants decrease in number because they cannot find the food they prefer.

■ Under fallow growth the soil is protected from sun and rain. Soil temperature is lower, moisture is higher. Microorganic life is less active than in heated soil. Organic waste decays slowly and is mixed more intimately with sand and clay. Fallow land create humus, improves the structure and fertility of the superficial soil layers.

Duration of fallow intervals

The duration of fallow intervals on farmlands varies a great deal.

- The duration varies with **the climate and the vegetation** found in the area. Forest, bush and herbaceous fallow land are terms commonly used.

- The duration of fallow intervals varies with **the quality, variety and abundance of plants** found there. When there are many plant species, the effects on soil fertility of leaving land fallow are rapid and significant. For instance, fallow land with legumes is more efficient than fallow land without these plants. Having different root systems is also important.

- The duration depends on **the quantity of seeds, stumps and roots of fallow plants** that remained alive during cultivation. The longer the cropping period, the longer it will take to establish good fallow conditions.

  As a general rule, a long period of cultivation must be followed by a long fallow interval in order to restore fertility to an acceptable level.

455

- The duration depends on **the amount of farmland available**. When land is in limited supply, the time when land is left fallow is correspondingly short. When fallow intervals are too short, soil fertility is not properly restored. In this case, farmers must turn to other ways of improving fertility (Lesson 36).

- The duration of fallow intervals depends on **the quality of the soil** and especially its composition in terms of sand, clay, gravel and mineral salts. Fallow vegetation thrives better on land rich in mineral salts than on land poor in these.

456

- Finally, the duration of the fallow period depends on **cultural methods**. On poorly tended soil or soil eroded by rain, fallow regeneration is handicapped. The wild plants that would normally provide fallow regrowth show little signs of activity on this badly eroded soil **(figure 455)**. It will take many years for fallow plants to provide proper soil cover and the decaying matter needed to restore soil structure and fertility. In contrast, fallow renewal is much easier on well-tended land where some tree stumps, trees and strips of grass have always been kept.

Variety in fallow vegetation

The forest plot **(figure 456)** was allowed to rest after cultivation lasting two years. Forest plants have grown again to replace the cleared trees. Nonetheless, forest fallow is quite

457

458

459

460

different from the original forest before cultivation because the variety of plants in forest fallow, as opposed to native forests, is not so great.

In the same way, the natural fertility of a given piece of land is lower after an interval of forest fallow than when first cleared.

Herbaceous fallow, composed of **ferns** and **Spear (sword) grass**, was photographed in the same forest region **(figure 457)**. These two plants growing after crops are a sign of soil exhaustion because cropping lasted too long. Trees in this forest fallow will have a struggle to survive and only savanna herbs will remain in the end.

Here is bush fallow in a region with fairly abundant rains **(figure 458)**. Residues are not as abundant as in forest fallow and, in addition, some residues are burnt and turn to ash. This is a young fallow plot with some cassava still in the ground.

The fallow land in **figure 459** is composed only of grasses in an area with heavy rainfall. The fallow state is three years old. It will soon be cleared and the trash buried. Residues in large quantities will rot on the ground but they will lack variety as they come exclusively from grasses.

A look at fallow land in a region with low rainfall is also instructive **(figure 460)**.

461

A harvested field lies in the background. Unlike the vegetation in the previous figure, the short herbs in the foreground only produce small quantities of trash and consequently return scant amounts of organic matter to the soil. Fortunately, the trees produce a good supply of residues and, like the herbs, help to improve soil fertility all the time.

196

462

How to make fallow land as productive as possible

■ **Respect the duration of fallow intervals.**

Good fallowing practices mean producing residues and humus equivalent in bulk to what was destroyed during the cropping period. This process takes time. Roots must grow in depth ; leaves, stems and trunks must develop, die, fall to the ground and decay. Plants must be present in sufficient quantities. Regeneration of forests takes many years, perhaps ten to fifteen. It is not usually so long in savanna.

If the fallow interval is too short, the soil layers exploited by cultivated plants after leaving the land fallow, will not be adequately enriched.

■ **Apply the right succession of crops (crop rotation) after clearing.**

Cultivators know which plants will thrive on which piece of fallow land. Cultivated plants are chosen in relation to soil fertility and the kind of vegetation growing on the fallow land. Fallow plots are often named with reference to the crops that will be sown after the fallow interval.

Siam weed dominates the fallow land in **figure 461**. This plant spreads rapidly not in forest but in untended fields in parts of Central Africa. Siam weed enriches the soil and grows in association with wild forest species.

Siam weed is native to Asia but is now firmly established in tropical rainbelts on the African continent. It sometimes invades land, becoming a serious problem, hard to control. However, this plant proves that the value of fallow periods for soil restoration is enhanced by the addition of enriching herbaceous plants.

Unfortunately, the fallow land in **figure 462** is seriously impoverished. The land was tilled by machines for many years with complete disregard for the soil, which is now exhausted and ruined by laterization. This land will never be cultivated again. A few plants tolerating poor soil grow here and there but they will not manage to restore soil fertility because they do not produce enough organic matter.

In some types of dryland farming, leaving land fallow means keeping land clear of all vegetation, including weeds and beneficial bush species, in order to conserve moisture as well as resting the land.

463

464

■ **Respect the organic matter produced by leaving land fallow.**

Turning organic matter into humus is the most important aspect of leaving fallow. Methods limiting the formation of humus must be ruled out, particularly the wanton destruction of organic matter by fire.

In most cases, clearing involves huge quantities of leaves, stems, branches and trunks. Fire is the only way to get rid of unwanted vegetation and open up fields. But there are ways of controlling bush fires so that only wood and plants of real disturbance to crops are destroyed while the remaining vegetation is left to rot on the ground or is heaped up out of the way.

The grass fallow **(figure 459)** is shown in detail in **figure 463**. Here, letting residues rot and then burying them would be beneficial, but this is difficult without the right tools. So, the farmer is forced to burn, against his better judgement.

In all events, the more superficial the fire, the better for soil life **(figure 464)**.

Intense, slow-burning fires overheat the soil, strip it bare and destroy soil life. Fires produce not humus but ashes that are blown away, leaving denuded soil exposed to erosion and laterization **(figure 465)**.

In regions with low rainfall, fallow plants are often smaller and thinner and, as a rule, can be easily buried by hand or, better still, ploughed into the ground.

In herbaceous fallow towards the end of the fallow interval, plants are sometimes cultivated so that wild species are quickly overgrown or at least brought under control. A woman farmer used this strategy in a herbaceous fallow **(figure 466)**. She first beat down the grassy stalks, did some rough hoeing and then sowed sweet potato. This plant makes herbage less active and more manageable for burying later on.

Agricultural implements for burying rather than burning trash during clearing operations

When vegetable trash is a nuisance in cultivated fields, machinery of various kinds can be used to crush and plough it into the soil. Short branches and stalks can be turned into the soil by using a mouldboard plough, but longer branches and grass stalks have to be chopped before burying.

Many implements, most of them powered by small engines, are designed for chopping up residues and debris.

■ The **disc plough** breaks up stubble and small branches before burying them in the ground **(figure 518**, Lesson 37). Tractor-drawn.

■ The **rotary crusher**, a big motor-powered blade that rotates very fast, chops up straw and fine stalks and spreads them on the ground. Drawn by tractor or mounted on a hand-steered mechanical cultivator.

- The **power saw** with a small blade 30 to 40 cm long, hand-controlled.

- The **rotary cultivator** used for cutting up trash left on the ground and for burying it **(figure 522, Lesson 37)**.

- The **chain saw** is another tool for handling organic matter produced by a fallow interval **(figure 468)**. Tree trunks and branches can be sawed and windrowed, i.e. swept into **windrows** as in **figure 467**. A windrow, in this instance, is a long row of stems, branches and tree trunks in a fallow clearing. Windrows can be set on fire without burning organic matter in other parts of the field. They can also be left to decay gradually. Decomposition is activated by termites, ants, rodents, worms, fungi, etc.

467

468

However, if windrows are burnt, the ashes should be spread on all the fields to be cultivated to benefit the crops.

Lesson 36

Cultural practices for producing humus

All over Africa, farmers use fallow intervals and rely on vigorous natural growth to restore soil fertility.

However, a whole series of other practices exist and achieve comparable or better results by the selective use of enriching plants rather than leaving natural vegetation to do the job on its own.

These interventionist practices are justified, for instance on farms where fallow intervals have to be curtailed because there is a shortage of land, or to speed up soil regeneration, or to make clearing easier because wild plants are sometimes hard to eradicate when clearing starts.

Here are some practices aimed at producing humus.

Cultivating plants specialized in soil improvement

Some plants improve soil structure more than others. Tropical kudzu is a particularly useful plant in this respect **(figure 469)**. It is a legume providing soil cover after two or three months, and may be cut

tropical kudzu

kudzu litter

469

470

tropical kudzu

back later on. Plant residues can be left on the soil but are more effective when ploughed in. Because kudzu is specialized in fertilizing the soil, it is called green manure.

straw

moist soil

471

472

Tropical kudzu can be added to herbaceous fallow as in **figure 470** where seeds were broadcast during the agricultural season immediately preceding the fallow interval. Legumes are highly recommended for this agricultural practice called fallow improvement.

Legumes commonly used as green manure crops are *Pueraria* (tropical kudzu), *Centrosema*, *Calopogonium*, *Crotalaria*, *Stylosanthes*, *Mucuna*.

Mulching the soil

Mulching is another way of improving soil structure and fertility, and of producing humus. A thick layer of mulch (straw, leaves, etc.) is spread on the ground. This litter retains soil moisture and prevents overheating. It decays slowly and is changed into humus.

Plant residues found on the spot or coming from other sources can be used for mulch.

On no account, however, should residues containing seeds be used. Seeds might germinate and cause weed infestation.

Here are two examples of mulching. Elephant grass leaves were used in **figure 471**. The soil underneath this cover is humid and fertile. Big banana leaves were used in the second case **(figure 472)**.

200

A young tree has just been planted **(figure 473)** and mulched round the base to maintain good soil structure. Mulching stops weeds from growing in proximity and from competing with the plantlet for food. The mulch cover rots gradually and, with decay, restores its mineral salts to

473

474 | *Tithonia*

475 | *Tithonia* hedge

476

Mulching material can also be obtained by growing or tending good cover plants in the fields or nearby.

Tithonia (Mexican sunflower) is such a plant. It is often found near villages in forest zones **(figure 474)**. It gives an abundant supply of humid stalks that are easily cut and spread to provide cover. *Tithonia* can form hedges for regular cutting whenever mulching is planned.

Making compost from organic waste

Spreading **compost** is another way of conditioning the soil. Compost is a mixture of all kinds of organic waste in various stages of decomposition. The waste is heaped up and usually heats and cools off after a few weeks. When the residues are thoroughly decomposed, they change into compost manure, containing humus, that is then spread on the fields. The greater the variety of compost matrial, the better the compost from the standpoint of soil fertility.

For example, plant material from the *Tithonia* hedge **(figure 475)** on the edge of a field is ready for cutting to make compost or mulch. Whatever happens, the plant must be used before the seeds reach maturity. Stems and leaves must be dried or heaped up to allow for heating so that budding is avoided when the waste is spread later on.

Compost is also made from straw and/or chopped branches. A **straw chopper** can be used for the purpose. This piece of equipment has a metal box with one or more rotating blades for chopping thin branches or straw.

The crushed waste is piled into a broad, high heap and left to rot. The heap should be sprinkled with water every day and turned over two or three times at intervals of six to eight days. Compost from vegetable waste can be enriched by adding animal waste products, ashes, lime and fertilizers. The composting materials are spread in layers.

A **compost crusher** fitted with a petrol engine is shown in **figure 476**. The branches are pushed into the bulk hopper and the crushed waste falls down a chute to the ground.

The straw chopper in **figure 477** is worked by hand and is only used on straw and grasses.

Making compost is a way of producing on the farm itself high-quality manure with humus and mineral admixtures (ashes, chemical fertilizers). **Regrettably, this agricultural method is not practised widely on African farms**, although its potential benefits are huge.

Points to remember when making compost are :

- the compost must be made out of **layers** of different materials ;

- the compost must be **turned** to promote beneficial decay and avoid putrefaction.

For example, the layers could be like this :

- a bottom layer of dung, 3 cm deep ;

- a middle layer of ash, 3 cm deep ;

- a top layer of vegetable matter, 15 cm deep.

- The pattern is then repeated - dung, ash, vegetable matter.

Planting trees

As a general rule, trees produce large quantities of organic matter and therefore of humus. Consequently, growing them in association with seasonal crops is always beneficial provided competition is avoided.

Tree roots are active in depth in the soil and every year their foliage produces large amounts of organic matter and humus. Since all trees are not able to make humus to the same extent, they should be planted on a selective basis with reference to their humus potential. *Leucaena* in a coffee plantation is a good choice from this point of view **(figure 478)**.

Associating complementary species

Associating complementary species in fields is a good way of improving soil structure, because many species growing together produce more organic matter than the same number cultivated in pure stands.

Such plant materials are also more varied and, when decomposed, provide better humus. Apart from producing humus, every plant modifies the soil.

blades

straw goes in here

handle

chopped straw falls out here

477 straw shopper

Leucaena

478 coffee

Notes

202

Sweet potato is associated with banana **(figure 479)**. The leaves of both plants will fall to the ground, provide thick cover, then rot and be transformed into humus while banana bunches and tubers are being harvested.

Use of night kraals (folds or pens)

Kraaling cattle means penning them at night in the same place for a given length of time. The soil is fertilized by dung and urine, always valuable adjuncts for producing humus.

The kraal can be moved round after a few days so that, by the end of the season, cattle will have manured many fields. Cultivators must be strict about burying the dung when the cattle move elsewhere. Otherwise, the dung is burnt by the sun, organic matter destroyed and nutritive value for the soil lost.

The kraal in **figure 480** is situated in the Sine Saloum territory of Senegal. The trees scattered in the background also help to fertilize the soil.

Some cultivators realize that kraaling is worthwhile, and they pay cattle owners to graze animals in their fields during the dry season. Others start their own herds and let them graze on the land.

Acacia albida

kraal

480

481

banana

479 sweet potato

Growing forage crops

Forage crops are excellent fertility builders and improve soil structure especially when regularly grazed by cattle **(figure 481)**. Their plant roots activate the soil, cattle feed on the leaves and manure the ground.

Associating forage crops with plants cultivated for human consumption

Wandering domestic animals are a common sight in many African settlements. Animals, mostly cattle, sheep, goats and pigs, stray and make do with any food they find such as wild plants and waste.

Stray animals are a source of endless trouble, e.g. crops destroyed, undernourished cattle, bad village hygiene, etc. A way of solving the problem of loose cattle and avoiding friction between farmers and cattle owners is by interplanting forage crops with plants for human consumption.

Palm planters in parts of Africa grow legumes under the trees or a mixture of legumes and grasses, thus providing pasture for livestock. This agricultural practice ensures food for cattle, and dung is restored to the soil to make humus for the palms.

In zones with lower rainfall, forage crops are grown with seasonal plants for human consumption. A legume called siratro *(Macroptilium)* is planted along with millet and acts as a fertilizer for the cereal. After the millet harvest, siratro is left in the ground and is grazed without too much damage from trampling. If preferred, it can be harvested, dried and stored **(figure 482)**.

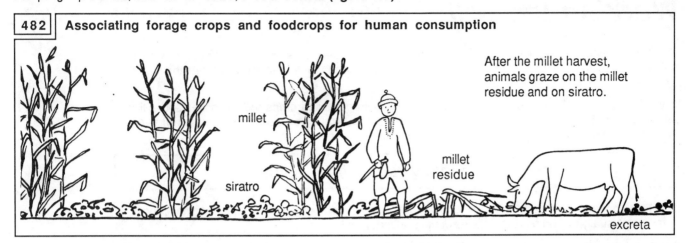

| 482 | **Associating forage crops and foodcrops for human consumption** |

After the millet harvest, animals graze on the millet residue and on siratro.

millet

siratro

millet residue

excreta

Siratro can be ploughed back into the soil at the next planting season. It is also used as a fallow plant, fertilizing the soil and improving fallow regeneration.

Trees producing forage can be interplanted with seasonal crops. This cultural method was discussed in Lesson 15 in connection with Serer settlements and the wide use of *Acacia albida*.

There is much to be gained by growing forage plants along with cultivated plants for human consumption :

■ this practice ensures a plentiful supply of varied vegetable waste that is changed into humus ;

■ farmers who intercrop forage plants and plants for human consumption boost farm economy. They solve the problem of wandering cattle, either directly by letting livestock into their own fields after harvesting, or indirectly by selling forage to cattle owners. Farmers growing their own fodder crops can feed stabled and tethered animals, and keep them away from cropping fields where cattle cause much damage ;

■ when fodder and other crops are grown in association, animals grazing in the fields, after the harvest, change plant waste into animal manure that, in turn, provides excellent humus.

Reminder

■ Because humus is needed to make soil fertile, farmers should aim at producing it all the time.

■ Only living soil is rich in humus. Only living beings are able to produce humus. Farm implements and machinery must be used with caution.

■ The number and activity of living organisms in the soil are directly related to the quantity and variety of plant and animal waste restored to the soil.

■ Soil life is very fragile. It can be destroyed by the wrong cultural methods as well as by fire, sun, erosion and needless felling of trees. Any products killing soil life must be used cautiously. Chemical fertilizers, herbicides, fungicides and pesticides can kill soil life without distinction. They are therefore dangerous for the soil if instructions for use are not followed scrupulously.

■ Increasing and protecting residues falling on the ground will increase and maintain food supplies for plants.

■ Setting fire unnecessarily to residues that could rot slowly in the soil is a waste of plant nutrients.

■ This is why it is always beneficial to let residues decay in the soil for some weeks before the planting season so that they have time to change into humus and to mix with sand and clay.

■ Various agricultural practices favour the formation of humus - leaving land fallow, using specialized plants, planting trees that enrich the soil, growing crops in mixture, mulching, making compost, etc.

■ Methods suitable for making humus in the soil are also good for retaining soil moisture.

■ Protecting and producing humus in fields guarantees the future of the land.

Lesson 37 🔲🔲🔲🔲🔲🔲🔲🔲🔲🔲🔲🔲🔲🔲🔲🔲🔲🔲🔲🔲🔲🔲🔲🔲🔲🔲🔲🔲🔲🔲🔲🔲🔲🔲🔲🔲

Working the soil

Weeding, hoeing and earthing up operations have already been described in connection with field maintenance. The aim of these operations is to control weeds, limit evaporation from the soil and improve the anchorage of cultivated plants.

In this lesson, all the land-forming work carried out before sowing crops will be discussed - tilling, earthing up, ridging, and bedding cultivation. Whatever the technique used, a layer of soil of variable depth is turned over to activate growth in the roots of cultivated plants.

There are many reasons for tilling

- **Tilling loosens the soil**, helps seed germination and root penetration. Good growth conditions for roots usually exist in proportion to the depth of inverted soil.

- **Air circulates more easily in tilled soil than in untilled soil. Air is good for soil life** and especially for microorganic life. Living soil is a favourable environment for cultivated plants.

- **Tilling is good for rainwater penetration.** Water runoff is reduced because the rain infiltrates the gaps between the clods of tilled earth. If the soil is clayey and water-saturated, **tilling helps drainage**.

- Good tilling methods help the farmer **to control unwanted plants** (weeds) on his land. Weeds that are properly turned over (roots up, leaves down) are killed and decompose in the soil.

- Proper tilling is **an efficient way of burying manure**, trash, ashes and fertilizers.

There are many ways of working the land

Tilling must take all these elements into account:

- **the kind of soil** to be cultivated. Light, sandy soil is not worked in the same way as heavy, clay soil or crumby soil rich in organic matter;

- **the depth of arable soil.** Shallow soil cannot be worked like deep soil clear of rock;

- land **on a slope** is worked differently from flat land;

- **soil moisture** is also a determining factor in the tilling method chosen. For example, a wet valley bottom must be tilled so that excess water flows into furrows whereas the objective in tilling dry soil is to capture all the rainwater;

- working the land depends on **the available tools**. Tilling by hand is completely different from mechanical tilling. But there are marked differences in

483

485

mechanical cultivation depending on the type of plough used. A disc plough does not give the same results as a mouldboard plough and share (see below) ;

- **the energy (force) used to make implements work** is also a factor influencing the quality of tillage operations. Manual work and work done by animals and mechanized equipment (with different kinds of engines) do not exert the same force, and therefore tilling results cannot be identical ;

- **the rooting systems of plants** differ from species to species so that tilling must be adapted to the plants to be cultivated ;

- tilling varies with the **bulk and shape of the organic matter** to be buried. Long stalks, for instance, are harder to bury than small leaves or animal manure.

Some common ways of working the land

484

tilling on the flat for maize

cassava on mounds

486

Tilling on the flat

Tilling on the flat means turning over the soil on the spot by hoeing or ploughing. Tilling is said to be on the flat because the surface of the field seems flat after tilling, in contrast to ridging, for example. **Figures 483 and 484** illustrate tilling on the flat done with a hoe, i.e. entirely by human energy. An ox-drawn plough is used for tilling on the flat in **figure 485**.

487

yam mounds

488

Tilling on the flat by hand is always shallow while animal-pulled and engine-powered machinery tills to a greater depth.

When deep tillage is desired but adequate implements and animal or mechanical power are not available, tillage involving **mounding, ridging, and bedding cultivation** can be used to concentrate richer topsoil with minimum effort.

These techniques all aim at giving the roots of cultivated plants a greater volume of loosened soil.

yam planted on ridges

489

organic waste

ridge

490

ridges made by a mechanical ridger

491

Mounding (earthing up)

Here are two kinds of mound prepared for a yam crop. In the first example, long, raised mounds (sometimes called hills) were earthed

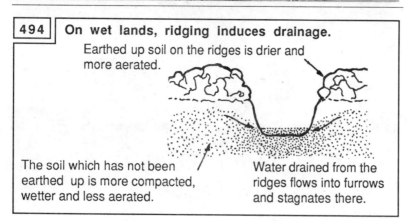

494   **On wet lands, ridging induces drainage.**

Earthed up soil on the ridges is drier and more aerated.

The soil which has not been earthed up is more compacted, wetter and less aerated.

Water drained from the ridges flows into furrows and stagnates there.

492

up under a tree **(figure 488)**. In the second example, the mounds are small and rounded off **(figure 487)**. Groundnut and yam have been interplanted on the mounds.

The fenced field **(figure 486)** was tilled on the flat for maize. Here and there the farmer earthed up and planted cassava cuttings.

Ridging

Ridging means forming the soil into raised lines called **ridges**. The hollowed-out rows between the ridges are called **furrows**.

Here are ridges made by hand to intercrop yam and sweet potato **(figure 489)**. Yam is planted on one side of the ridge, potato on the other. To the right, stakes can be seen planted in the ridge tops to train the climbing yam stalks as they grow upwards in contrast to the sweet potato which is a creeper.

The ridging technique photographed in **figure 490** is used in the north-west of Cameroon. The ridges come above the knee to a height of between 60 cm and 70 cm. Here and there, planters heap up large quantities of trash and organic waste and cover them with earth.

The land under these trash heaps will be well fertilized when waste has decayed. Trash is sometimes set on fire under the covering layer of earth. It burns very slowly releasing its own mineral salts in the form of ashes. As it heats the soil, it also releases the mineral salts contained in the surrounding clay.

Ridging can be done mechanically **(figure 491)** but ridges made in this way are low because ploughs cannot raise earth as high as hand tools.

Ridging can be established on fields with or without tree plantations. The ridges in the banana grove **(figure 492)** have been intercropped with all kinds of plants - maize, potato, cocoyam, bean, groundnut and a choice of vegetables.

In another field, ridging was used to drain water saturated soil **(figures 493 and 494)**.

Ridging in Senoufu country

Senoufu farmers in the Ivory Coast, Burkina Faso and Mali use a hoe with a concave blade and bent-back handle **(figure 497)**.

Tilling is carried out in two stages. The farmer starts by digging the furrow **(figure 496)**. The soil is broken into clods and turned over

sugar cane

493

first stage : opening of the furrow

496

second stage : deepening the furrow, raising the ridge

495

onto the ridging row. As the farmer goes along, he makes sure to turn over the weeds uprooted from the furrow and to cover and choke the weeds on the ridging row. By the time he has worked up and down both sides of the row, the ridge is well formed. Parallel ridges are established all over the field.

The second step in this ridging method is to deepen the furrows and raise the ridging. Soil dug from the furrows is cast onto the ridging giving it more height **(figure 495)**. In this way, weeds still growing on the top of the ridges are suffocated.

If the farmer thinks it advisable in view of the cropping plan, he may dig the furrows a second time and raise the ridges still more.

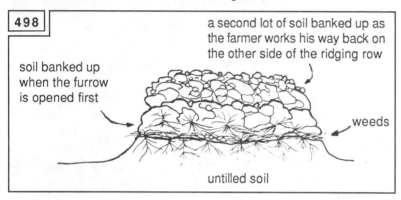

498

a second lot of soil banked up as the farmer works his way back on the other side of the ridging row

soil banked up when the furrow is opened first

weeds

untilled soil

497

ridging in a rice field

ridge

kayendo (hoe)

earth bank

499

If we were to make a soil profile in the ridges in this field, this is what we would find **(figure 498)**.

Next year, the ridges will be opened and the soil will be earthed up to form new ridges covering the old furrow slices. This agricultural method ensures that every year the soil is thoroughly mixed to a depth of 60 cm. This is deep tillage with total burial of weeds.

How rice fields are tilled in Casamance and Guinea

In Casamance (Senegal) and in the neighbouring regions of Guinea Bissau and the Gambia, farmers grow rice in mangrove swamps. They work the soil with a tool called the **kayendo (figure 499)** which has a long handle and a wooden blade with a reinforced iron tip.

The kayendo cuts and forms the side of the ridges and digs the furrows. The soil is thrown onto the ridges as in the Senoufu method just mentioned. By using the kayendo, the soil is tilled in depth.

In order to lower the water level and stop formation of a salty soil crust, furrows are sometimes dug very deep. The land in Guinea Bissau **(figure 500)** was tilled in depth to grow rice. The ridges are more than 60 cm high. During the dry season, they will start

rice field

anti-erosive banks

ridges

furrows dug out with a kayendo (hoe)

500

501 | Rotation of ridges

first cultural season

fallow herbs

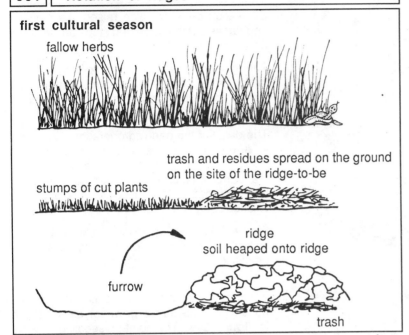

trash and residues spread on the ground on the site of the ridge-to-be

stumps of cut plants

ridge
soil heaped onto ridge

furrow

trash

second cultural season

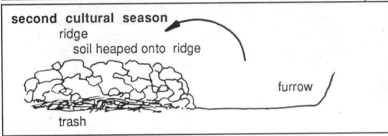

ridge
soil heaped onto ridge

furrow

trash

third cultural season

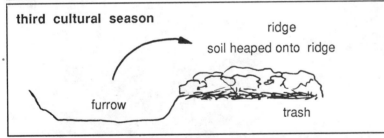

ridge
soil heaped onto ridge

furrow

trash

cut ridge

502

sinking as a result of erosion and the furrows will begin to fill up.

Cutting and rotating ridges

Figure 501 shows how ridges are rotated from one year to the next.

Ridges are being cut in two in **figure 502**. Trash can be seen in the furrow. It is lying under the first layer of earth thrown down from the cut ridge.

Notes

503 banana / bedding cultivation

504 earth was cast onto the bed and covers the litter / edge of bed

505

Bedding cultivation

Cultivation beds are also raised like mounds and ridges but the cultivated strips are wider as pictured in **figure 503**.

The ground was hoed and spread with tall fallow herbs. These were then covered with soil scraped up all round the surface about to be cultivated.

The detail of the bedding operation stands out in **figure 504**. In the foreground, the soil is covered with trash. On the side to the right, the surface is bare because the earth has been scraped up and cast onto the bed. The finished operation is seen front left.

Figure 505 gives a general view of bedding cultivation. In this field, bedding rotation is practised as with the ridging rotation in **figure 501**.

Bedding cultivation is often used for horticulture (vegetable growing) as illustrated in **figure 506**. The tomato seedlings in the bed on the right will soon be transplanted to the bed in the centre which is ready to receive them. Eggplant has just been sown in the bed in the foreground. The bed is covered with straw to maintain the moisture needed for germination.

506 horticultural beds

507

508

509

Bedding cultivation involves a lot of hard work because large quantities of earth have to be transported. However, when machinery is not available, bedding cultivation is an effective manual method of increasing the volume of earth exploited by the roots of cultivated plants.

Planting between mounds, ridges and beds

When mounds, ridges and beds are ready, the furrows can also be cultivated if the soil is deep enough.

This cultural practice is displayed in **figure 507**. On both sides of the picture, there are large mounds where maize and yam are growing in mixture along with gourds and other vegetables. In the wide furrow in the centre, the soil was worked and formed into low mounds where cassava cuttings were then planted.

When maize and yam have all been harvested, the farmer will start breaking the old mounds and filling the centre of the furrow to earth up the small mounds planted with cassava. Before transporting the soil to the new mounds, he will flatten the maize stalks and other trash to the ground. He has begun the task and is using a special type of local hoe **(figure 508)**.

In order to increase the volume of earth exploited by roots, the farmer can practise simultaneously more than one method of using the way roots occupy the soil. Two examples make this point clear.

Trees line a cultivated bed in the first example **(figures 509 and 510)**.

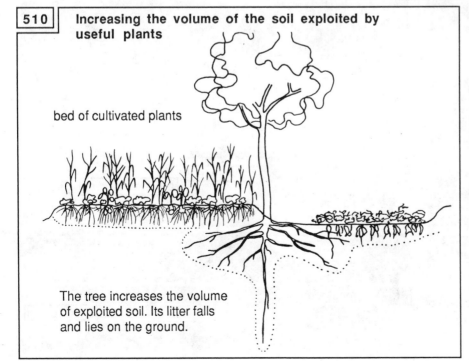

510   **Increasing the volume of the soil exploited by useful plants**

bed of cultivated plants

The tree increases the volume of exploited soil. Its litter falls and lies on the ground.

212

511

cocoyam

512

weeding-mounding
of bean ridge

Their roots penetrate deeply but do not compete with the roots of the cultivated plants which remain confined to the volume of loose soil in the bed itself.

The second example shows beans growing on ridges **(figure 511)**. Trees scattered over the field root in depth, and residues from them fall

weeding and
mounding
groundnut by hand

513

514

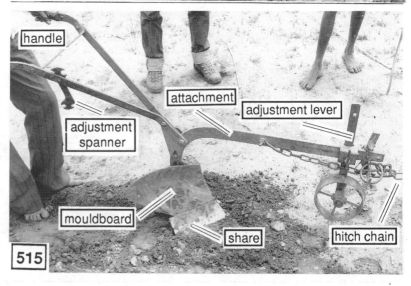

handle

adjustment
spanner

attachment

adjustment lever

mouldboard

share

hitch chain

515

to the ground and act as fertilizer. In this kind of association, trouble may be caused by tree density. If there is too much competition from the trees, they will have to be pruned.

Weeding ridges and mounds

Earthing up and ridging are sometimes combined with field maintenance. Lesson 31, **figure 382**, showed how useful it was to earth up maize to help the growth of adventitious roots. Mounding reinforces stems at the base and improves anchorage.

Weeding of bean ridges is under way (**figure 512**). Soil in the furrow is scraped and cast onto the ridge. Weeds are uprooted. The woman farmer is careful to cover with earth the weeds that sprang up with the beans and could cause competition. This is a weeding-mounding operation that aims at eliminating weeds and building up mounds and ridges. It is also a way of covering up any bean plantlets with roots exposed through rain erosion.

The groundnut field (**figure 513**) has just been weeded. The soil was scraped between the rows and earthed up round the young plants at the flowering stage. This is also a weeding-mounding operation. Because the earth has been loosened, the groundnut pegs which later form pods can easily enter the soil. The operation could be carried out because cleaning was early and the woman did not have to cope with tenacious weeds.

Implements for mechanical tilling

Ploughs

Tilling is much easier when animals, motor-powered cultivators or tractors are available. Instead of sheer elbow grease, animal and machine power is harnessed and makes deep ploughing easier in a single operation (**figure 514**). A tractor can till to a depth of 30 cm. A team of oxen ploughs 20 cm deep, while donkey power will not go deeper than 10 or 15 cm as a general rule. These depths obviously depend on soil type and the sort of tool being drawn.

There are all sorts of ploughs. One of the commonest kinds is illustrated in **figures 515 and 516**. The main working parts can be distinguished clearly - attachment, handle, share, mouldboard and heel. The share cuts the soil, the mouldboard turns it over, the heel slides along the ground. In front, there is a depth wheel and adjustment lever to regulate the depth of penetration. There is also a hitch chain which is linked to the ox yokes.

516

animal-drawn plough with reversible share

517

disc plough tractor-drawn

518

There are many other kinds of plough. The one in **figure 517** has two shares and two mouldboards. These parts are reversible so that the plough can go up and down the furrows, always inverting the soil to the side already ploughed. This is not possible with the model in **figure 515**.

519

520

521

The disc plough (**figure 518**) must be tractor-drawn. Concave discs are mounted on a heavy metal chassis. The use of such ploughs is limited in tropical Africa because they are expensive to use.

Two ploughing operations are shown in **figures 519 and 520**. In the first case, the plough was drawn by two oxen. The soil is fairly sandy and, as it was inverted, it separated into smallish clods. The result is shallow ploughing to about 15 cm.

The heavy clay soil in the second example was tilled in depth by a tractor-drawn plough. The plough reached a depth of 30 cm but the clods are large and the soil is barely broken up. Other implements will be needed to go over the field and break up the clods because the soil has not been sufficiently loosened to allow seed germination.

Cultivators

The implement in **figure 521** is a **cultivator**. It is mounted on a metal chassis and consists of a wheel, a handle and three tines which dig into the soil with animal traction.

A cultivator works differently to a plough because the soil is not inverted. The tines break up the surface and uproot weeds that are left lying on the ground.

Cultivators are useful for light, easily-workable soils.

Rotary cultivators

Hand-controlled and tractor-powered cultivators are available (**figures 522 and 523**). Curved tines (**figure 522**) are attached to an axle rotated by the engine . The tines break up and mix the soil to a depth of 15 to 20 cm. These cultivators loosen the soil well but weeds are not completely buried. This implement does a good job on light sandy soils, but is not so successful on heavy soils.

Harrows

The soil photographed in **figure 520** was harrowed with the result shown in **figure 524**.

Harrows are large rakes mounted on a metal or wooden frame to which tines are attached. Harrows are used to break up large clods of soil on ground already tilled and to prepare a loose refined bed for seeding.

The metal harrow in **figure 525** was made by village blacksmiths. Another type that can

522

523

524

525

easily be made by local craftsmen is illustrated in **figure 526**. Four rows of tines are attached to a strong metal frame. The harrow can be drawn by hand or by an animal. To make sure that the tines penetrate the soil, the tiller must stand on the frame or place a heavy weight on it.

Notes

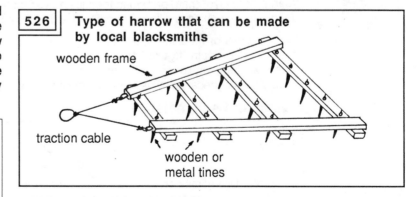

526 **Type of harrow that can be made by local blacksmiths**

wooden frame

traction cable

wooden or metal tines

Distinctive signs of good tillage

The aims of tilling were explained at the beginning of the lesson - to loosen the soil, promote air and water circulation, control weeds, bury manure. These criteria should be applied to every field to judge the standard of tillage. Two concrete examples are discussed in **figures 527 and 528**.

527

528

Light soil was tilled by a donkey-drawn plough **(figure 527)**. The clods were properly inverted but the amount of earth turned over was not enough to cover surface weeds completely. They will recover quickly and will compete with cultivated plants.

Little mounds were formed when the field in **figure 528** was tilled. Soil clods were inverted with a **daba** (type of hoe) and cast onto low mounds. Furrows were not dug out. The weeds were buried properly but the mounds are too low. This is an example of shallow ploughing. The roots of the plants sown on the mounds will quickly grow down to the untilled, harder, soil layer.

Tilling standards do not depend only on the implements used, be they hand or mechanical. They depend above all on the way the cultivator uses and sets the implements in order to achieve his aims.

Good tillage depends on correct timing

Good tillage needs to be done at the very moment when the soil is neither too dry nor too wet. Overdry soil is hard to till. It is difficult to break the soil crust and to plough deeply.

Saturated soil is no better because it is heavy to move. Clods are large and with few cracks allowing air to circulate. Tools flatten the clods that become shiny on the surface. Flattening stops air from penetrating the soil.

Shallow and deep tillage

It is a good idea to vary tillage depth each year. When a field is tilled to the same depth every year, a hard layer (hard pan) may form just under the ploughed earth and this stops plants from rooting deeply. By ploughing to different depths, a hard pan is not allowed to form.

When animal or machine-drawn ploughs are used, the implement can be set correctly to obtain the desired tilling depth **(figure 515)**. It is much harder to reach the right depth with hand tools because manpower only equipped with a hoe cannot penetrate deeply. It is worthwhile making beds, earthing up and ridging, because by using any of these methods the soil can be raised to the desired height.

Tilling and soil moisture

Tilling methods greatly influence soil moisture. One of the aims of tilling is to improve soil structure and we have already seen (Lesson 21) that good soil structure is favourable to the retention of moisture.

- Correct tilling limits erosion. Water is trapped in the furrows and between the mounds. It infiltrates into the countless soil cracks caused by tilling instead of running off on the surface.

- When organic matter is buried, it decays and forms humus which retains water well.

- On eroded land, water runoff removes clay and silt. When runoff is reduced, these soil elements are deposited on the soil surface and form a thin, impermeable layer. Tillage breaks up this thin layer and mixes the clay and silt once again with the other soil fractions.

- When land suffers from excess water, tilling helps drainage. Water can drain away more easily from the soil clods through the cracks caused by tilling.

Dangers of tilling

As a general rule, tilling is needed to make seasonal crops grow, but it may be dangerous. Why is this so ?

■ **Tilling strongly activates soil life and sometimes to such an extent that organic matter decays much faster than it can be used.** This situation may arise if land is tilled straight after forest clearing.

In this particular case it is often advisable not to till after clearing but to sow one or many plants able to benefit from the field in its unworked state. The gourd mentioned in Lesson 7 is a case in point. Here we saw that tilling was initiated gradually :

□ no tillage during the first cultural season, only gentle soil maintenance and scraping where the gourds were sown ;

□ light shallow tillage for the maize seedbed in the second cultural season ;

□ deeper tillage and thorough cleaning before groundnut was sown in the third season ;

□ after that, tillage was limited in this plot to the places where pineapple, cassava and trees were to be planted.

■ **Tillage exposes the soil to rain, wind and direct sunlight.** As it is well known, these elements may constitute a threat to soil life. There is the danger of erosion by water and wind, and of desiccation by sun and wind.

529 | valley | eroded furrow | beans and potatoes | slope

530 | slope

■ **In fields lying on sloping ground, the wrong tillage can contribute significantly to the formation of eroded gullies.** When the ploughing method used results in the formation of down-the-slope rills, water runs off and flows with increasing speed **(figure 529)**.

531 | slope

The plots in **figure 529** were ridged downwards with the slope. This type of ridging contributes enormously to erosion. The results can be observed on this land where earth is carried away by runoff, rocks are apparent and soil is removed.

The landscape in **figure 530** is also sloping and much the same as in **figure 529**. But here tillage is perpendicular to the slope. Every ridge bars the way to runoff water which can then be trapped in the furrows and used for the benefit of the crops (refer back to **figure 248**, Lesson 22).

On hill farms in Rwanda

On hilly ground in Rwanda, women farmers use a heavy metal hoe with a long handle. They begin tilling at the bottom of the cultivated patch, so that the inverted soil falls down the slope (**figures 531, 532 and 533**).

This way of tilling by hand is often practised in mountainous country because it is less arduous to let the soil fall down the slope than to raise it up. Unfortunately, this method has many disadvantages.

■ As the years go by, the soil falling down the slope may leave the field, and rocks may break through the surface. This agricultural practice means that soil losses, already high due to erosion, are even greater.

■ Up the slope at the top of the plots where tillage stops every year, high vertical earthbanks are formed. These banks separate the lower plots from those cultivated higher up, but they may collapse, being swept away by the force of water runoff and water infiltration.

If the women farmers in Rwanda had more physical strength with which to manipulate their tools, or if they had animal-drawn ploughs, they could then work the soil by raising it up rather than letting it fall downslope. They could also try to hold the falling ground by anti-erosive strips which would lead

top limit of plot

eroding gully

| 532 | flat tillage on a steep slope |

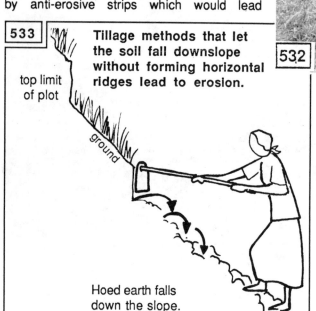

533

top limit of plot

ground

Tillage methods that let the soil fall downslope without forming horizontal ridges lead to erosion.

Hoed earth falls down the slope.

gradually to the formation of terraces (Lesson 22). Everything depends on the power and means available.

Correct tilling on sloping ground traps water and stops runoff.

However, it must be remembered that only well-sealed furrows retain water. If furrows are open at the ends, water is not trapped. Instead, it flows away down the slope and forms gullies that grow deeper with every rainfall (**figure 534**).

Plants also help to till the soil

Lesson 34 described how plant roots transform the soil. Consequently, properly selected plants help to work the soil.

As a rule, tubers work the soil well. They penetrate, sometimes deeply, and push back the soil at the time of tuber

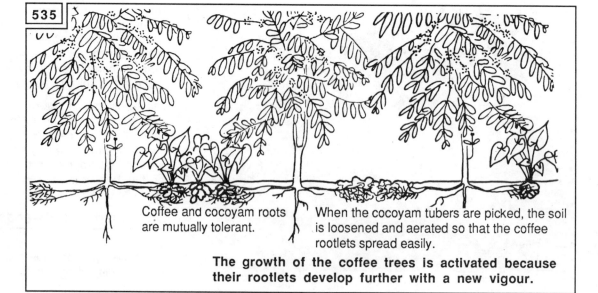

535

Coffee and cocoyam roots are mutually tolerant.

When the cocoyam tubers are picked, the soil is loosened and aerated so that the coffee rootlets spread easily.

The growth of the coffee trees is activated because their rootlets develop further with a new vigour.

enlargement. When pulled, they leave large openings in the ground. Such 'tilling' plants can be used in two ways : in association with other crops (coffee and cocoyam, for instance), or in pure stands but in rotation.

Coffee and cocoyam are growing in mixture **(figure 535)**. On the left, the roots of the two plants are entwined between the rows of trees. When the cocoyam tubers are pulled (on the right), the soil is loosened and water and air penetrate easily after tuber picking. The coffee roots have a favourable growth environment thanks to the soil loosening accomplished by the tubers.

534

Summing up of cultural practices connected with soils

Cultural practices that affect the mineral fraction of soils :

- burning and spreading ashes. Both operations feed mineral salts into the soil ;

- spreading chemical fertilizers because they are also rich in mineral salts ;

- hoeing, tilling, weeding and harrowing that turn over the soil and break up the clods ;

- measures to stop leaching (loss of soluble nutrients) ;

- measures to prevent erosion.

Cultural practices that affect the organic fraction of soils and activate soil life :

- letting trash rot on the ground and/or burying it ;

- spreading compost and/or animal manure ;

- mulching ;

- sowing plants that fertilize the soil. They are called green manure or cover plants ;

- in the fields, planting and protecting trees that give plenty of litter ;

- respecting fallow intervals, improving fallow methods ;

- cultivating legume plants in fields and fallow plots ;

- kraaling livestock ;

- increasing the amount of organic matter produced in fields by growing the right crops in mixture.

Cultural practices that help the circulation of water and air in the soil :

- tilling, hoeing, harrowing, weeding, mounding, ridging, bedding ;

- working the soil by mixing plants with different root systems (taproot, fibrous, tuberous, rhizomatous) ;

- subsoil tillage and drainage.

Cultural practices that make the most of water resources :

- providing soil cover (mulching and cover plants) to prevent erosion and runoff ;

- associating plants that take up water at different times and/or at different depths ;

- establishing correct plant density ;

- increasing the amount of organic matter in the soil ;

- planting trees and hedges that intercept wind and help water infiltration in the soil ;

- weeding soil surfaces to reduce evaporation ;

- watering and irrigation ;

- stopping water runoff on slopes.

The cultural practices designed to improve the arable soil layer all help to establish and reinforce good soil structure. They set off intense biological activity.

But these practices are not enough on their own.

Good farmers take care to avoid :

- overgrazing and trampling by livestock ;

- overheating the soil by exposure to sun and fire ;

- direct exposure of the soil to the beating action of the rain ;

Better management of fields and settlements should concentrate on :

- fighting erosion caused by wind, rain and sun ;

- introducing a wide variety of cultivated plants ;

- choosing species that suit land forms and soil properties ;

- adjusting agricultural calendars.

Lessons 38 and 39

Cutting the farmer's risks, increasing his chances of success

Farming is always a hazardous occupation because, to some extent, it faces risks. Farming is risky because results depend on factors over which the farmer has no control, that is to say, factors which he can only influence to a minimal degree. For instance, he does not determine rainfall, he cannot contain a sudden attack of locusts or stop the outbreak of a serious disease fatal to his crops.

With what risks and perils is farming confronted ?

Risks fall under four major headings : weather risks, biological, economic and technical risks.

- **Weather risks**

  Weather hazards make farming chancy. If rainfall is inadequate or untimely, plants dry up, yields are in jeopardy. A sudden hurricane flattens plants to the ground. A spell of exceptionally cold weather destroys flowers and young fruit. There are many varied ways of combatting weather risks but there is no foolproof way of protecting crops from bad weather.

- **Biological risks**

  Plant health is also subject to certain hazards. Plants may be attacked by microorganisms carrying disease, they may be eaten by caterpillars, slugs, insects and rodents. Monkeys, loose animals and elephants can devastate crops. People also cause devastation, for example, by lighting uncontrolled fires. Other biological risks include pest attacks, rotting and fermentation of stored products, damage caused by wandering livestock.

- **Economic and social risks**

  Economic risks spring from the economic and social environment of farms (Lesson 5). The farmer is not sure of finding buyers for his produce after the harvest, or market prices may have slumped when his produce is up for sale.

  Production factors such as manpower, fertilizers, seeds, tools, machinery may be in short supply when they are needed for the planting season (Lesson 2). The cost of production factors may shoot up or workers fall ill.

- **Technical risks**

  A mechanical breakdown may occur at a critical time - a plough part breaks, an engine stops working. If the spare part is not available, agricultural production may suffer a serious setback because the farmer will be forced to halt cultural operations.

  Admittedly, uncertainty about the future works both ways - farmers can win or lose depending on the outcome of events. These lessons describe ways in which smallholders try to cut their risks and increase their chances of success.

Lesson 38

Weather and biological risks in farming

The farmer is confronted by numerous hazards due to weather - from too much wind, cold, heat or rain (**figure 29**, Lesson 4). Hazards connected with rainfall will be examined more closely.

Circumventing weather risks to rainfed farming

Rain calendars were studied in detail in Lesson 26 and two particular aspects of weather hazard connected with rainfall were underlined :

- rains may be inadequate for a cultural season taken as a whole ;

- they may be unreliable and, above all, they may be totally lacking at crucial times in plant life.

Drought inevitably follows lack of rain and, with it, damage to plants that are not at the end of their life cycle (Lesson 27). There are many ways of cutting drought hazards for rainfed agriculture.

In Burkina Faso, associating rice and maize in wet valley bottoms

The rain calendar in this particular region of Burkina Faso is similar to the one in **figure 311**, Lesson 26 (Sudan climate). The rainy season lasts between five and six months.

The field in **figure 536** is low-lying and is usually wet enough to grow rice. Planting takes place when the soil is very moist, about the end of June. However, the soil is already moistened in April with the onset of the first rains. Rainfall between the months of April and June is unreliable, so it would be risky to plant rice before June. On the other hand, rainfall often suffices for sowing maize, a plant with greater adaptability during growth. When the farmer decided to go ahead with the maize seeding, this is how he reasoned :

- If I sow maize with the first rains in April and it goes on raining, I stand a chance of a maize harvest in July. The maize will come on top of the rice crop I hope to harvest in October. if it doesn't rain enough in May and June, maize yields will be poor, but at the worst I'll only have lost the seeds.

- The rice I'm going to sow in June, in the middle of the rainy season, will surely get enough rain at the early growth stage. But will there be enough rain in September and October to let it mature fully ? If the rice crop in October is disappointing for lack of rain, maybe I'll be lucky and at least have the maize harvested in July.

Several reasons underlie the farmer's decision

- By acting as he does, **he cuts his economic risks**. In other words, when planting the two crops, he acts in the hope that by growing two crops rather than one he is sure of a higher yield from his field. The argument is valid because :

- The farmer intercrops two species (maize and rice) with different life cycles, with different water requirements and different planting seasons. **If there is drought, it will hit the plants at different times in their cycle.** Drought in May lowers maize yields but does not affect the rice that is not yet sown. Again, the rice may suffer drought in August, but not the maize already harvested.

- The farmer also provides better growth conditions for the rice crop. During the planting season and emergence, **the rice plantlets are protected by the maize leaves** from the harmful effects of hot sun and wind.

By acting in this way, the farmer strikes a balance between his chances of success and the hazards for which he must allow even though they may not occur.

This example of intercropping rice and maize is not limited to valley bottom cultivation in Burkina Faso. It is found in many regions and on many types of land.

Association of maize and rainfed rice in Cameroon

Maize and rice are also associated in Cameroon, 2000 kilometres away from the farm in Burkina Faso. Here again, we shall see how this cultural practice helps to limit farming risks, but for other reasons and in other weather conditions.

The rain calendar of this region can be compared to the one in **figure 310**, Lesson 26. There are two rainy seasons with a short intervening dry season. The first rainy season runs from March to June, the second from September to November. Rainfall is less in July and August with a marked risk of drought periods for many consecutive days.

With the same variety of rice, farmers have two different ways of spreading harvests and income over the year.

In March, rice is sown in valley bottoms, i.e. wet lands, and then harvested in July and August during the short dry season. The valley bottoms are still moist at that time and the grains are able to swell. The situation is straightforward on wet lands where there is little risk of serious drought in July and August.

Things are different, however, on drier plateau lands where there is a strong risk of drought besetting the rice during the short dry season. In order to forestall this eventuality, farmers grow rice in association with maize. Their cropping plan is like this.

Maize is sown with the onset of the first rains in March while rice is planted a little later, in April and May. In July and August, the maize crop is harvested but the dried stalks are not flattened to the ground because they provide shade for the rice.

Now, during the short dry season in July and August, rice is at the stage of emergence. It has not yet flowered and because the weather is dry, it becomes inactive, shaded by the dried maize stalks, until the rains begin again in earnest. The rice plants then resume growth and the ears mature in October.

We infer from this example that farmers in Cameroon associate rice and maize for the same reasons as their counterparts in Burkina Faso, but other reasons also underlie their crop management.

By associating maize and rice, farmers are able to have two rice harvests at different times in the year. If maize and rice were not intercropped on drier plateau lands, there would only be the one rice harvest in July/August from valley bottom fields. Thanks to the crop association, a second rice harvest is possible in October and November. Staggering the rice crops is a boon for family subsistence, for labour requirements (spread of work) and for monetary income.

This crop management is also advantageous from another point of view. By spreading harvests over a period of time, conservation and storage of produce on farms take up less room and extend over longer periods.

How to combat weather hazards especially those associated with rain

There are many ways of coping with weather risks such as :

- **Mixing many seed varieties of the same plant species.** Because some are early and others late, they react differently to drought and soil moisture at any given time in the agricultural calendar.

- **Mixing two or more complementary species.** The plants, being different, will not need water at the same time in the cultural season. When two or more species are mixed, varieties of these species can be added.

- **Sowing as early as possible** when the first rains have moistened the soil, even if it means sowing or transplanting later on if the seedlings have suffered from drought.

- **Staggering the planting season.** This means seeding a given variety, then seeding a second time two or three weeks later with seeds of another variety. A first plant species can also be sown, and another species two or three weeks later.

- If the dry season is short and seeds fail, it may be better to **transplant** rather than sow all over again. The reason for this practice is that transplants are more advanced and have a chance, however slight, of reaching maturity before the end of the dry season, whereas there is no hope for seedlings. This is why millet was transplanted onto little mounds in **figure 537**.

As illustrated by the dried-up millet field in **figure 537**, it is important to choose cultural practices to fit the events of the season. Adaptability helps to cut farming risks and increases the likelihood of good results.

537

- **All the cultural practices that increase and retain soil moisture limit weather hazards.** They include mulching, cover plants, admixture of organic manures, etc.

- **Associating trees and hedges in fields** in order to lessen the effects of drying winds and provide shade.

- Protecting and planting in fields **drought-tolerant species** such as trees.

- **Irrigating plants** by drawing water from rivers, lakes, bore holes or wells (see below).

However, it is important to remember that the climatic dangers lying in wait for tropical farms are not only due to drought. They may originate in torrential rains and rainfall intensity whose effects can be checked by the methods discussed in Lesson 22.

Taming water resources

In regions with low rainfall, farmers face a serious problem when rains, already too infrequent, flow swiftly away into gullies and rivers **(figure 538)**. Rainwater is lost for good and crops suffer from drought. Farmers can stop this by storing water so that it can be used to meet crop needs.

Water can be stored on the surface of the ground by building dams. The weir of a dam is seen in **figure 539**. The lake that formed when the dam was built supplies the inhabitants and livestock of a big village with water throughout the dry season and gardens can be watered at the same time.

538

stone wall acts as water barrier

weir

Water flows in this direction.

540

dam

weir

539

building gabions

stones

wire netting cages

541

Water can also be forced to infiltrate the ground rather than flow away on the surface (Lesson 22). When this happens, water is stored in the soil, especially in the water tables.

Small farmers in Cape Verde undertook land management **(figure 540)** in order to bring about water infiltration. The stone wall barring the valley is seen on the left. When it rains, water flows from

224

542

direction of water flow

544

545

the right and is partly stopped by the stone wall and stagnates. The rest flows over the wall. The stagnant water deposits the silt and clay removed from the slopes. The soil that builds up at the foot of the stone wall is thoroughly moistened and enriched.

This stone barrier makes the water infiltrate rather than letting it flow further away down the valley. Some of it percolates to the water table from where it can be drawn.

Figure 541 illustrates another type of water barrier. This is the **gablon** technique. Barrage stones are packed evenly into strong wire-mesh baskets.

Water stored during the rains can be used later when the dry season sets in and plants are in need of moisture. This is why irrigation lessens farming risks.

If there is an adequate supply of water, irrigation or watering of plants takes place at the very time when plants need water, and the required quantity can be drawn from dams, wells and bore holes **(figure 542)**.

Water can also be brought to where it is needed by building small channels. Two types can be seen in **figure 543,** a cement inlet rill on the left and plastic piping (PVC) on the right.

cement inlet rill for water

543

PVC piping

Plants can be watered regularly if cultivated land is managed properly. Here is a field planned to let irrigation water flow evenly along the path traced **(figure 544)**.

Management of irrigated plots must be such that not a drop of water is wasted and infiltration is maximal. Maize and okra in **figure 545** are being irrigated. The water channelled there is trapped by earth banks. A small amount evaporates, but the greater part penetrates the soil.

546

motor pump · well

548

water for sprinklers

banana plants

Young vegetable gardeners use a small **motor pump** to draw water from a well and water their plots **(figure 546)**.

A seedbed for irrigated rice is being prepared **(figure 547)**. Irrigation water comes from a dam. Land below the dam was levelled and then divided into plots by building low dykes.

547

earth bank

Irrigation by sprinkling has been installed in the banana grove **(figure 548)**. The water is pumped from a river, channelled through pipes and forced into the air so as to fall like rain on the bananas.

One can also irrigate by using a permeable earthen jar. The pot is buried in the ground, filled with water and covered. The water is topped up every day. A gourd is being irrigated in **figure 549**.

Anticipating water shortages by planning irrigation systems requires :

■ **water reserves ;**

■ **the means of supplying water**

　□ **where** plants need it ;

　□ **when** plants need it ;

　□ **in the exact quantities** needed to avoid waste.

Biological risks : plant diseases and pests

Plant diseases are caused by a wide range of pests - insects, worms, fungi, bacteria, viruses, birds, rodents and sometimes other mammals as well.

Some of these pests attack all plants indiscriminately. Monkeys eat bananas, heads of cereals, tubers and many other plants. Some birds eat any and every sort of seed. Some aphids pierce and suck the sap of different plant species. In the same way, hoppers **(figure 550)** and some caterpillars eat any foliage available.

549

earthen jar filled with water

gourd

cover

550

damaged
fruit

551 okra beetles

Most pests, however, are specialized :

■ They live on one or more particular plants.

■ They attack specific plant parts : leaf and flower buds, young leaves, flowers, young or ripe fruit, stems, roots.

For instance, a maize pest will not attack millet or sorghum or nearby trees. A cowpea pest cannot live on sorghum. The insect that pierces and sucks cassava leaves and transmits mosaic virus (**figure 354**, Lesson 29) is not attracted by groundnut leaves. Another kind of piercing-sucking insect that avoids cassava carries rosette virus to groundnut.

There are many ways of lessening the risk of plant diseases and pests. Here are possible measures that can be taken :

■ **Sources of infection** should be wiped out. By sources of infection we mean places, usually their nests, where pests are lying in wait to attack crops.

Figure 551 illustrates another common source of infection. Insects (in this case, **beetles**) pierce the young okra fruit and lay their eggs on the plant. Infected okra plants then become sources of infection for other okra plants growing close by. Burning the infected stems is a way of limiting the risk of beetles spreading to healthy plants.

Another disease and source of infection shows up in **figure 552** where many cocoa pods on the same stem have contracted black pod. The fungus that transmits this brown-coloured rot is quite likely to carry it to healthy pods. This is why the diseased pods, source of infection, must be eliminated by **plant sanitation** which involves picking and burning infected pods and the fungi on them. During the picking/burning operation, great care must be taken not to touch the healthy pods because they might be contaminated.

infected pod :
source of
contamination

healthy
pod

552

■ **Choosing disease-resistant plants** for seed propagation also helps lessen the risk of pest attacks (Lesson 28).

■ **Practising good crop rotations and plant associations** lessens the risk of pest depredation. The beneficial effects of this method can be explained easily. When the same plant is cultivated for consecutive seasons on the same plot, specialized parasites remain there from one season to the next. The danger of parasitic infestation is increased because the specialized pests have kept the same habitat. But when crops are rotated, parasites are adversely affected by the absence of the plants they are accustomed to eat. The pests either decrease in number or disappear entirely. It is clear therefore that crop rotation is a way of fighting sources of infestation.

Witchweed *(Striga)* fixes on sorghum roots. Because witchweed seeds germinate in the presence of young developing roots, the roots of sorghum, groundnut, cotton, pea, castor and other plants stimulate witchweed germination. But at the same time, witchweed seedlings can only settle on sorghum. Without it, they die.

So, if a plot of sorghum is infested by witchweed, crop rotation will help eradicate the weed to a great extent. Growing the other crops mentioned eliminates witchweed seeds so that sorghum can be grown there again at a later stage.

Crop rotation and growing crops in mixture are important aspects of weed, pest and disease control and by their impact, they decrease farming risks.

■ **Seed density influences the chances of disease.** The relationship between seed rate and the rosette virus in groundnut illustrates this point. Rosette virus is caused by an aphid that pierces the underside of the leaves and injects the disease. However, this particular aphid avoids a humid, shaded environment **(figure 553)**. Consequently when groundnut plantlets are closely spaced and/or shaded, pest activity is restricted. On the other hand, wide spacing provides aphids with foliage and the fairly dry environment they prefer.

Methods of cultivation that maintain moisture under leaves lessen infestation of groundnut by rosette virus.

| 553 | **Seed density affects plant health** |

Under closely spaced groundnut, conditions are very humid and therefore unfavourable for aphid insects.

Under widely spaced groundnut, conditions are favourable for aphid insects, and these pierce the leaves and transmit the rosette virus.

■ **The use of chemical products** - if the farmer can afford them - is a means of fighting disease.

Pesticides are poisons designed to kill the insects (**insecticides**), fungi (**fungicides**), and bacteria (**bactericides**) that cause damage in fields and plantations.

Their use produces immediate effects - the danger of crop infestation by disease or destruction by pests is averted. The use of these products therefore reduces farming risks and increases production.

Manufactured chemical products for pest control are often necessary but they must be used with extreme caution. It is advisable, where at all possible, to use natural products from the soil or from plants because they are more adapted to the living environment.

The danger of these manufactured pesticides is two-fold. They can have harmful effects on human life and on animal life. They are often non-selective, they make no distinction between useful and harmful insects, fungi and bacteria, so that they are just as likely to kill useful living beings as kill the pests.

■ When farmers cannot afford pesticides, cheaper methods must be found. **Growing plants in mixture** is one inexpensive way of tackling the problem. Researchers at the International Institute for Tropical Agriculture (I.I.T.A.), others at Morogoro in Tanzania and elsewhere have carried out successful experiments on these lines.

They have shown that the aphid that carries rosette virus in groundnut does not tolerate maize or any other cereal in the field. **Hunting spiders** that eat aphids and other maize pests are found in larger numbers and are more active on maize stalks when this plant is intercropped with groundnut.

The maize stalk-borer, an insect at the larval stage of a moth, lays its eggs on the plant stalks. The eggs develop into larvae that bite and chew the stalks, and so are called **borers**. It has been observed that when maize grows in mixture with cassava, there are fewer borers than when maize is grown in pure stand. This decrease in population is explained by the fact that the flight of the borer is confused by the cassava foliage.

There are many examples of this kind, typical of each locality. While it is not easy to ascertain the exact effects of one plant on another where pest attacks are concerned, it is evident that such effects exist.

Hence the importance of observing them in every region and of experimenting to find the best plant associations for pest control.

Controlling the cowpea beetle by intercropping cowpea and a cereal

In Tanzania, in the region of Morogoro, an insect from the beetle family pierces the flowers and leaves of cowpea, causing stunted growth of pods and fall of leaves.

The insect moves from one cowpea plant to the next, but does not like flying or walking on the ground. When cowpea plants overlap, the insect moves freely from plant to plant, causing much damage. The insect's movements can be seen in **figure 554 (a)**. When the plants barely touch each other, the insect becomes sedentary. By intercropping cowpea with a cereal such as maize, millet or sorghum, the cereal acts as a barrier and prevents the insect from travelling through the cowpea as in **figure 554 (b)**. In conclusion, the cereal stops the pest from spreading and so protects the cowpea, while the latter is useful to the cereal by enriching the soil (Lesson 49). The maize growing on rich soil is stronger and more resistant.

| 554 | Careful plant associations help control pests |

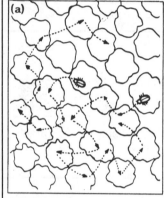

(a) (b)

When rows of cowpea are interplanted with rows of maize, the maize forms barriers, inhibiting insect movement, thereby controlling the pest.

| 555 | Two cultural methods of controlling pests in cropping fields : growing crops in mixture and choosing the right seed density and row arrangement |

Maize in pure stand

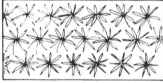

16 plants out of 100 are attacked.

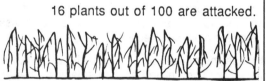

Maize and cowpea in mixture, with row arrangement as shown below

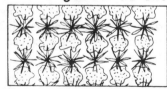

10 plants out of 100 are attacked.

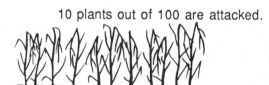

Maize and cowpea in mixture, with row arrangement as shown below

8 plants out of 100 are attacked.

In this case, the cereal is a useful weapon against the beetle pest. However, even if this precaution fails and the cowpea crop is damaged or destroyed, the field will not be a total loss because the maize harvest is intact.

How does cowpea affect maize ? The presence of cowpea limits the attacks of the maize stalk-borer. Researchers have observed the results drawn in **figure 555**.

Three points based on their observations are worth remembering :

■ Associating maize and cowpea is profitable because grain yields are higher in association than in pure stand.

■ The risk of pest attacks increases or decreases, depending on the circumstances. Some pests are inhibited by associated plants while others, on the contrary, find a more suitable environment for multiplication.

- But the likelihood of an attack on all the crops is reduced, except where non-specialized insects, for example, locusts that eat everything in their path, are concerned.

Seizing opportunities

In some parts of West Africa, farmers try to make the most of early rains.

Millet is usually sown in July after two or three rainshowers when the soil can be easily worked.

Occasionally, it rains heavily some weeks earlier, say in May. When this happens, the women hurry to the fields and broadcast millet seed or sow it after scraping the soil lightly.

What happens then ?

If it does not rain again until July, the seeds will have germinated and the parched plantlets will dry up with the resultant loss of seed.

On the other hand, if it rains again in June, the millet plants will develop satisfactorily and will produce two or three times as much as plants sown later.

By resorting to this cultural method, women farmers seize a passing opportunity in the form of early rains. If the opportunity is not available, they will sow in July once the rains are well established.

Controlling pests with natural products available locally

The leaves, seeds and roots of certain familiar plants can provide effective insecticides. They are used in two ways. They are either dried and ground into powder, or they are left to infuse in water. The infusion is then applied to the diseased plants.

*It always pays farmers to cultivate **insecticidal and insect repellent plants**. Here are some of these useful plants : neem leaves and seeds in powder-form, peppers and chillies (their crushed seeds are mixed with harvested produce), some cacti dried and powdered down, shea butter, certain Graminaea (leaves of finger millet, Andropogon, Spear grass, Citronella), some eucalypts, Annona, derris, some species of Lantana, tobacco, mint, pepper, arrowroot, Siam weed.*

Pyrethrum leaves and flowers are effective in controlling insects. Many oils afford good protection for seeds and pulses. Oils from palm, maize, cotton, soya bean, coconut and grountnut are particularly efficient admixtures. Small quantities of oil are mixed in, enough to cover the stored grain with a thin film of oil.

Many mineral products are used to protect crops and harvests, especially ashes and charcoal which, no matter what their origin, act as disinfectants. Some, however, are more powerful than others, for instance, ashes from mango wood, tamarind, babul (Acacia nilotica), Casuarina and Eucalyptus. Ashes from rice trash are also used in some parts.

Clean, dry sand and clay (especially kaolin) are frequent admixtures for harvested produce and work well.

Smoke is a preservative agent used in all village kitchens. Smoke is also an insect repellent because it drives insects away but does not kill them.

Important points

- *As we have seen, there are many methods of controlling plant pests. Some of these methods and means of applying them are available in fields and rural settlements themselves. Others must be bought commercially.*

- *Priority must be given to applying agricultural methods that reduce the risk of pest attacks - crop rotation, cultural associations, combinations of plant varieties, seed density.*

- *The second task is to feed plants properly. Nourished and watered, plants are much more likely to resist disease than poorly fed plants growing on exhausted soils. Fertilizing the soil, improving soil structure, using varieties known to be resistant to disease, these and similar measures all favour the growth of healthy plants.*

- *Hygiene and prevention are also very important - clean tools, clean water for watering, plant sanitation, eradication of sources of infestation.*

- *Before thinking of buying chemical products for pest control (pesticides), farmers should consider using natural means that are more easily available at village level, are cheaper and, above all, not so dangerous as chemical products.*

- *Chemical control comes as a last resort. However, this will not produce results unless the conditions listed above are in force. It is pointless spending money on stunted plants which, even if sprayed, will have negligible yields. But if chemical methods are used, pesticides must be chosen with care so that they kill selectively and do not blindly destroy useful living beings that are necessary for the ecological balance of the living environment (Lesson 3).*

Lesson 39

Economic and technical risks

Farmers are confronted by many economic and technical risks - these concern production factors, marketing, prices, labour and so on (**table 23** in Lesson 2 deals fully with production factors). A farmer falls ill, prices plummet, unforeseeable expenses have to be covered, land and money are short, transport services inadequate.

Whether on farms, in villages, over a whole region or nation-wide, there are many ways of facing up to economic risks:

- adapting and improving agricultural methods;

- setting up reserves of food and money;

- promoting mutual help schemes;

- making farm and village activities as varied as possible.

Adapting and improving agricultural methods

Agricultural methods that reduce climatic and biological risks have been examined in detail. The use of the same methods also lessens economic risks.

The fields in **figures 556 and 557** illustrate what is meant by an economic risk. The cotton, field 1, is in full growth and the first flowers are in bloom. The outlook is promising provided good husbandry practices, including spraying to prevent disease, are carried out and everything goes smoothly in the coming weeks.

Dressings and sprayers must be available on time and spare parts available in case of a breakdown.

But supposing a sprayer breaks down, parts are not delivered or the farmer cannot afford them, the hope of a good yield is dashed and the money spent on fertilizers and dressings will have served no purpose.

The owner of field 2 reasons on different lines. He is growing cotton and maize in mixture, four rows of cotton to one of maize.

556   cotton in pure stand

maize   557   cotton

Of course, he will try to give the cotton plants all the applications they need in order to maximize yields. Still, if a machine breaks down or dressings run short, the farmer will not be empty-handed at harvest time. Cotton yields may be poor but maize will compensate the loss to some extent.

If, on the other hand, there is no hitch with machinery, and materials reach the farm on time, the farmer can perhaps save a little money by managing without one spraying operation thanks to the maize crop. In any case, the maize will supplement the cotton yield. It is obvious that by associating maize with cotton, the second farmer has lessened his risks and increased his chances of success.

Building up reserves

There are many ways of storing consumer goods and production factors.

- Setting goods aside is called **storage**. The term is used for consumer products (cereals, flour, cloth, oils, foodstuffs) as well as production factors (seeds, fertilizers, plant dressings, and so on). Storage is provided in **granaries, shops and warehouses**.

- Setting sums of money aside is called **saving money**. It can be on an individual or collective basis.

- Setting aside production factors such as land, machinery, buildings and money is called the **accumulation of capital**. For example, when title deeds to land are accumulated, this is the accumulation of land capital. The purchase of housing property is the accumulation of housing capital.

The saving of goods and money can be organized **individually** or **by groups**. Collective undertakings may take the form of putting aside consumer and production means to ensure mutual aid among individuals. This is what banks, cooperatives and insurance agencies do.

Reserves are sometimes established under the authority of a headman charged with the welfare of his community. This is, in fact, what often happens when a family or group headman assigns tasks to people under his authority in order to store the fruit of their collective work. In this case, the allocation of tasks is decided by the headman. Every member of the group must execute the assigned tasks, but he also has the right to benefit from collective measures when he is in need.

The tradition of working in villages with a view to establishing collective stores is on the decline. Yet, it is often the only way of guaranteeing the necessities of life in sickness, infirmity and old age.

Usual ways of storing consumer goods and production factors include such strategies as :

earthen granary

opening

558

- **Individual and/or collective granaries (figure 558)** where foodstuffs and seeds can be stored for periods of one or more years. In some parts of Africa, granaries are big enough to take supplies lasting seven to ten years.

- **Seed banks** whose usefulness was discussed in Lesson 30. Farmers club together to store seed after the harvest. When the next planting season comes round, they obtain supplies of seed from the bank as required.

  Risk of seed shortage is less likely because, although every member can or must deposit a fixed quantity of seed, contributors will not all withdraw seed at the same time. The person in charge must keep a record of deposits and withdrawals.

- **Mutual savings and credit banks.** Members with money and no immediate use for it deposit their savings. Other members in need of money are able to borrow. The bank can give them credit thanks to the savings handed in. After a few months or years, loans are paid back with interest which is then divided among the lenders.

- **Tontines and kitties** also counteract the shortage of funds. Members meet monthly to pool money in order to cover the heavy financial outlay planned by one of the members. The following month, another member receives the contents of the kitty after fresh monthly contributions have been paid.

- **Insurances.** Members subscribe for insurance against economic risks. A typical example is insurance against accidents to oxen. Many ox owners subscribe, whereas only a few have the misfortune to lose oxen. Some of the money paid in by all the members is then used to replace the dead animals. It can be argued that those who do not lose livestock pay for the rest. However, later on, it may be the turn of the lucky members to get paid insurance money, unless they are fortunate enough never to lose a beast.

- Steps to reduce economic risks by storing consumer goods and saving money can also be taken on a regional or national scale. For example, State-controlled banks or marketing boards are set up to **stabilize** the price of agricultural products. In this way, huge amounts of consumer goods and money are managed by government bodies with a view to stabilizing the price paid to farmers for their produce.

For example, the marketing board fixes the price of coffee paid to the producer at 300 francs per kg. One year, the coffee is exported for 400 francs per kg. The difference between the cost of buying and selling, 100 francs per kg, is allowed to accumulate. The following year, the country exports coffee for 220 francs per kg and the money saved from the previous year is used to pay the guaranteed price of 300 francs. First, the board earned and saved 100 francs per kg of coffee, then it spent 80 francs per kg, while the producer received the fixed price of 300 francs per kg for both harvests.

Promoting mutual aid

Mutual aid among farmers is another way of limiting economic, technical and social risks. It is frequently found at village level, in family circles, within the framework of more or less structured associations and cooperatives.

Cooperative activities limit risks for farmers because these bodies represent the combined force of all the members rather than individual effort. By working together, cooperative members can undertake operations beyond the scope of the individual. For instance, they market and transport goods without relying on middlemen, make bulk purchases without waiting for suppliers to turn up, and defend common interests.

Mutual help is also of paramount importance in the case of illness. Without help from neighbours, a farmer who falls sick even for a few days may lose his entire harvest. With all these joint ventures, success demands mutual trust, honesty and properly audited accounts.

Diversifying economic activities on farms and in villages

Diversifying activities on farms and in villages takes the edge off economic risks :

- Farmers cultivating 10 or 15 kinds of produce at different times of the year have a more regular cash income than farmers who go in for monocropping (one crop only). The income from multicrop farming is spread out over the year.

- Diversifying farm products helps improve the family diet (Lesson 47). Good food reduces the risk of ill health for workers and their families.

- Expanding farm activities by combining crops and livestock (mixed farming) is another way of reducing economic risks.

- Promoting trade in villages by the exchange of work, money and goods confines the farmer's economic risks. A farmer who buys a tool from the local blacksmith helps him to earn a living. But the blacksmith in turn helps the farmer by buying farm produce. Trade at village and regional level is a form of mutual help.

Suppose the same farmer decides to buy an imported tool. He at once becomes dependent on people whom he does not know personally - the city businessman, the transporter, the shipping line, middlemen

559

560

and factory workers abroad. If one of the go-betweens drops out, the tool will not be delivered on time. Being dependent on others is always riskier when there is no way of putting pressure on the slack link in the supply chain.

■ All the goods needed for family use cannot be produced on farms, in villages, in the region nor perhaps in the country. Yet, local people who go to the trouble of learning to produce goods that are usually manufactured abroad make farmers' lives more secure.

The example given below illustrates the importance of village production.

The donkey-drawn seeder in **figure 559** was made in Senegal. It can easily be repaired by village artisans. Money paid for repairs is used by local people. When artisans buy food produced on the spot, they help farmers earn their living. By buying seeders from local artisans, farmers encourage them to learn and practise their trade, thus developing the local economy.

The seeder in **figure 560** is tractor-drawn. It works fast and well. But what happens if a main part breaks or the tractor does not work ? If the village artisan can repair the machine, everything is fine. But if the owner has to wait for a repairman, or if the spare part must be ordered from abroad, the planting season may be over before the machine is ready again. Such machinery also, eventually, displaces people from farm jobs. Methods of tillage also illustrate the point just discussed.

A farmer using oxen for cultivation loses an animal for one reason or another. A second farmer with foreign machines has a breakdown. In all probability, the first man will manage to replace the animal quickly from a source in the neighbourhood even if only on loan. The second will depend on a dealer in town or on foreign industries. The repair may be expensive and will have to be paid for immediately in cash.

When buying machinery for mechanized farming, the potential risk should be reckoned as well as the work the implement can perform.

Why is rural economy based on local or regional trade safer than an economic system based on international trade ?

The answer lies in the comparison between the two means of transport, a motor car and a buggy photographed in Senegal **(figure 561).**

561

The motorcar

The motorcar was manufactured in Europe. European workers earned their living by making it. They buy their food from European farmers.

The motorcar runs on petrol imported from foreign countries. Petrol is shipped by sea and distributed to petrol pumps in Senegal.

Some of the money spent on buying petrol pays Senegalese who buy their food in Senegal. Some of the money is spent to pay foreign workers and middlemen who buy nothing from Senegalese farmers or buy it cheaply (groundnut, for example).

The motorcar is a fast, efficient means of transport but it is much more expensive than the buggy. The owner of the motorcar pays a lot of money to people he does not know. He cannot bargain with them over the price. How can you bargain with a petrol pump ?

The buggy

The buggy was made by Senegalese workers. Only the wheels and axle were made abroad (they were recovered from a derelict truck). The men who made the buggy buy their food from Senegalese farmers or from neighbouring countries.

The buggy is drawn by a horse who eats grasses and grain. The horse's fodder can be grown in the country. In this case, horse feed is supplied by Senegalese farmers who make money from the sale and are then able to buy consumer goods or pay for personal transport.

The buggy is a much slower means of transport than the motorcar, but its manufacture and use depend solely on Senegalese artisans and cultivators. A hard bargain can be struck over the cost of transport and the amount charged will always be less than the cost of travelling by motorcar. Sometimes, instead of paying for the buggy ride, an arrangement is made to pay by an exchange of services.

234

Of course, it would be wrong to decide, on the basis of this example, that the use of motorcars should be banned in Senegal. If it were so, how would people travel long distances and move quickly from one town to the next ? However, the suggestion frequently heard that buggies should be done away with would be an even greater mistake. In actual fact, it would amount to a decline in job opportunities and trade among the Senegalese. Such a step would benefit employment and trade in distant countries, as least until such time as the Senegalese began to manufacture motorcars themselves.

The example of the motorcar and the buggy can be applied to farming. Every time a farmer gives work to members of his family or community, he reduces his own economic and technical risks.

Establishing sound farming systems, which are safer because they are less exposed to risk, means primarily counting on mutual economic and social support and believing in the value of local trade. Autonomy is a source of development ; dependency is a source of danger.

We can say, by way of conclusion, that when choosing farm equipment the following points must be considered :

- **what the farmer wants to do with it** - till, weed, earth up, harvest, transport, grind, process produce ;

- **the means at his disposal** - labour, money, land ;

- **the degree of risk engendered by the equipment** - parts easily repaired or not, repairs expensive or not, money to cover running costs available or not ;

- **mutual economic and social help** - by choosing this piece of equipment, am I helping local people, or am I helping people in foreign countries ? Am I promoting trade between the men and women of my own country ? When we help a worker from our own region, we indirectly help ourselves.

It must never be forgotten that developing trade and local jobs is the best way of improving one's own security.

562

The farmer who gives custom to the village blacksmith can reason like this : 'When I buy from the forge **(figures 562 and 563)**, I help the old blacksmith to earn a living **(figure 563)**. Because I have helped him to make ends meet, his son has had time to learn his father's craft.'

Thanks to what he learned in the forge, the blacksmith's son is skilled in his trade and makes equipment such as pumps for drawing well water **(figure 565)**, traps for game **(figure 566)**, wheelchairs for the handicapped **(figure 564)**.

564

563

565

566

Are the plastic or wire hen-houses imported from Europe any better than those made by village artisans **(figure 569)** ? The imported ones are dearer, and what happends if the local workers lose their livelihood ?

Another example of useful local trading might run on the following lines : 'I bought milk from the shepherd **(figure 567)**. I boiled it and my child drank it. The shepherd bought millet and forage from me. We have exchanged money. If I had bought a tin of powdered milk, my money would have left the country' **(figure 568)**.

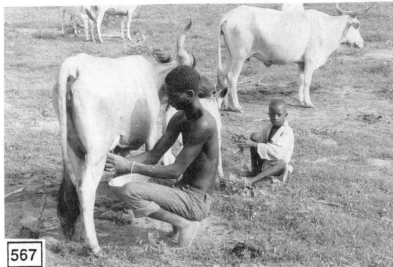

567

568

569

236

570

Making beer from millet is a traditonal village industry **(figure 570)**.

Is bottled beer, delivered in truckloads, tastier than local beer ? What happens to the money spent on it ? Does the money drained away with the empty beer crates contribute more to the welfare of villagers than the money exchanged in the locality or region ?

We can compare this with rainwater. Infiltrating water is beneficial for crops, whereas runoff water is useless. Infiltrating water emerging at springs flows slowly but surely. Water that runs off instead of infiltrating leaves settlements and removes soil as it goes.

Notes

Lessons 40 to 43

Feeding the soil with fertilizers

Plants take up some of the food they need through their roots while the leaves absorb more food from the air (Lesson 23). The next four lessons deal with **fertilizers** that can be used to enable the soil to feed plants better.

Fertilizers are often called by other names with more or less the same meaning - **manures, topdressings, soil conditioners** and **improvers.**

Fertilizers fall into two main categories :

- **Organic fertilizers** come from living matter and are natural. Lesson 41 covers this subject ;

- **Mineral fertilizers** come from lifeless (inorganic) substances. These fertilizers, often called **chemical fertilizers,** can be **natural** or **artificial.** They are natural when found in nature, for example, in mines and quarries. They are artificial when manufactured from other products such as oil and air. They are discussed in Lesson 42.

However, we must first discover exactly what plant nutrients are taken up from the soil.

Lesson 40

Plant nutrients

Plants are nourished by many different kinds of food, each of which has a specific role to play rather like the various relishes combined to make a tasty dish (Lesson 23).

These nutrients are mainly composed of **mineral salts** of which the major components are :

- **Nitrogen** found in large quantities in the atmosphere, in organic matter and residues. Nitrogen is essential for making stems, leaves and roots, as well as animal and human flesh. Being a significant component of all proteins, new growth depends on it ;

- **Carbon** also found in the atmosphere and in organic matter. It combines with nitrogen to make wood and straw, and with water and air to make sugars and fats. Coal and diamonds are formed from pure carbon ;

- **Phosphorus** found in the ground is a component of human and animal bones. Plants need phosphate to flower and bear fruit, and to develop strong roots ;

- **Potassium** is an important nutrient for tuber and fruit enlargement. This element maintains healthy activity of all plant tissues ;

- **Calcium** and **magnesium**. Magnesium is a component of chlorophyll, the substance that makes plants green and enables them to use light energy. Calcium is present in all plant cell walls. It builds bones and teeth in men and animals.

Nitrogen, phosphorus, potassium, calcium and magnesium are taken up by the underground roots of plants. Carbon is absorbed in the air through the stomata (Lesson 23). These foods, along with water, are the basis of plant nutrition.

Other nourishing substances taken up in tiny quantities and contributing to plant growth are called **trace elements**. There are many trace elements such as **sodium, copper, zinc, boron** and **molybdenum**. We can liken them to oil in machinery. Very little is needed but without it, machines will not run smoothly.

All these plant nutrients are found, to a greater or lesser extent, in the soil, but, like water, they travel a lot and the pattern of their movements has to be described.

Cycle of organic matter and mineral elements

Figure 571 represents the cycle of organic matter and mineral elements.

Plant roots take up nutrients disssolved in water. Nitrogen, phosphorus, potassium, calcium, magnesium and trace elements taken up by the roots through the root hairs are combined with water. This mixture forms the unelaborated sap ascending through the stem and branches towards the leaves.

In the leaves, the salts contained in the unelaborated sap are combined with carbon present in the atmosphere and breathed in by the plant.

In response to light, all these components unite to form the elaborated sap that goes to make organic matter, i.e. proteins, sugars, fats, oils, wood, straw.

When it dies, organic matter changes, being gradually decomposed by the action of countless animals and microorganisms in the soil.

With decay and the formation of humus, carbon is released into the atmosphere while the mineral salts - nitrogen, phosphate, potassium, calcium, magnesium - along with trace elements are mixed again with sand, silt and clay. At this stage, plant roots can absorb them to start producing organic matter once more.

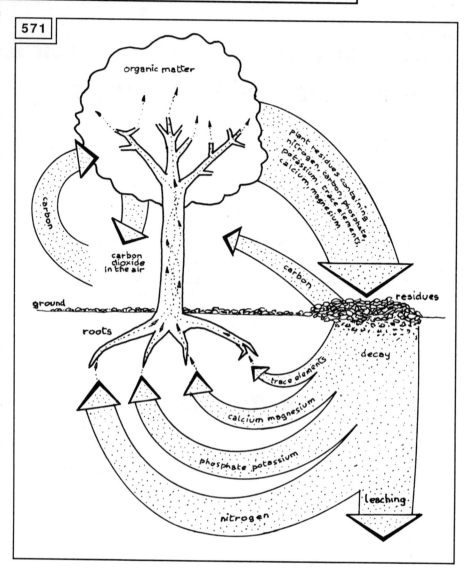

There are therefore two phases in the cycle of organic matter and mineral salts. There is the phase when mineral elements are present in the soil or in the atmosphere, and the phase when they are present in plants in the form of organic matter. Leaching carries some minerals deep into the soil so that they cannot take part again in the cycle of organic matter.

Decomposition of vegetable waste

Vegetable waste decays in two ways :

- it may rot as a result of **biological activity**, that is, the activity of insects, worms, other small animals, fungi and microorganisms that break down these residues. This is the process known as the **formation of humus (humification)**. Decay is gradual. The minerals are restored, little by little, to the soil and to the atmosphere over a period of weeks, or maybe many months in cool weather conditions. Organic decay, i.e. the formation of humus, adds nitrogen and other minerals into the soil ;

- decomposition can be caused by **fire** that acts violently. Fire releases carbon and nitrogen into the atmosphere. Phosphorus, potassium, calcium, magnesium and trace elements are restored to the soil in the form of ashes without being changed into humus.

Slow breakdown by humification is usually the best way of improving soil structure. It means that plants receive a regular supply of food for two reasons. First, the mineral elements are put back gradually into the soil. Secondly, they are used up by soil inhabitants that retain them in their living matter for some time before getting rid of them in the form of waste or dead remains which, in turn, are food for other living beings.

At times, organic decay is too slow. This is so with large tree trunks and branches after forest clearing. This kind of wood is better burnt and their ashes spread on the fields, but light quick-rotting residues should not be destroyed by fire.

The cycle of organic matter and the removal of mineral elements from farms

---

As we have just seen, when all plant residues are returned to the soil and rot there, reserves of mineral nutrients are built up again from year to year. This is what happens in natural forest and on fallow land.

However, the process of renewal is altered when some of the organic matter leaves the field. Less waste is returned to rot in the soil with the result that the amount of nutritive elements restored to the soil is less than the amount taken away **(figure 572)**.

What happens, for instance, in a coffee plantation ? On leaving the field, the coffee berries carry off all the minerals they contain. Or, to give another example, millet grains and legumes taken away from the field also remove the minerals taken up from the soil.

Agricultural produce leaving fields is said to **export** (carry away) the mineral elements taken from the soil. When produce is exported in large quantities and at regular intervals, the soil becomes

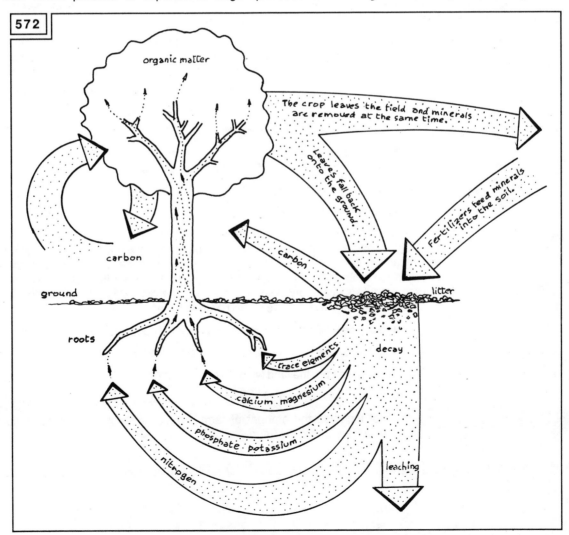

572

organic matter

The crop leaves the field and minerals are removed at the same time.

Leaves fall back onto the ground

Fertilizers feed minerals into the soil

carbon

carbon

ground

litter

roots

decay

trace elements

calcium · magnesium

phosphate · potassium

nitrogen

leaching

exhausted. It contains fewer and fewer minerals. Plants on this kind of soil are badly nourished. Their leaves turn yellow and the plants themselves are much more susceptible to disease.

Exporting mineral elements always harms the soil layers exploited by the roots.

The aim of agriculture is to find in the earth and take from the earth what is needed for a livelihood, but removal without restitution inflicts slow death on the land.

The cultivator exporting mineral salts from his farm must always be careful to replace them. Fertilizers enable the farmer to make up for the minerals removed when harvests are taken away.

Fertilizers are used in fields for many reasons :

- **they contain the minerals** needed to nourish plants ;

- **they compensate for the loss of minerals** removed when crops are harvested or lost through leaching ;

- **they affect soil structure.** Generally speaking, organic manures give better results than mineral fertilizers ;

- **organic matter and humus retain soil minerals, while organic manure improves the effectiveness of mineral fertilizers.**

Some thoughts on farm economy

We have stressed that harvested crops leaving the land export the minerals that make soil productive. We know that minerals removed from the land must be replaced in the same proportion and that fertilizers enable the farmer to make this restitution.

However, in order to make this restitution possible, **the price obtained for agricultural produce sold off-farm must include the cost of the fertilizers** the farmer has to use. If the cost of fertilizers is not completely covered by the sale of farm produce, the farmer is better off selling less. If he goes on exporting minerals without replacing them, he is fast exhausting the soil.

Calculating the sale price of a bag of produce means adding up :

- the cost of fertilizer per bag sold ;

- the cost of seeds and other production factors (pesticides, fungicides, tools and so on) per bag sold ;

- cost of farm workers per bag ;

- farmer's work valued in money per bag.

Unfortunately, manufactured fertilizers are very expensive at the present time. The cost is so high that often it is not covered by the sale of the particular crop involved. African governments are worried by the comparison between the price obtained for produce and the cost of fertilizers, and are negotiating the matter with those countries buying African produce and selling fertilizers. Meanwhile, until such time as satisfactory solutions are found where African smallholders are concerned, what should the farmer do ? He should be guided by these rules :

- produce his own fertilizers as far as humanly possible ;

- not waste the fertilizers available on his own land ;

- only take from the land what is strictly needed ;

- give back to the land all available residues, dung, ashes, especially as compost ;

- not increase the amount of farm produce for sale unless he is sure of being able to enrich his land in the future.

Lesson 41 ▨▨▨▨▨▨▨▨▨▨▨▨▨▨▨▨▨▨▨▨▨▨▨▨▨▨▨▨▨▨▨▨▨▨

Organic fertilizers

Organic fertilizers are substances of plant, animal or human origin. When spread on the soil or buried, they increase the organic fraction of the soil (Lesson 18) and can be transformed into humus. A distinction is made between these organic fertilizers :

- **Farmyard manure**, a mixture of animal dung, poultry droppings and bedding ;
- **Compost**, a mixture of vegetable waste heaped up and left to rot ;
- **Green manures**, plants grown to be buried in the soil and left to rot there ;
- **Slurry (liquid manure)** from animal urine.

In other words, all decomposable materials of plant, animal or human origin constitute organic manure upon decay. Mineral and chemical fertilizers are different. They do not rot, they dissolve slowly like grains of salt in water.

Human faeces can be collected from pit latrines and composted, as long as this is done thoroughly and carefully so that it does not encourage diseases such as dysentry. Alternatively, bananas and other fruit crops can be planted over old latrine sites. Human urine can be applied to compost heaps as an additional nutrient source.

The role of organic fertilizers

Organic fertilizers feed plants

Like all organic waste, organic fertilizers rot in the soil and thus give back to the land the minerals they contain.

Organic fertilizers make plants more resistant

Organic fertilizers also supply the soil with substances that help plants develop and increase their resistance to disease. These substances are like remedies taken in small quantities and good for one's health. Plants growing on soil rich in organic matter stand up better to disease, pest attacks and drought. When diseased, they recover faster than plants growing in poor soil.

Organic and inorganic (mineral) fertilizers can be compared by picturing two tables, both set for a meal. On the first table, the one with organic fertilizers, there are many nourishing dishes and relishes. The guests choose what they need, depending on how healthy and how hungry they are. On the second table, the one with chemical fertilizers, there are three dishes, one relish and some salt. There is not much choice. The guests make do with what is on hand even though they might prefer other food for health reasons.

Organic fertilizers are more nutritious than mineral fertilizers ; they are conducive to plant health and resistance.

Organic fertilizers improve soil structure

As was explained in Lesson 19, soil structure is improved by increasing its organic fraction. When soil structure is good, plant roots develop vigorously, moisture is increased, the soil can be worked more easily and is more resistant to wind and rain.

Organic fertilizers combat soil impoverishment

Organic matter is like clay. It holds mineral salts, reserving and storing them. This is an important function because when mineral salts are not held in this way, they may be carried deep into the soil by infiltrating water and there they are of no use to plants (leaching).

Organic fertilizers promote soil moisture

All organic matter contains water. Organic matter therefore retains water until decomposition is completed. Microorganisms that break down this organic matter are also rich in water. This is why living soil is moister than soil with no organic matter.

Organic fertilizers activate soil life

Good soil is always living soil. Treating soil with organic fertilizers gives more food to soil inhabitants that change these organic manures into organic matter which decays and is, in turn, changed into humus releasing mineral nutrients.

Some organic fertilizers are more useful than others

Some organic matter rots quickly, for example, leaves and fruit pulp, but certain types, such as straw, wood and fruit shells, decay much more slowly. Some kinds of wood, e.g. the African fan palm, do not rot at all. (Straw is made of a substance called **cellulose** ; wood is made from **lignin**).

Good organic fertilizer contains a mixture of easily decomposable materials (they rot quickly) and of materials that do not decay so fast. Manure composed of animal excreta and chopped straw is an example of good organic fertilizer. Good compost is composed of green materials - leaves and soft stems - as well as thin branches and chopped straw.

The woman working in the rice plot **(figure 573)** is spreading dried cattle dung collected from her kraal and cowshed. She will bury the dung along with trash lying in the field.

Mixing two kinds of organic matter gives quality manure that plants can use at once. Animal excreta containing much water and nitrogen is quickly broken down by soil microorganisms. Trash, on the other hand, is drier, has a higher carbon content and decays gradually.

The woman has collected large quantities of straw and spread it in one corner of the field **(figure 574)**. Straw is rich in cellulose and poor in materials that rot quickly. The farmer will have to wait at least a year before being able to cultivate this part of the field.

Moreover, burying too much straw and wood is not good for crops. Better results are obtained with a mixture of organic matter rich in nitrogen - excreta, leaves, fruit and so on - and materials richer in carbon like wood and straw.

Crushing and chopping straw by machine speeds up decomposition (Lesson 36). Wood will decay thoroughly if it is first cut and eaten up by termites, ants and fungi. It is then available for microorganisms. The food chain is longer for wood and straw than for moister material with less cellulose and lignin or, for example, for animal excreta (Lesson 4).

573   dry dung

574   straw and other waste   groundnut

Green manures

Green manures are plants grown specifically to obtain organic fertilizer. These plants take up large quantities of minerals found naturally in the soil or added by mineral fertilizers, and restore them in the form of organic matter that is gradually changed into humus. In fact, they have the same effect as fallow plants, the difference being that they are cultivated on purpose. Green manure plants are grown for their organic manure rather than for grain, fruit or leaves for human consumption. Most green manure plants belong to the legume family and some of them were named in Lesson 9.

Actually, all plants act to some extent like green manure when their leaves and stems fall to the ground. However, it is worthwhile producing organic matter systematically and not leaving things to chance.

Figure 575 gives the main qualities one would expect to find in good green manures.

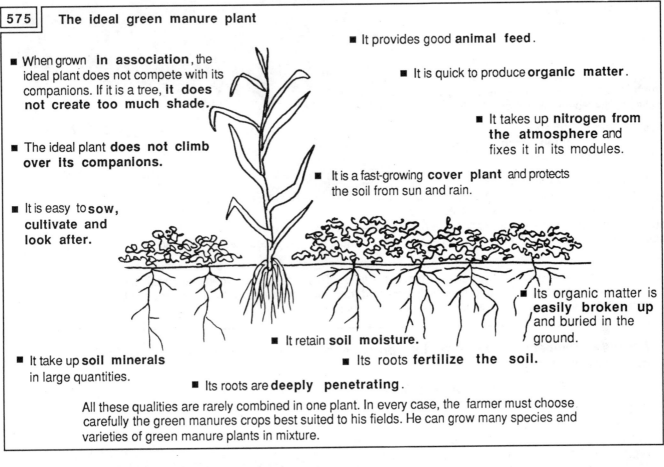

| 575 | **The ideal green manure plant** |

- It provides good **animal feed**.

- When grown **in association**, the ideal plant does not compete with its companions. If it is a tree, **it does not create too much shade.**

- It is quick to produce **organic matter**.

- It takes up **nitrogen from the atmosphere** and fixes it in its modules.

- The ideal plant **does not climb over its companions.**

- It is a fast-growing **cover plant** and protects the soil from sun and rain.

- It is easy to **sow, cultivate and look after.**

- Its organic matter is **easily broken up** and buried in the ground.

- It retain **soil moisture.**

- Its roots **fertilize the soil.**

- It take up **soil minerals** in large quantities.

- Its roots are **deeply penetrating**.

All these qualities are rarely combined in one plant. In every case, the farmer must choose carefully the green manures crops best suited to his fields. He can grow many species and varieties of green manure plants in mixture.

Ways of growing green manure crops

When trees, like the *Leucaena* mentioned many times in this book, are planted , they are grown in orchards, care being taken to prevent the green manure trees from overshading or from invading the root zones of companion crops.

Green manure crops are often herbaceous plants. **They can be sown in pure stand for one or more consecutive seasons** in which case they are part of the cycle of crop rotation just like natural fallow.

The field in **figure 576** is planted with the green manure called tropical kudzu. (See Lesson 36, **figure 470**, for a close-up of this plant.) Five months after planting, the stems were cut and will be dug in, once they start rotting in the rain. At the same time, some maize and cocoyam plantlets grew up through the kudzu. Note how vigorous they are. The abundance of organic matter that will be ploughed into the soil, will nourish it and restore good structure.

Green manure crops are tended in the usual way - by tilling, weeding, cutting, battening down climbing stems.

Green manures can also be grown in mixture with edible plants. Siratro sown with millet (Lesson 36) is an example of this

576

kind. Planting can be staggered, the green manure being sown when cereals are already in full growth.

The tropical kudzu in the *Hevea* plantation **(figure 577)** shows how useful it is to choose a green manure plant that can be used for animal feed.

Green manure crops are sometimes sown during the agricultural season preceding the natural fallow interval. Their fertilizing action makes fallow regeneration more effective.

Associating different species and varieties of green manure crops is also worthwhile.

Lastly, farmers may get better value for money by spreading artificial fertilizers on green manure crops rather than straight onto other

577 | rubber | tropical kudzu

cultivated plants. These costly fertilizers are assimilated quickly by green manures and then released slowly to cultivated plants. Farmers should experiment and note the results obtained by this practice.

When plants have actually produced organic fertilizers, there still remains the task of using these manures to the best advantage by burying them in the fields. This is hard work when plant trash such as stems and dry leaves are long and fibrous. If tools and machinery for crushing trash are not available, the only alternative is to spread residues on the ground and wait until they rot.

Possible competition between the production of organic fertilizers and the need for animal feed

The farmer has just harvested groundnut on sandy soil **(figure 578)** *and heaped up the residues that will be sold for animal feed.*

Perhaps the sale of plant trash for fodder will imperil the land. Will there be any organic matter left to fertilize the soil ? Understandably, the farmer wants to earn money but at the same time he is depriving his land of the food it needs.

Surely it should be possible to manage fodder crops in such a way that they benefit existing and future crops.

As this example shows, it is not easy to strike a balance between three aspects of agricultural production : the production of food for human consumption, the production of animal feed, and the production of green manure crops. Yet, this balance is vital to achieve at one and the same time the aims of self-sufficiency, sale of produce and soil conservation.

578

Lesson 42
Mineral fertilizers

Mineral fertilizers are substances containing mineral salts. When they are mixed into the soil, they can be taken up by plant roots. It is important to remember that mineral salts (and therefore mineral fertilizers) must be dissolved by soil water before they can be used, just as salt must be dissolved in cooking.

Some mineral fertilizers, like urea and sulphate of ammonia, dissolve quickly in soil water and become rapidly available to plants. **Others, such as phosphate and lime, are slower to dissolve** and thus affect the soil for longer periods.

Dissolved mineral salts are carried away by water. When water infiltrates deeply, fertilizers are transported in depth and are leached off. On the other hand, when water is retained by clay or organic matter in the soil, salts remain in the root zone.

It is understandable that the destination and effectiveness of mineral fertilizers depend on soil capacity to retain water and minerals. Applying fertilizer on poor soil is a complete waste of money.

Straight, compound and complete fertilizers

Manufactured fertilizers are divided into major groups depending on the concentration of minerals they contain.

Straight fertilizers contain one mineral element : nitrogen, phosphorus, potassium, calcium, etc.

- **Nitrogenous fertilizers** contain nitrogen as their major component, e.g. **ammonium nitrate, ammonium sulphate** and **urea.**

- **Phosphatic fertilizers** contain phosphate, examples being **natural phosphates** extracted from mines, **phosphatic** and **superphosphatic fertilizers.**

- **Potassic fertilizers** supply the soil with potassium of which two common forms are **potassium chloride** and **potassium sulphate.**

- **Calcic and magnesic fertilizers** are mostly found in **lime** rich in calcium or in

Meaning of fertilizer notations

579

The bags of fertilizer in **figure 579** have the following notations :

Complete fertilizer
N.P.K. SH
14.23.15 + 6 S + 1H₂O₃
Fertilizer 1982
50 kg

The figures 14.23.15 are read in the order N.P.K. They mean that 100 kg contain 14 kg of nitrogen (symbol N), 23 kg of phosphorus (P) and 15 kg of potassium (K). The fertilizer also contains 6 kg of sulphur (+ 6 S), hydrogen (H) and oxygen (O). However, the hydrogen and oxygen are not taken into account here to define the fertilizer. These are the minerals composing water. They are also found in air.

The figures give the amounts of N, P, and K contained in 100 kg of fertilizer, but the bags photographed only contain 50 kg each and therefore half the major elements indicated, i.e. 7 kg of nitrogen, 11.5 kg of phosphorus, 7.5 of kg of potassium and 3 kg of sulphur.

The figures indicate the major mineral elements contained in a given fertilizer. Here are examples :

N.P.K. 10.10.20., major element is potassium ;

N.P.K. 30.10.10., major element is nitrogen ;

N.P.K. 20.30.5. has high phosphorus and nitrogen content, little potassium.

Because fertilizers differ widely in composition and the effects they produce, they must be exactly right for the soil on which they are spread and fit plant requirements.

Often, the symbol P is written P₂O₅ (as phosphate), and the symbol K is written K₂O (as potash).

magnesic lime containing magnesium and calcium. **Limestone**, a rock rich in calcium, can be crushed and used as a calcic fertilizer.

In regions where, for example, oyster shells and fish bones are available, they are heated on metal sheets and broken down into a greyish-white powder that is used as calcic fertilizer.

Compound fertilizers are a mixture of several minerals. **Phosphate** of **ammonia** contains phosphate and nitrogen, **potassium nitrate** has nitrogen and potassium.

Lime is a mineral salt of great importance for soil structure and is often found in compound fertilizers. **Bicalcic phosphate** contains phosphate and lime. **Dolomite** contains calcium and magnesium.

'Complete' fertilizers form the third type of inorganic manures and are called N.P.K. fertilizers because they contain the three major plant nutrients - nitrogen, phosphate and potassium. In fact, they are not complete because plants need many other elements.

Some fertilizers are found in a natural state, often in the form of crushed stone. They are referred to as **natural fertilizers**. Natural phosphates and lime fall into this category. Others are manufactured, hence the term **artificial fertilizers**.

Fertilizers must be adapted to meet plant requirements

Because the farmer removes mineral elements from the land along with harvested products, he has to replace them by spreading organic or mineral fertilizers. However, as each plant carries off different nutrients, the fertilizers needed vary from crop to crop.

Three crops are compared and losses in nitrogen (N), phosphate (P_2O_5), potash (K_2O) and lime (Ca) are stated in each case.

- 14 000 kg (14 tonnes) of cassava tubers are picked in field 1 and remove 25 kg of nitrogen (N), 10 kg of phosphorus (P), 20 kg of potassium (K) and 5 kg of lime (Ca).

- The maize crop in field 2 yields 500 kg of dry grain (about 7 bags). The cereal contains, and therefore carries away, 47 kg of nitrogen (N), 1.5 kg phosphorus (P), 1.5 kg potassium (K) and 0.2 kg lime (Ca).

- Field 3 yields 500 kg of fresh groundnut pods (about 8 bags) that remove 26 kg of nitrogen (N), 2 kg of phosphorus (P), 6 kg of potassium (K) and 3.5 kg of lime (Ca).

Here are the relevant figures for plantations.

- 500 kg of dry coffee berries remove 7 kg of nitrogen (N), 1 kg of phosphate (P_2O_5), 12 kg of potash (K_2O) and 1.5 kg of lime (Ca).

- A banana harvest of 300 bunches causes a loss of 25 kg of nitrogen (N), 7 kg of phosphate (P_2O_5), 90 kg of potash (K_2O) and 5 kg of lime (Ca).

Losses are even greater when the entire crop is removed. When groundnut is harvested and the haulms are also removed for animal feed, the minerals removed are the sum of the pods and haulms. Thus, 500 kg of groundnut haulms, removed from the field, represent 7 kg of nitrogen (N), 0.4 kg of phosphate (P_2O_5), 6 kg of potash (K_2O) and 5 kg of lime (Ca).

580	Mineral elements, in kilograms, removed with harvested crops			
	N nitrogen	**P (of P_2O_5)** phosphorus (of phosphate)	**K (of K_2O)** potassium (of potash)	**Ca** lime
14 000 kg cassava tubers	25	10	20	5
500 kg maize/dry grain	47	1.5	1.5	0.2
500 fresh groundnut pods	26	2	6	3.5
500 kg groundnut haulms	7	0.4	6	5
500 kg dried coffee berries	7	1	12	1.5
Bananas - 300 bunches	25	7	90	5

Comparative figures are shown in **table 580 and lead to the conclusion that the minerals removed vary in quantity from crop to crop. It is therefore obvious that, in order to restore the minerals removed from the soil, fertilizers must be chosen to fit the needs of each individual plant.**

Plant needs vary with every stage of the vegetative cycle

Lesson 24 described how the light requirements of plants vary with each stage of the life cycle. Then water requirements were examined stage by stage, especially for yam and sorghum (Lesson 27). Nutrient requirements also vary with every phase in plant growth. The nutrient requirements of groundnut are set out in **table 581** (page 248). Each column deals with a phase of the cycle, first with plant life and growth at that particular time and then with the food needed.

Leguminous crops and fertilizers

Leguminous crops feed heavily on nitrogen. Nitrogen is found in large quantities in the atmosphere, and legumes fix it in their root nodules through the nitrogen-fixing bacteria, *Rhizobium*. The plant must work hard to fix atmospheric nitrogen.

Legumes can be compared to a greedy man who likes sweetmeats so much that he makes them himself with all the appropriate ingredients at his disposal. However, if the sweetmeats are ready-made and there for the asking, why should the man bother to make them himself?

Nitrogen is the sweetmeat of legume crops. This is why fertilizers for legume crops must never be too rich in nitrogen. Supplied with too much nitrogen, legumes fix atmospheric nitrogen less efficiently **(figure 582)**.

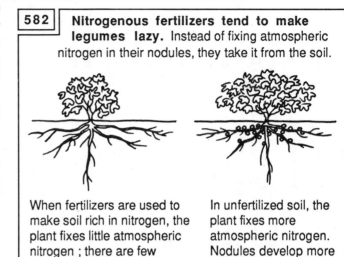

582 | **Nitrogenous fertilizers tend to make legumes lazy.** Instead of fixing atmospheric nitrogen in their nodules, they take it from the soil.

When fertilizers are used to make soil rich in nitrogen, the plant fixes little atmospheric nitrogen ; there are few nodules.

In unfertilized soil, the plant fixes more atmospheric nitrogen. Nodules develop more abundantly.

The quantity of fertilizer needed for any given crop depends on soil structure

Let us suppose that a field growing maize yields 1000 kg grain that remove 95 kg of nitrogen, 2.5 kg of phosphate, 3 kg of potash and 0.3 kg of lime. The soil layers exploited by the roots must be replenished with the same quantities of mineral elements.

At the same time, the possible leaching of mineral salts must be kept in mind. Minerals may be carried away into deep soil layers beyond the reach of plant roots. Leaching is all the greater when soil is sandy and poor in

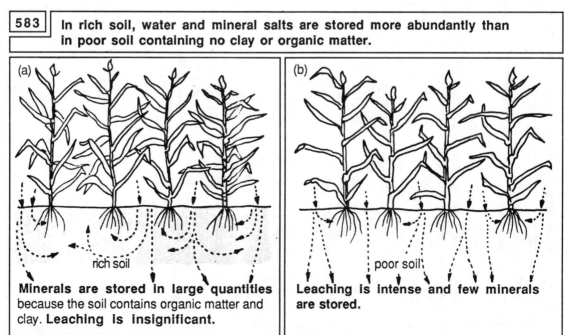

583 | **In rich soil, water and mineral salts are stored more abundantly than in poor soil containing no clay or organic matter.**

(a) rich soil

Minerals are stored in large quantities because the soil contains organic matter and clay. **Leaching is insignificant.**

(b) poor soil

Leaching is intense and few minerals are stored.

581 | Groundnut : nutritive requirements at each stage of the vegetative cycle

Germination and emergence	Vegetative growth	Flowering	Fruit enlargement	Fruiting and ripening
Groundnut seed is dry and dormant. It is inactive. After sowing and as soon as the first rains have moistened the soil well, the seed swells and is imbibed with water. It starts breathing in soil air. First the radicle, then the plumule appear. Soon the first two leaves and short stem are seen on the soil surface.	Once the seedling is well established, it grows fast. New stems and leaves develop. At this stage, the groundnut grows very fast underground. The roots penetrate deeply at first, then spread extensively. Nodules show on the rootlets.	Leaves are still growing when the first flowers appear. Full flowering takes place two or three weeks later, by which time foliage is more or less complete. The abundant leaves are busy manufacturing food reserves to ensure fruit enlargement.	Flowering is almost over. Ovary enlargement takes place, fruit forms at the end of the carpophores or pegs that elongate and grow downwards penetrating the soil. The pegs are the pod-bearing stalks.	The pods, now in the ground, swell gradually, fed by the nutrients in the leaves and by those absorbed from the soil. When the pods are ripe, the leaves turn yellow and the roots die off. It is almost harvest time.
At this early stage of life, the seedling relies entirely on the food stored in the cotyledons. **The cotyledon reserves contain the mineral elements the plant needs**, such as nitrogen, phosphate, potassium, etc.	Food reserves are now exhausted. **Roots and stems must develop. Mineral requirements are high.** In good soil, the plant thrives and responds to mineral fertilizers. In poor soil, plant-building nutrients are missing and growth is stunted. **Phosphate** is now effective. It contributes to root extension and nodulation. The nodules take up nitrogen from the air and use it for plant growth.	**Much food is needed** especially nitrogen. The nodules are busy because leaf and flower development calls for large quantities of nitrogen. Supplying this from a fertilizer bag will depress natural nitrogen fixation in the nodules. The activity of the nodules can be checked by crushing them between the fingers ; red flesh is a sign that they are working hard. Groundnut drains other soil minerals but they are not all used up at the same time. **Some are used at once, others are stored in the stems and leaves.** These reserves are used later for pod enlargement.	At this stage, groundnut feeds on large quantities of **nitrogen and phosphorus** in order to ensure fruit enlargement. **Some of these minerals are absorbed by the roots straight from the soil ; others are drawn from the reserves mentioned in the previous column**, and are transported from the leaves and stems towards the pods.	The change in roots and leaves takes place because **the pods must have minerals** to ensure enlargement, and exhaust the supplies contained in the leaves and roots, causing them to wither. The pods contain three-quarters of the phosphate taken up from the soil during the growing season, two-thirds of the nitrogen and half the potash. These minerals leave the field when the crop is removed. Other minerals, for instance, calcium, magnesium and most of the trace elements, remain in the withered stems and roots.
It is too early for fertilizer. However, if any is used, it should be one that dissolves slowly so that the plant takes up the minerals later on.	**Fertilizers are recommended at this stage.**		**Fertilizers are most beneficial at this time because they promote the build-up of mineral reserves.**	**There is no point in applying fertilizers at this stage because the roots are dying.**

organic matter because, as we have seen, organic matter and clay hold and store up water and mineral salts.

Now, let us compare leaching in depth in two fields, one rich in organic matter and with good soil structure, the other poor in organic matter and with bad soil structure.

What are the fertilizer requirements, in both cases, if a N.P.K. mixture appropriate for maize is used ? On fields of the same size, we shall need :

- about 200 kg, or four 50 kg bags, in field 1 **(figure 583a)**. If the fertilizer costs 70 francs per kg, fertilizer input is 14 000 francs ;

- in order to obtain the same yield from field 2 **(figure 583b)**, 300 kg will be needed, or six 50 kg bags, costing 21 000 francs, the additional outlay coming to 7000 francs.

The amount of fertilizer required for the same crop growing in fields of the same size depends on soil quality.

Lesson 43
The proper use of mineral fertilizers

Putting mineral fertilizers in the right place

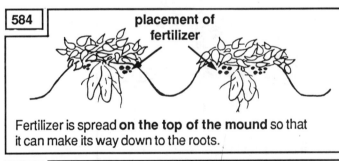

584

placement of fertilizer

Fertilizer is spread **on the top of the mound** so that it can make its way down to the roots.

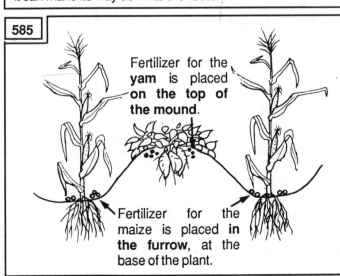

585

Fertilizer for the **yam** is placed **on the top of the mound.**

Fertilizer for the maize is placed **in the furrow,** at the base of the plant.

There are many ways of applying fertilizer :
- it can be spread over the whole field. This is called broadcasting ;
- it can be spread on the rows of cultivated plants or between the rows (band application) ;
- it can be applied in a ring round the plant (ring method) ;
- sometimes, it is dropped into each planting hole along with the seeds.

One general rule holds whatever the method used. Mineral fertilizer must be spread where the roots, and particularly the root hairs, are located (Lesson 31). To ensure accurate applications, the root systems of the plants to be manured and their exact location in the soil must be known. Here are examples to illustrate this basic point.

Yam **(figure 584)** is growing on hills or high mounds. Fertilizers must be applied on the top of the mounds and not in the furrows where they would not benefit the plant.

Here is an example of mounded yam associated with maize planted in the furrows **(figure 585)**. Fertilizer for the yam plant must be spread on the mound, whereas fertilizer for the maize is applied in the furrow near the maize roots.

Figure 586 traces three phases in the life cycle of coffee trees. The location of roots and root hairs changes at every phase with the result that fertilizer placement also changes.

| 586 | **Fertilizer placement depends on the age of the trees.** |

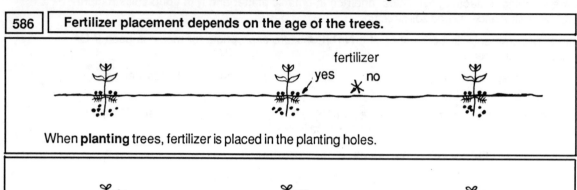

When **planting** trees, fertilizer is placed in the planting holes.

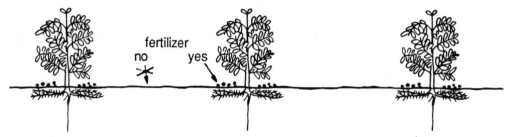

During growth and when roots have not yet spread, fertilizer is applied round the foot of the tree.

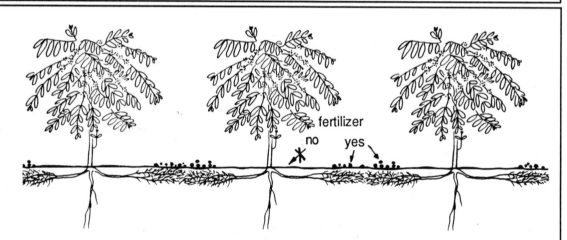

The root system of **mature trees** is fully developed and root hairs are some distance from the trunks. At this stage, fertilizer is spread between the rows **where the roots hairs are located.**

Before spreading fertilizer in orchards, rootlets and root hairs must be located accurately.

Incorporating fertilizers

Wherever possible, fertilizers should be buried lightly. This is advisable because roots are drawn to areas where nutrients are stored (Lesson 31). Thus, if fertilizer remains on the ground, roots might work their way towards the surface. Some fertilizers are lost by prolonged exposure to the air.

Roots above ground tend to shrivel. It is best, therefore, to dig in manures using a hoe, or a tilling implement like the ridger that moves up and down between the rows after the girl has dropped fertilizer at the base of each cotton plantlet **(figures 587 and 588).**

Fertilizers must be incorporated at all costs if there is a risk of water runoff, because runoff carries away mineral fertilizers without any difficulty.

587

588

All the same, it may be advisable to leave fertilizer on the surface if the soil is very sandy and rainwater infiltrates easily. This practice slows down leaching at least to some extent.

How to reduce fertilizer leaching

We say that fertilizer is leached when it is carried deep into the ground by infiltrating water. Then there is no fertilizer in the soil layers exploited by the roots of cultivated plants.

Steps can be taken to reduce the leaching of fertilizers (figure 589) and such measures include :

- **manuring the soil with organic matter** that retains water and mineral salts ;

- **helping roots to grow vigorously.** As the volume of rootage increases, plants are better able to recover mineral salts which, once taken up by plants, can no longer travel deep into the ground. Fertilizers turned carefully into the soil contribute to root growth in the best possible way ;

- **mixing plants with different root systems.** When water infiltrates, plants with shallow roots capture fertilizers first, and any mineral salts not taken up by these superficial roots will be retained by deeper roots ;

- **spreading mineral fertilizers** exactly when the nutrient uptake of cultivated plants is highest. Correct timing is all important ;

- **practising split applications near the roots**, i.e. applying fertilizers in small successive doses that plants can absorb at once. This method reduces the leaching of minerals which occurs when uptake by plants is not immediate.

589

Three major barriers to prevent the leaching of fertilizers

- vitality of the environment

- abundance of organic matter in the soil
- maximum occupation of the soil by the plant roots

Applying fertilizers in splits

In temperate climates, such as in Europe, rainfall is not so intense and soils retain mineral salts quite satisfactorily. There is some leaching but it is fairly slow.

The opposite is true of the tropical regions of Africa. Rainfall is often torrential, especially at the onset of the rainy season, and soil leaching is a familiar feature of these zones.

Imagine what happens if fertilizer application is followed by a downpour. The enormous mass of infiltrating water carries down to the water table half or maybe two-thirds of the fertilizer that has just been spread, with great loss to the farmer.

Spreading fertilizers in **split applications** is always recommended in order to reduce the risk of leaching. This method involves applying small doses of fertilizer at intervals of several weeks, rather than spreading them in large quantities all at once. For instance, 300 kg of fertilizer are not spread on a field in one operation. They are applied in three or four operations, using 75 kg or 100 kg each time.

Applying fertilizers in splits therefore lessens the risk of disappointing returns for fertilizer input.

Split applications are also a way of adapting the amount of fertilizer used to suit plant requirements. These requirements change during the cultural season, just like water and light needs. Each application can be adjusted to fit the corresponding phase in the cycle. **Figure 590** presents the first of two examples of split applications. It deals with sorghum.

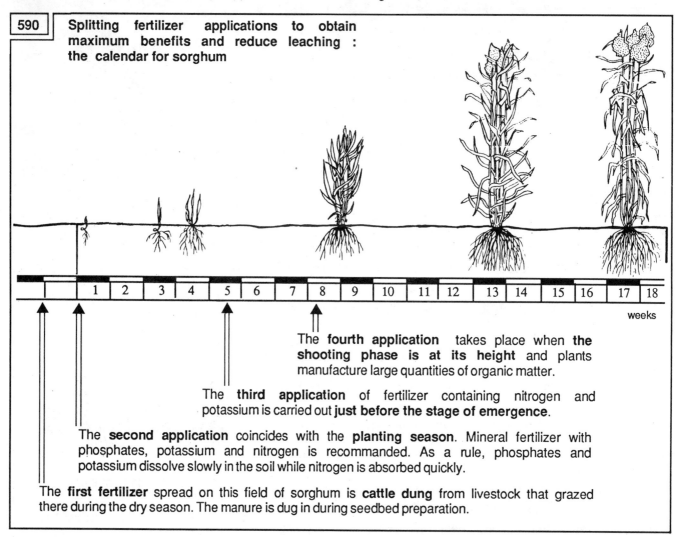

590 **Splitting fertilizer applications to obtain maximum benefits and reduce leaching : the calendar for sorghum**

weeks

The **fourth application** takes place when **the shooting phase is at its height** and plants manufacture large quantities of organic matter.

The **third application** of fertilizer containing nitrogen and potassium is carried out **just before the stage of emergence.**

The **second application** coincides with the **planting season**. Mineral fertilizer with phosphates, potassium and nitrogen is recommanded. As a rule, phosphates and potassium dissolve slowly in the soil while nitrogen is absorbed quickly.

The **first fertilizer** spread on this field of sorghum is **cattle dung** from livestock that grazed there during the dry season. The manure is dug in during seedbed preparation.

The second example of split applications is practised in banana plantations in a very wet region of the Ivory Coast.

■ Phosphatic fertilizer is placed in the planting holes of the banana seedlings. This fertilizer dissolves very slowly.

■ Nitrogen and potassium are applied in six splits between planting and fruiting, and are spread near the plants.

■ Lime and a mixture of trace elements are applied twice - first when the crop is planted and a second time about six months later.

This makes a total of nine splits.

Applying fertilizers in splits means that small quantities of certain kinds of fertilizer are spread each time to meet plant needs as closely as possible and to lessen the risk of mineral leaching in heavy rainfall.

253

Weeds thrive on fertilizers

Fertilizers benefit weeds just as much as cultivated plants. After fertilizer was applied **(figure 591)**, unwanted grasses began to grow up through the cotton crop.

Weeds compete with cultivated plants for the nutrients supplied by fertilizers. Hence, when fertilizers are used, severe weed control is essential so that undesired plants do not overrun cultivated species.

It is now too late to weed this cotton field where the crop has already been invaded. A cleaning operation should have been carried out before the weeds were established and before their roots had spread extensively. The aim of weeding is to leave the field and fertilizers at the sole disposal of cultivated crops.

591

There are other ways of increasing the return from every bag of fertilizer used. One such method associates useful plants, some erect, some creeping. Creeping species and plants providing thick soil cover are called cleaning plants because they are capable of controlling weeds. These cleaning plants keep weeds down and at the same time take up the minerals contained in fertilizers for the good of their own produce. Just like weeds, cleaning plants may compete with other cultivated plants, the difference being that farmers organize this competition on purpose in order to increase production.

Some cleaning plants belong to the legume family. Crops can be interplanted with cleaning plants that manure and clean the soil while also keeping weeds under control. **Figure 592** explains this cultural pattern.

---

592 | **Using mineral fertilizers and green manure plants**

Oil palms have been interplanted with tropical kudzu that **covers and cleans** the soil efficiently. Kudzu is also suitable for grazing. Palm branch can be lopped if they overshade the green manure plant and disturb its growth. **Trimming the overhead branches** lets light reach the cover plant.

Trimmed palm branches

Tropical kudzu

Fertilizer spread in the plantation is taken up quickly by tropical kudzu and changed into organic manure. The process is like this. The cover plant uses mineral fertilizer to produce more organic matter that finally rots on the ground. **As it decays, this organic waste feeds the palms slowly with useful minerals.** Kudzu fixes atmospheric nitrogen and does so more efficiently when the right fertilizer (with low nitrogen content) is applied. By carefull maintenance of the kudzu crop, the farmer increases the nitrogen-fixing capacity of the plant rather than buying nitrogen fertilizer itself.

Why mineral fertilizers can be dangerous

Farmers using mineral fertilizers for years on end may notice signs of soil exhaustion.

Sometimes too, fertilized plants are seen to be more susceptible to disease and pest attacks. Maybe produce is softer and rots faster than produce from land that received no mineral fertilizer. At times, the edible produce tastes of the fertilizer used. When inappropriate fertilizers are applied, plants start producing too many leaves, to the detriment of fruit and grain yields.

What causes all these harmful effects ?

■ **Fertilizer applications activate soil life.** Microorganisms feeding on the soil transform organic matter. Now, by making microorganisms more active, fertilizers speed up the breakdown of organic matter and a chain reaction is started. Water and minerals contained in the organic matter are leached off to deep soil layers and soil structure is impoverished. This is the first reason for possible soil exhaustion resulting from the use of inappropriate fertilizers.

■ A second danger lies in the fact that **manufactured mineral fertilizers contain large quantities of nitrogen, phosphorus and potassium with negligible quantities of trace elements.** Because N.P.K. fertilizers make plants more active, the uptake of trace elements is greatly increased. These elements are removed with harvested crops and are not replaced because N.P.K. fertilizers do not contain any. Trace elements in the soil are gradually exhausted and plants are left without these important nutrients. Plants are like people who sit down to a meal and find there is only a heap of staple left and no relish. Plenty of nutrients are fed into the soil but plants are not able to eat and use the food because an essential trace element is missing. (There is a nutrient **deficiency.**) Trace elements can be compared to the kola nut. Its seeds do not really feed the chewer but, when he is deprived of them, he feels listless and has no energy.

■ Sportsmen are a case in point. They must look after their health by eating a varied diet. They chew, salivate, eat staples, meat and relishes. They spend their energy by taking a lot of physical exercise. But maybe these athletes get lazy and cannot be bothered to make the effort to keep fit. They fall back onto a diet of soft foods that do not need chewing (staples perhaps) and stop taking exercise. In this kind of situation, health and energy deteriorate.

Plants react in more or less the same way when, without effort on their part, they receive plenty of fertilizers, and no longer have to fight for food in the soil. Their flesh goes soft and they become more prone to disease.

In conclusion, spreading mineral fertilizers greatly increases crop yields, but often has adverse effects on plant health and the quality of produce. This is why farmers using large quantities of chemical fertilizers must also resort to pesticides because plants supplied with chemical nutrients usually forfeit their inborn resistance to disease and pests.

■ Sometimes, **fertilizers stimulate plants to overproduce foliage.** Leaves transpire and lose a lot of water. In areas of low rainfall, the use of fertilizers may be one of the causes of soil desiccation.

There are many ways of counteracting the adverse effects of mineral fertilizers. They include :

■ **mixing mineral fertilizers** with **organic manure ;**

■ **making many plant species work the soil rather than one,** i.e. **multiple** as opposed to **sole cropping ;**

■ **not over-fertilizing ;**

■ **splitting fertilizer applications and adjusting the amounts** to fit each phase of the life cycle ;

■ **using natural manures** in preference to artificial fertilizers.

A final point on the subject of fertilizers. Some plant varieties are hardy, meaning they tolerate the natural environment where they have lived for a long time. When fertilizer is applied, the plant is likely to respond badly. Thus farmers who use fertilizers intensively are sometimes forced to buy new seeds, specially selected for use on fertilized fields. In this situation, farmers must ask themselves the all-

Fields must be in good shape before fertilizer application.

It is pointless to think of spreading fertilizer on this field (figure 593). It would be leached off at once by water runoff.

593

important question, 'Can we afford to buy selected seeds and all the products needed to protect our harvests ?' **Figure 368**, Lesson 30, may be consulted in this connection.

Before learning how to calculate returns from fertilizer applications (this important subject is covered in Lesson 48), we must first know how to estimate crop yields and maximize them by good planting practices.

Notes

Reminder

Many factors must be considered in order to ensure the proper use of fertilizers. These factors are :

■ the food, especially the mineral elements, that every plant needs ;

■ timing, that is to say, the moment when plants need these nutrients ;

■ soil characteristics ;

■ the composition of the fertilizers on sale ;

■ the amount of water infiltrating the deep soil layers with consequent leaching of fertilizers.

The only sure way of making mineral fertilizers available to plants is to maintain or establish good soil structure. Good structure implies the presence of clay and organic matter that retain water rich in minerals in the root zone. Spreading chemical fertilizers on poor soil is simply throwing money away.

Prudence and caution save money. This is done by seeking counsel from qualified technicians and by following their advice scrupulously. Farmers also achieve results by observing the effects of fertilizers on small experimental plots.

Lesson 44

Making the most of field space

This lesson examines spatial arrangement - the way cultivated plants are arranged and field space is occupied.

Good plant arrangement **(figure 594)** means trying to imagine and reconcile four aspects of plants :

■ the way plants **occupy aerial (aboveground) space (figure 595)** ;

■ the way plants **occupy soil and soil volume (figure 596)**. This subject was covered in Lessons 31 and 32 ;

■ the way plants **occupy the soil extensively above and below ground**, that is to say, the interplant arrangement of root zones and leaf canopy **(figure 597)**,

■ **the time element**, considering what the spatial occupation is like at any particular time in the life of the plant, since plant occupation evolves with each phase of the life cycle.

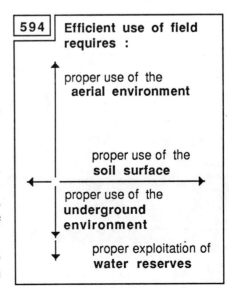

594 | **Efficient use of field requires :**

proper use of the **aerial environment**

proper use of the **soil surface**

proper use of the **underground environment**

proper exploitation of **water reserves**

It would be impossible to review the whole range of planting and sowing arrangements. What is important is to grasp the need to experiment with arrangements, adjusted to suit every cultural activity. **Arrangements depend on a series of factors :**

- **the lie of cultivated land** and land relief ;

- **soil characteristics** - depth, structure, moisture ;

- the plants that are going to be cultivated and **their morphological characteristics**, meaning the shape of roots, stems, leaves ;

- **the life cycle** of cultivated plants ;

- **available tools** ;

- **weather hazards** (flooding, drought, cold, tornadoes), risk of water runoff and erosion, biological risks (pest attacks), technical and economic risks ;

- **method of tillage** - on the flat, mounding (hilling), ridging, bedding.

In order to describe possible sowing and planting arrangements, we must take into account :

- **the shape** of plants ;

- the method of **sowing or planting** ;

- **seed or planting rate**, and spacing ;

- **the way the land is formed** for sowing and planting.

What exactly do these terms mean ?

The shape of plants

The expression **shape of plants** is used to describe the way plants stand above ground. The following plant shapes are illustrated in **figures 598 to 601** :

- **arborescent shape** - mango, avocado, African locust bean, palm, baobab, shea butter, breadfruit, kola, characterized by their trunk and branches ;

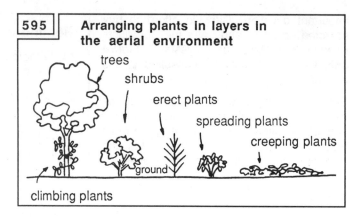

595 | **Arranging plants in layers in the aerial environment**

trees
shrubs
erect plants
spreading plants
creeping plants
ground
climbing plants

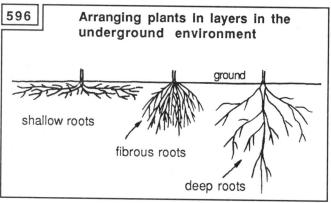

596 | **Arranging plants in layers in the underground environment**

ground
shallow roots
fibrous roots
deep roots

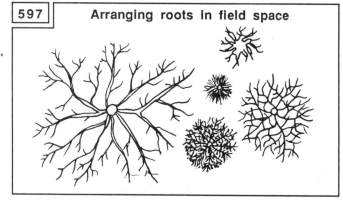

597 | **Arranging roots in field space**

tree shape

shrub shape

598 | bush shape

- **shrub shape** - small, low trees such as lemon, cocoa, wild coffee, pawpaw, cashew ;

- **bush shape** - plants with many low stems growing from the base, examples being cassava, guava, castor, wild cotton, pea, pruned dwarf coffee.

Arborescent, shrub and bush shapes refer to plants with **ligneous** or woody stems.

Herbaceous plants do not have woody stems. Characteristic shapes are erect, spreading, creeping and climbing :

- **erect shape** of maize, sugar cane, rice, millet, okra ;

- **spreading shape** of taro, groundnut, earthpea, potato, pineapple, sisal, tobacco ;

- **creeping shape** of sweet potato and gourd, for example, with stems spreading along the ground.

- **climbing shape** of yam, pepper, some beans, peas, giant granadilla.

Shape also depends on agricultural methods. Coffee is a shrub in its natural state, but it can be shaped into a bush by pruning. A climbing

tree shape

shape shape

bush shape

599

erect shape of maize

creeping shape of potato

601

spreading shape of cocoyam

creeping shape of gourd

erect shape of sugar cane

600

yam will creep along the ground if it is not trained up a stake.

The way in which plants occupy aerial space is an important aspect of farming for various reasons.

- When plants of similar shape are sown too close together, they may compete for air and light. But plants with different shapes occupy different spaces and are complementary.

- Plant shape and its effect on soil cover must be understood. Sweet potato and cowpea, for instance, are far better than millet, sorghum or maize at protecting the soil from the beating action of the rain.

- Harvesting operations must not be forgotten. It is easier to pick fruit from bushes and shrubs than from tall trees.

- Farm work is also affected by plant shape. Creeping plants, like gourds, stop weeds from thriving so fewer weeding operations are needed where gourds are present. But if only erect species are cultivated, the field has to be weeded many times.

Ways of sowing and planting

There are many ways of sowing and planting. Seeds can be sown one by one or dibbled into planting holes (Lesson 32). Seeding one by one can be done at random or in rows. The same goes for seeding in planting holes. These may be spaced haphazardly or arranged in rows. Dibbling and seeding in rows has a distinct advantage - plant development can be checked and crop maintenance

fonio
(Hungry rice or acha)

602

pea

bean

603

is easier. There is room between the rows and weeding is made easier.

When seeding follows a definite pattern, seed density and pest attacks are more easily controlled. The need for transplanting can be spotted. Moreover, farmers using machinery must sow in rows to allow passage for animals and machines.

Some examples of more or less organized sowing

Fonio **(figure 602)** was sown **by broadcasting**. The seeds were tossed into the air and fell in a haphazard fashion on the tilled soil. Seed rate is not even so that plants are closely spaced in some parts and there are none in others. There is no uniform sowing pattern in this field.

In the field photographed in **figure 603**, bean and pea were sown at random, seed by seed. However, unlike the fonio crop, every seed was placed in a hollow in the ground and covered with a light layer of soil.

Maize seeds were sown one by one in rows **(figure 604)**. The rows are wide apart but spacing between plantlets in the rows is much closer.

The millet crop **(figure 605)** was sown in planting holes spaced unsystematically.

Mechanized cultivation, done here by a team of oxen **(figure 606)**, calls for row planting. Would oxen be able to tread their way through a field with planting holes dug at random or between plantlets distributed without any spacing arrangement ?

Figure 607 presents different ways of sowing and planting. These patterns are also suitable for herbs and trees.

maize sown
in rows

604

dibbled millet

605

606

Seed rate and plant spacing

The **seed** or **planting rate** gives the number of plants per hectare or per are, for example, 200 mango plantlets per hectare (or 2 per are), or 5000 yams per hectare (50 per are), or 160 000 groundnut plants per hectare (1600 per are).

Small plants, such as groundnut, can be close seeded, but larger plants, cassava for example and trees, must be more widely spaced.

Seed and planting rates depend on the characteristics of the cultivated plants. The farmers also must remember that crops need to find food in the soil without undue competition, be free to thrive normally above ground and enjoy light.

Spacing refers to the distance between two cultivated plantlets of the same species. To give an example, the correct spacing of a groundnut crop is 25 cm x 25 cm. This means that there is a distance of 25 cm between every plantlet **(figure 608b)**. Lemon trees in plantation must be spaced 5 m x 5 m so that foliage does not overlap and that root zones do not invade each other. Bananas are spaced 3 m x 3 m **(figure 608a)**. For any one species, actual spacing may be altered according to the variety (for example, tall or dwarf) and according to the soil fertility.

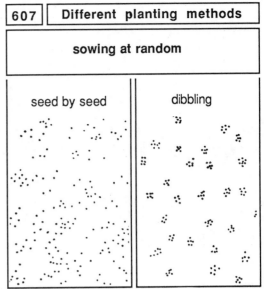

607 | Different planting methods

sowing at random

seed by seed | dibbling

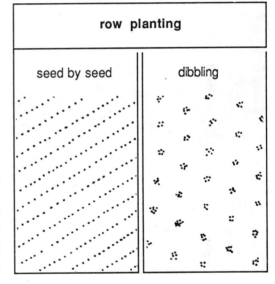

row planting

seed by seed | dibbling

608 | Plant spacing and seed rate

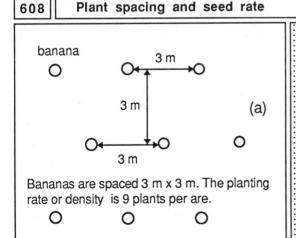

banana — 3 m, 3 m, 3 m (a)

Bananas are spaced 3 m x 3 m. The planting rate or density is 9 plants per are.

groundnut (b)

Groundnut is spaced 25 cm x 25 cm. The seed rate or density is 1600 seedlings per are.

609

610

611

Plenty of air and light penetrate between the double rows

pineapple

tropical kudzu

612

Spacing and seed rate are terms that go hand in hand, because seed rate depends on the spacing chosen for any particular seed or plant. The closer the spacing, the higher the seed rate.

Land forming

Land forms are the contours or configurations of land formed for sowing and planting. Land forms depend entirely on the tillage method decided upon - on the flat, earthing up, ridging, bed cultivation. Lesson 37 dealt with these land forms.

Arranging plants and exploiting field space

We shall look at a wide range of fields and discover plant arrangements starting with simple patterns and moving onto more complicated models. All patterns are acceptable provided adjustments are made between :

■ the space occupied by plants with different shapes and root systems ;

■ the method of planting ;

■ plant rate or density ;

■ land forms.

Cultural layers or storeys

The field of irrigated rice **(figure 609)** is in pure stand. The planting rate of the transplanted rice seedlings is very high. Aerial space is occupied to a height of 60 cm and roots have penetrated to a depth of 20 cm to 30 cm.

Growing rice in pure stand is justified in this case because soil structure is suitable and land forming was carried out to control water supplies. This swampy land only produces rice crops and their trash, but yields are high because the management of water and the organic matter cycle are satisfactory.

Here is a pineapple plantation, also in pure stand **(figure 610)**. The suckers were planted in double rows to facilitate the passage of workers and animals. Although spacing is close, there is little competition because the plantlets have access to air and light by spreading sideways to the empty spaces between the double rows **(figure 611)**.

Sweet potato and tropical kudzu provide good soil cover between the pineapple plants.

The sweet potato stems creep along the ground but do not scramble over the pineapple, which gets enough light and heat to ensure proper fruiting. Sweet potato not only covers the soil but also keeps weeds in check.

By growing these plants in mixture, the farmer harvests the first crop (potato) after four to six months and does not have to wait sixteen to twenty-four months until the pineapple is ripe for harvesting.

The organic matter from the sweet potato and its residues retain soil moisture. When the trash rots, after the harvest, it will produce good manure for the pineapple.

Tropical kudzu is also an excellent cover plant and it enriches the soil like most legumes. But this plant tends to climb and, if the farmer does not flatten it down, it may scramble over the pineapple stems and hinder growth. When kudzu foliage invades pineapple, as in **figure 612**, it may cause disease and malformation, or simply overshade the fruit crop.

Observation leads us to conclude that, while tropical kudzu is most useful as a fertilizing cover plant, it should not be associated with low plants, such as pineapple, because it may interfere with them in the aerial environment. Tropical kudzu is much more beneficial in tree plantations, or as a fallow plant, because of its climbing shape and its capacity to overrun untilled land and control other plants with lower fertilizing potential.

Two layers of cultivation on the flat

This field of vigorous, erect maize and creeping potato **(figure 613)** was featured in Lesson 22. Here and there, peas form a halfway storey or layer of vegetation, but they do not show up in the photograph.

This field yields four kinds of agricultural produce:

- maize grain;
- tubers;
- legume seeds;
- abundant organic matter that will return to the soil and improve its structure.

The value of this kind of fourfold association from the point of view of nutrition will be discussed in Lesson 47.

The field in **figure 614** has been arranged in two layers. Millet forms the lower, seasonal layer. Mango and locust bean form the arborescent, perennial layer.

613

614

615

The millet growing near the mango trees (see the centre of the photograph) is retarded because mangoes are shade trees and dry up the soil but their fruit is a valuable food.

The arborescent layer in **figure 615** is composed of many tree species and will soon be sown with sorghum and cowpea. When the crops reach maturity, there will be three layers :

■ an arborescent layer composed of many trees yielding varied produce. *Acacia albida* produces wood, and animal feed during the dry season. The locust bean supplies seeds for human consumption. Shea butter fruit is eaten and the nuts are used to make butter. All these trees restore vegetable waste to the soil and fertilize it. Their roots penetrate much deeper than those of seasonal species ;

■ two herbaceous layers formed by sorghum and cowpea. Sorghum is an erect cereal plant. Cowpea, spreading or creeping, yields seed rich in proteins (Lesson 47).

In **figure 616**, coffee trees were spaced to leave room for seasonal crops like groundnut, pineapple and cocoyam. The farmer decided on wide spacing per hectare. As a result, the annual coffee yield per hectare will be less than if the trees had been more closely spaced but, in addition to coffee, the farmer will have other crops for family consumption or for sale.

Of course, if the farmer had not spaced out the trees properly in the first place, he would not have been able to cultivate light-demanding plants, e.g. groundnut and pineapple.

616

The coffee trees in the next illustration **(figure 617)** were planted close together. The cocoyam manages to thrive, whereas groundnut and pineapple are doing poorly.

617

618

It is obvious that this second farmer is interested in producing coffee, a cash crop for export, and only that. The first farmer is keen to produce food crops along with coffee for export.

Figures 618 and 619 present two different planting arrangements for cotton intercropped with maize. Maize in **figure 618** is widely spaced. It was planted one row in five at the same time as the cotton. The rows of maize are oriented east-west and follow the sun so that shade from the maize barely interferes with the cotton except along the maize row.

In the second example **(figure 619)**, maize was planted much closer. Cotton was planted on its own every second row, the other row being planted with maize or maize and cotton together.

The shrubby cassava dominates the mixed seedbed of maize and bean **(figure 620)**. The casssava has reached maturity. The maize/bean

619

620

seed rate is so high that it may be necessary to pull out some plants to prevent them from choking each other.

During the season, the midway storey now occupied by cassava will be taken over by erect maize stems, whereas the bean crop with spreading leaves will provide good soil cover and protection from the heavy rains.

Here is yet another spatial arrangement and another way of layering cultivated crops **(figure 621)**. Fruit trees were planted in rows 20 m or 30 m apart with a view to planting

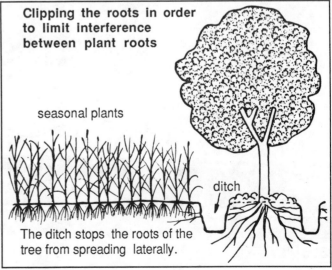

621

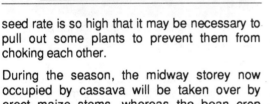

622 | **Pollarding trees (cutting back the top branches) in order to limit shading**

branches cut back

seasonal plants

Clipping the roots in order to limit interference between plant roots

seasonal plants

ditch

The ditch stops the roots of the tree from spreading laterally.

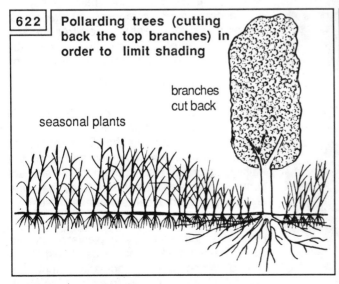

arborescent layer :
fan palm, shea butter,
locust bean

shrub layer :
cassava

herbaceous layer :
millet, earthpea

623

624

625

seasonal crops in between the rows. When the photograph was taken, beans were being cultivated mechanically. This is a good example of how well-arranged tree plantations can be managed with the mechanized cultivation of seasonal crops.

Many layers of trees, bushes and herbs

Many layers or storeys of vegetation are to be found in this field **(figure 623)** already mentioned in Lesson 14 :

■ herbaceous plants - millet and cowpea ;

■ shrubby plant - cassava ;

■ arborescent plants, mostly shea butter and fan palm.

Note the great variety of produce from this field and the way production is staggered (spread out in time) :

■ cereals (millet) and legumes (cowpea) ;

■ tubers (cassava) ;

■ fruit and oil seeds (shea butter) ;

■ caterpillars on the shea butter tree (Lesson 14) ;

■ fruit, matting and wood from the fan palm.

Note too that plants which last one cultural season and others with longer life cycles are growing together. Mixing plants with different life spans establishes permanent field occupation.

Some aspects of **figures 624 and 625** were discussed at the beginning of this lesson **(figures 598 and 599)**. **Figure 624** illustrates a careful plant arrangement. Tall trees growing on the edge of the plantation form the upper storey and act as a shelterbelt against the wind. In the background, an arborescent layer, composed of *Leucaena*, shades the coffee trees and fertilizes the soil. In the foreground, two layers can be distinguished - the shrubby layer provided by the coffee trees and the herbaceous layer consisting of *Paspalum*.

The layers were planned quite differently in the plantation in **figure 625**. The shrub-shaped coffee trees are the dominant feature growing all over the plantation and forming a halfway storey. The uppermost layers are composed of banana and fruit trees that were planted unsystematically, wherever it seemed best.

Here is a coffee and palm plantation **(figure 626)**. Palms and some large forest trees spared during clearing operations form the uppermost storey. Coffee trees form the shrub layer while the understorey is composed of herbaceous plants.

This arrangement was slow to establish. Palms were planted first and when they had reached a considerable height, coffee shrubs were added. If the two species had been planted at the same time, the palms would not have left enough room for the coffee trees to spread their branches in the normal way.

The planter's main task is to seed the right number of plantlets per species so that each of them, and especially those with produce fit for human and animal consumption, are able to enjoy the best growth conditions.

The layering of crops and the role of land forming

The layering of cultivated plants and their distribution in fields can be helped by land forming. A land form is the configuration or outline of the land after ploughing, earthing up, ridging or bedding.

Here are two examples of layered cultures on formed land (**figures 627 and 628**). Yam hills were formed to take a variety of yam that grows in full sun. The yam is climbing on props made from branches and therefore gets better exposure. The tubers thrive well in the loose soil of the mounds.

In the first example, maize, an erect plant, is growing in association with yam (**figure 629**). The maize seedlings have taken root at the foot of the mounds where there is more moisture.

In the second field, groundnut was sown on the sides of the yam hills (**figure 630**).

626

627

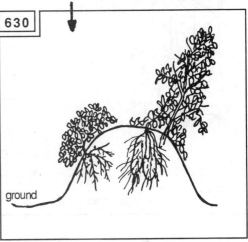

628

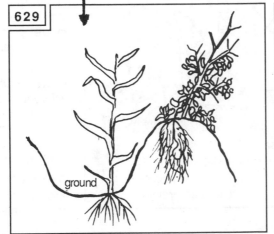

629

630

Again, there are two cultural layers - the staked yam and the spreading groundnut. However, in contrast to the maize in the drills, the groundnut was sown in the loose soil of the hills because groundnut pegs would never be able to penetrate the harder soil at the base of the mounds.

It is interesting to note that the yam stakes were placed on the top of the hills but just off-centre and at a slight angle to give more light to the groundnut underneath.

A way of improving this cropping pattern would be to sow more groundnut up the sides of the yam hills and maybe to till the furrows and cultivate a creeping plant that would give a third cultural layer.

Arranging field space into different cultural layers

This rice field in Casamance (Senegal) is dotted with palms and can be flooded in the rainy season **(figure 631)**. The soil has been ridged inside the squares bounded by low earth banks.

Near the rice paddy but on land that cannot be flooded, seasonal crops are cultivated under palms **(figure 632)**. Here, the farmer is establishing rows of bananas to divide the field into plots. The bananas will form a layer midway between the herbaceous plants and the storey of tall trees. Tillage is on the flat, not in ridges as in the previous case.

The wet valley bottom illustrated in **figures 229 and 230**, Lesson 21, is good for flooded rice. By forming the land, the planter was able to establish two cultural layers **(figure 633)**.

631   palm   rice paddy

632   palm   banana

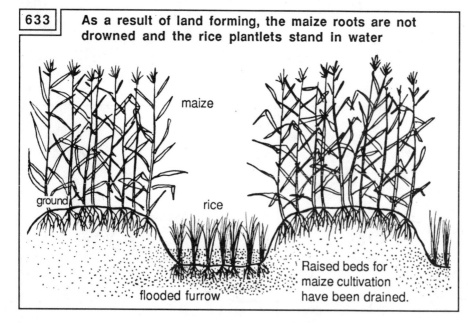

633   **As a result of land forming, the maize roots are not drowned and the rice plantlets stand in water**

maize

ground

rice

Raised beds for maize cultivation have been drained.

flooded furrow

The rice plantlets are standing in the flooded furrows between the raised beds. The maize is on dry soil but is in no danger of drought because of the water nearby.

In the same region, it is not unusual to find sweet potato growing between maize plants and forming a third cultural layer.

Field management, as illustrated in **figure 634**, is such that three different crops are produced. Groundnut was sown in four rows out of five with maize in the fifth row. Yam is planted under the tree on hills that are seen in detail in **figure 635**.

Figure 636 is another example of a field managed in three cultural layers - mango, maize and groundnut. Seed

634 maize groundnut maize

rate and plant distribution were carefully adjusted to suit each plant.

In this case, the seasonal crops will be phased out gradually as the mango trees take up more and more aerial space. As was mentioned earlier, mature mangoes deprive other plants of light and are not beneficial companions.

The sowing/planting arrangement of these fields **(figures 634 and 636)** is sketched below in **figures 637 and 638**.

The valley bottom **(figure 639)** is managed to grow flooded rice, the low-lying land being devoted exclusively to that plant. Maize and okra are cultivated in drier areas in the background. Dikes and some earth banks are wide enough to grow sweet potato, maize, chillies and other vegetables, care being taken that the roots of these plants do not burrow into the dam and lead to seepage.

635 shea butter yam hills

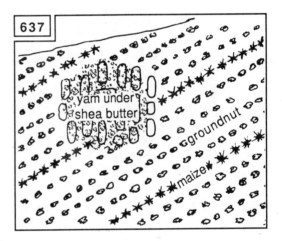

637 yam under shea butter groundnut maize

Figure 640 is a transect of this valley bottom (Lesson 10). It shows how plants are distributed with regard to the level of the plot and the degree of soil moisture. High ground is used for plants whose roots do not tolerate water, but the low, flooded ground is reserved for water-loving rice.

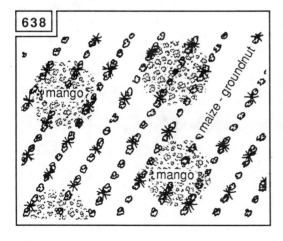

638 mango maize - groundnut mango

636 mango groundnut maize

This plot is poor in clay and is not flooded regularly. Rainfed rice and vegetables are grown here.

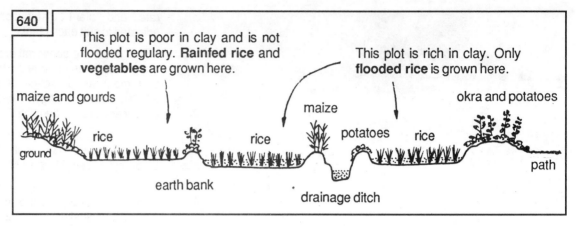

> 640

This plot is poor in clay and is not flooded regulary. **Rainfed rice** and **vegetables** are grown here.

This plot is rich in clay. Only **flooded rice** is grown here.

maize and gourds

okra and potatoes

maize

ground

rice

potatoes

rice

rice

path

earth bank

drainage ditch

> 639

We may conclude by saying that there are no hard and fast rules for plant distribution. It is certainly worthwhile trying out arrangements with associated crops and experimenting with cultural layers everywhere in the tropics. It is stressed that what is best for one particular farm depends on all the factors discussed at the beginning of this lesson.

Lessons 45 to 49

Agricultural returns

Many technical aspects of farming have been studied so far and it would seem that while some agricultural methods are efficient, others are inefficient or even dangerous. The efficiency or inefficiency of farming methods are gauged by comparing them with the results obtained in terms of the farmer's goals. He wants to produce food and guarantee a secure livelihood for his family, earn money, conserve and improve the resources of the farm and the settlement at large.

All farmers want to improve output gradually. But how can improved output be measured and how can a particular farmer compare his output with that of others ? How can he compare the yields from fields season after season ?

We shall try to answer these questions one by one under the following headings :

■ units of measurement and farm output (Lesson 45) ;

■ production and return (Lesson 46) ;

■ the value of the food we eat (Lesson 47) ;

■ measuring returns from fertilizers (Lesson 48) ;

■ returns from intercropping (Lesson 49).

Lesson 45

Units of measurement and farm output

Standard units of measurement are needed in order to determine farm output accurately. What are these basic units of linear, land, volume and weight measures ?

Linear measures

An easy way of measuring length is to count the number of feet or steps. This method is handy for rough calculations. But obviously some people have bigger feet and take longer strides than others, with the result that a distance of 50 steps taken by one person does not necessarily measure the same as 50 steps taken by another. Hence the advantage of choosing a unit measure used all over the world. This unit measure is the **metre**.

There are many units of linear measure based on the metre and obtained by simply dividing the metre into smaller units or multiplying it to obtain larger linear units. **Table 641** gives the linear measures based on the metre.

The people standing on the metre rule **(figure 642)** act as guide marks to help us visualize linear measures. The two men on the left are between 1.80 and 1.90 m tall. The step taken by the first man on the left, meaning the distance between his back right heel and his front left heel, comes to about one metre, . The short step taken by the second man measures about half a metre or 50 cm. The second child on the right is one metre tall.

People who do not have their own metre rule should make a careful note of the length of their foot or their step to be able to make fairly accurate measurements in metres. For example, if a man's step measures 0.70 m or 70 cm, then he knows that if he takes 20 steps, he covers a distance of 20 x 0.70 m or 14 m.

641	**Units of linear measure**

- by dividing one metre into 100 equal parts, we obtain 100 **centimetres** (cm) ;
- by dividing one metre into 1000 equal parts, we obtain 1000 **millimetres** (mm) ;
- by multiplying one metre by 10, we obtain a unit length of 1 **decametre** ;
- by multiplying one metre by 100, we obtain a unit length of 1 **hectometre** ;
- by multiplying one metre by 1000, we obtain a unit linear measure or distance of 1 **kilometre** (km).

The ruler on the righthand side of this page measures 25 cm. If we multiply it by four, we obtain a ruler one metre long.

642 1 m | 50 cm | 0.5 m | 1 m

Capacity measures

Here are various containers - two drums, a jerrican, a bottle and a glass. The capacity or volume of each one has been measured in litres.

A litre is the volume (space occupied) of a cube whose sides all measure 10 cm **(figure 645)**.

The contents of the containers illustrated in **figure 643** were measured accurately in litres. But most of the containers and vessels on the table **(figure 644)** have no standard volume or capacity. However, volume can be determined by filling up each vessel and counting the number of litres of water is contains. The one-litre bottle on the edge of the table on the left can be used as a unit measure. Ten one-litre bottles of water can be poured into the metal bucket on the table and 15 litres into the bucket on the floor.

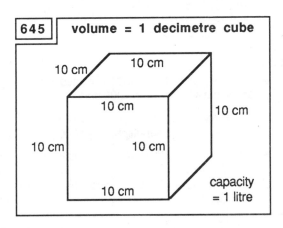

645	**volume = 1 decimetre cube**

10 cm · 10 cm · 10 cm · 10 cm · 10 cm · 10 cm · 10 cm

capacity = 1 litre

643

644

Weight measures

The unit measure used for weight is the **kilogram**. A litre of water weighs 1 kg, that is to say, the weight of the water contained in the one-litre cube illustrated in **figure 645** is 1 kg. However, items of the same volume do not necessarily have the same weight. For instance, a bag of sand is much heavier than a bag of the same size full of straw or groundnuts. A litre of sand weighs more than a kilogram ; a litre of groundnut or straw weighs less than a kilogram.

Figure 646 gives an idea of what everyday items weigh. The little girl is pointing to a packet of sugar weighing one kilogram. The one-litre bottle contains one kilogram of water, but the packet of salt, although smaller than the packet of sugar, also weighs a kilogram. On the right are some items of which the smallest

646

weighs 30 g. The tin of sardines weighs about 200 g, contents and tin together. The bucket containing 15 litres of water weighs 15 kg plus the weight of the metal.

The 200-litre drum in **figure 643** can contain 200 kg of water or 180 kg of oil. Oil weighs less than water. When oil and water are mixed, the oil rises to the surface.

Sometimes one needs units that are smaller or bigger than the kilogram to be able to weigh goods. The weight measures used for agricultural produce are given in **table 647**.

A final point about weights. Only the weight of contents has been given here ; the weight of the container was not taken into account. The weight of a container and/or wrapping is called the **tare**. Tare must be weighed separately and then deducted in order to determine the exact weight of the contents. For example, if a bag of groundnut weighs 30 kg and the tare (empty bag) weighs 400 g, the groundnut weighs 29.6 kg.

Land measures

Land measuring is essential in order to estimate farm production and compare outputs. The three most common units of measurement are the square metre, the are and the hectare.

647

- A **gram** (g) is the weight obtained by dividing a kilogram into 1000 parts.

- A **tonne** (t) is the weight obtained when a kilogram is multiplied by 1000.

- A **quintal** is the weight obtained when a kilogram is multiplied by 100. Farmers often weigh produce in quintals.

- A **square metre** (m²) is the surface enclosed in a square 1 m x 1 m. A shape with an area of a square metre may take the shape of a circle, a triangle, a rectangle or some other shape. A shape such as a rhombus or a triangle with equal sides of length one metre has an area less than one square metre.

- An **are** is the surface enclosed in a square 10 m x 10 m. The are **(a)**, is a surface equal to 100 m².

- A **hectare** is the surface enclosed in a square 100 m x 100 m. A hectare **(ha)** covers 100 ares, or 10 000 square metres (100 m x 100 m = 10 000 m²).

How to measure one are of land

The simplest way is to take a piece of rope or strong twine 35 metres long divided by three knots into four parts. Knots 1 and 2 must be 10 m apart, knots 2 and 3 are 14.15 m apart, knots 3 and 4 are 10 m apart. The square measuring 1 is marked out as follows.

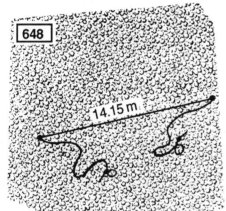

First step

The piece of twine in the middle measuring 14.15 m is fully stretched and staked down firmly through the knots at the both ends. The two end pieces of twine are left loose (**figure 648**).

Second step

When the middle piece of twine has been tied to the first two stakes, the two loose ends are stretched so that the end knots touch one another. A third stake is pushed into the ground where the knots meet (**figure 649**).

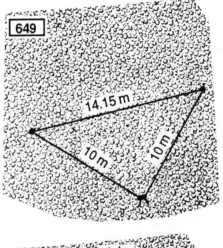

Third step

In order to locate the fourth corner of the square, leave the piece measuring 14.50 m where it is. Stretch the other two pieces to the opposite side so that the ends meet (**figure 650**) and place the fourth stake here. The twine can now be removed.

The surface bounded by the four pegs measures one are (**figure 651**).

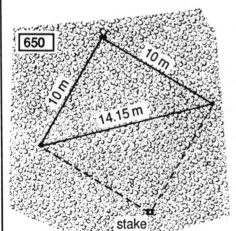

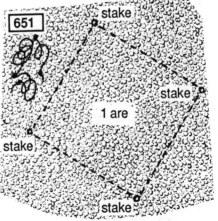

In West Africa, cotton planters often use a land measure called a cord. It is the equivalent of 0.50 hectares or 50 ares. The method described above is used to delimit the plots, but the middle piece of string is 100m long, the two side pieces being 70.7 m each.

Notes

All the linear, capacity, weight and land units mentioned make it possible to estimate agricultural returns. They are summarized in **table 652**.

| 652 | **Useful units of measurement** |

Linear measure

1 kilometre (km)	=	1000	m
1 hectometre (hm)	=	100	m
1 decametre (dm)	=	10	m
1 metre (m)	=	1	m
1 decimetre (dm)	=	$\frac{1}{10}$	m
1 centimetre (cm)	=	$\frac{1}{100}$	m
1 millimetre (mm)	=	$\frac{1}{1000}$	m

Land measure

1 hectare (ha)	=	10 000 m²
1 are (a)	=	100 m²
1 square metre (m²)	=	1 m²

Weight measure

1 tonne (t)	=	1000	kg
1 quintal (q)	=	100	kg
1 kilogram (kg)	=	1	kg
1 gram (g)	=	$\frac{1}{1000}$	kg

Capacity measure

1 hectolitre (hl)	=	100 l		1 centilitre (cl)	=	$\frac{1}{100}$ l
1 decilitre (dl)	=	$\frac{1}{10}$ l		1 millilitre (ml)	=	$\frac{1}{1000}$ l
1 litre (l)	=	1 l

Lesson 46
Production and return

| 653 | **Overall production on two farms** |

(a) Farm of 1.5 ha with two workers

Production

2	bags of beans
10	bags of groundnut
20	bags of maize
50	bunches of yam
20	baskets of oranges
100	pineapple
20	bunches of banana
3	baskets of okra

(b) Farm of 4 ha with five workers

Production

2	bags of beans
10	bags of groundnut
20	bags of maize
50	bunches of yam
20	baskets of oranges
100	pineapple
20	bunches of bananas
3	baskets of okra

The efficiency of farming systems can only be estimated when three basic aspects of farm management have been weighed up. They are production, return and cash income.

The farmer is interested in the overall production of his farm, that is all the products he obtains. Secondly, **he is interested in the return from production factors** (Lesson 2), that is the production from every unit - land, work, water, head of cattle and money. Thirdly, **he is interested in the cash income** earned by his work on the farm, that is the money available after the sale of farm produce.

Overall farm production

The produce from a farm can be itemized in many ways. For instance, a farm has produced 10 bags of groundnut, 20 cans of millet, 3 granaries of sorghum, 20 bunches of yam, 25 baskets of fruit, 15 bunches of bananas, and so on.

Farmers describing their produce in this way know exactly how much food is available for family consumption. They can also calculate the cash income that can be earned by selling this produce, by multiplyiung the sale price of each item by the quantities produced.

However, when farm yields are expressed in such terms, there is no way of finding out whether farm management is efficient or not. The amount of land cultivated, the number of farm workers and the money spent are not mentioned. The production described above may relate to a small farm occupied by two people, or to a large farm with five people. The overall production of two or more farms are in fact hard to compare. **Table 653** describes production on two farms where overall production was the same but production factors differed in quantity.

By comparing the yields obtained on the two farms, we see that every worker on the second farm (b) produced less than every worker on the first farm (a). In concrete terms, a worker on farm (b) produced four bags of maize, but his counterpart on farm (a) produced ten bags. The second farm is obviously less efficient than on the first farm.

The same remark can be made about land use. On farm (b), a hectare of land produced an average of five bags of maize per hectare (20 bags per 4 ha), but on farm (a), yield reached just over 13 bags per ha (20 bags per 1.5 ha).

The table demonstrates that production figures are not enough in themselves to describe the efficiency of farm management on a given holding.

Return from production factors

When speaking of farming, return is an idea that helps determine the efficiency of farming systems and particularly the efficiency of production factors. The idea allows us to make comparisons and such statements as :

■ That plot produced more beans this year than last year ;

■ Plot 1 was less productive than Plot 2, yet both are the same size.

Comparisons of this kind are especially important when farmers want to compare the results of innovations with the results of usual farming practices.

This idea of return from production factors is set out in detail in **table 654.**

In conclusion, if we want to find out how efficient farming is, we must first establish the efficiency of every production factor taken separately. After that, we proceed to work out the comprehensive efficiency of the combined production factors. We need to find out the extent to which farming goals are achieved, the goals of family self-sufficiency, cash income for consumer goods and social spending, security and the conservation of farm and settlement resources.

These are the questions that should be asked to get a clear picture of the efficiency of a farming system :

■ Is return from the land satisfactory ? Can it be improved ?

■ Is return from labour satisfactory ? Can it be improved ?

■ Is return from money acceptable ? Is expenditure justified ? Is money wasted ? Is the money available used to boost production or is it used solely on consumer items or for social spending ?

■ Are water resources used properly ? What about rainwater and irrigation water ?

■ Is return from livestock satisfactory ? Is the security it offers maximized ? Is livestock used to the best advantage to improve agricultural methods and production ?

Bearing in mind the way in which the production factors are used and their degree of efficiency :

■ What about family self-sufficiency ? Is it satisfactory ?

■ Is cash income on the farm sufficient ?

■ Does the farm guarantee a secure livelihood for everyone living on the farm, workers and families alike ?

■ Is conservation of all the resources on the settlement, especially its soil, actively pursued ? Are these resources improved?

The answers to these questions not only give an accurate picture of farm and settlement economy. They may also uncover situations like these :

■ On a particular farm, return from the land is high, but return from labour is poor because there are many farm workers.

■ Perhaps the opposite is true. Return from the land is bad but farm workers are scarce and return from labour is high.

■ Sometimes, return from the land is excellent but return from money is disappointing because production costs are high.

- Perhaps return from water resources is poor because water is badly used and wasted, but at the same time return from the land and from labour is good. If water has to be paid for and is not used efficiently, this production factor will cost a lot, maybe to no avail, in which case, return from money is also bad.

| **654** | **Return from production factors** |

Return from the land

Return from the land is **the production from an are or a hectare of land during a cultural season.** It is expressed like this :

- 4 cans of sorghum per are ;
- 10 bags of unshelled groundnut per hectare.

Return from the land is calculated like this :

$$\text{Return per are} = \frac{\text{number of kilograms produced}}{\text{number of ares cultivated}}$$

$$\text{Return per hectare} = \frac{\text{number of kilograms produced}}{\text{number of hectares cultivated}}$$

If three hectares of land produce 1500 kg cereal, the return or yield per hectare is

$$\frac{1500 \text{ kg}}{3 \text{ ha}} = 500 \text{ kg per ha.}$$

Return from money

Return from money is **the amount of money earned as a result of spending a certain sum.** For example, a farmer spends 10 000 francs on a field. He sells the produce for 25 000 francs making a profit of 15 000 francs.

For every franc spent, one franc was paid back plus 1.5 francs profit. Return on the money is said to be 2.5 francs per franc spent. The concepts return and profit must be kept separate.

Return from livestock

Return on livestock is expressed in **the annual production per animal.** For example, a cow yields 1200 litres of milk per year. 50 dairy cows yield 58 000 litres of milk per year with a return of 1160 litres per cow per year ($\frac{58\,000 \text{ litres}}{50 \text{ cows}}$).

Return from labour

Return from labour is **the production obtained from a day's work.** For example, by working 20 days in a field during a cultural season, a man harvested 10 cans of cereal. Labour return is 10 cans divided by 20 days' work = 0.5 cans per workday per season.

The idea return for work is very important in farming because it shows how much the worker is paid for his effort. Two examples illustrate this.

Because of drought and pest attacks, one worker only harvested two bags of millet as a result of all his labour during the season. This means that he has only two bags of millet with which to feed his family for a whole year. Obviously this is far too little.

Another worker was luckier and harvested 20 bags of millet at the end of the season. He and his family have more than enough millet for the year and he can even sell some produce. If the farmer worked 50 days during the year, labour return is 0.4 bags of millet per workday ($\frac{20 \text{ bags}}{50 \text{ days}}$). The return obtained by the first farmer is ten times less ($\frac{2 \text{ bags}}{50 \text{ days}} = 0.04$ bags per day).

If a bag of millet is worth 10 000 francs, the second farmer is said to earn 4000 francs per workday (0.4 bags x 10 000 francs per bag) whereas the first worker earns only 400 francs per day.

Return from water

When water is scarce and expensive, it is essential to put every drop to the best possible use. **Return from water is expressed by the quantity of produce obtained from a given quantity of water used for irrigating plants.**

Return is always the quantity of produce obtained for every unit of a production factor spent. Return is calculated separately for every production factor. Here again, units of measurement must be fixed because without them no comparisons are possible (Lesson 45).

The improvement of efficiency standards on farms always begins with the production factor which is the greatest obstacle to farm work. For example, the farmer with only a little land will do everything in his power to improve return from the land. The farmer who is short of labour will be particularly concerned to improve labour return. When water is scarce and expensive, the first task will be to improve water efficiency.

How to estimate returns from cultivated fields

Measuring these returns allows the farmer to make two sets of comparisons.

- **He can compare the performance of many fields during the same season** - fields situated in different parts of the farm, fields where different cultural methods were put into practice, fields with seeds of different varieties, etc. He can judge the efficiency of experiments, with a view to modernizing cultural methods. He can compare his performance with those of other local farmers.

- **He can compare the yield from the same field season after season.** He can see for himself improvements or deterioration resulting from his cultural methods.

Comparisons between seasonal returns are only valid if units of measurement are comparable.

How to compare returns from fields in the same settlement

In the map of the hamlet **(figure 655)**, dwellings, trees, hedges and field boundaries are marked.

Because fields vary in size, it is not easy to compare returns without fixing a unit of land measure. It was therefore decided to estimate return per are in four fields, and four square plots measuring one are were established at A, B, C and D. All the seasonal or yearly production of all the crops growing on these four one-are squares will be weighed at harvest time. Nothing may leave the field before or after the harvest without being noted.

If the four plots were planted with sorghum, returns will be given as in **table 656**.

But if the total area and production of the four fields are taken into account, the returns will be estimated as in **table 657**.

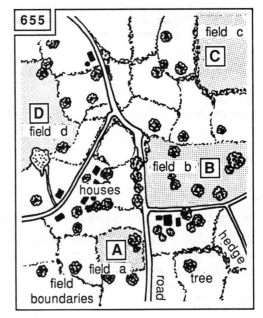

656	size of plot	quantity of sorghum harvested	return per are
plot A	1 are	5.40 kg	5.40 kg per are
plot B	1 are	2.60 kg	2.60 kg per are
plot C	1 are	1.90 kg	1.90 kg per are
plot D	1 are	6.80 kg	6.80 kg per are

657	size of whole field	total production of field	return from the field
field a	5 ares	27 kg	5.40 kg per are
field b	18 ares	47 kg	2.60 kg per are
field c	60 ares	114 kg	1.90 kg per are
field d	38 ares	258 kg	6.80 kg per are

The return for each field is calculated by dividing the total production in kilograms by field area in ares. For example, in field b, return comes to 47 kg for a total area of 18 ares, or 47 kg divided by 18 ares = 2.60 kg per are.

Table 657 reveals that the return from field d is higher than those of the other three fields. Field c has the poorest return. The third column shows that while field b produces more in total quantity than field a, return is lower. This table brings out the vast difference between production and return.

Returns in quantity and quality from intercropped fields

Where plants are grown in mixture, many different crops are harvested. The way to calculate returns from the field is the same as explained above, but it is more difficult to make comparisons with other fields.

A field on Lake Kivu on the borders of Zaire and Rwanda will serve as an example. This is a mountainous region where the equatorial climate is favourably affected by altitude. Rains are abundant and regular throughout the year.

Five crops were growing in mixture when the field was visited - maize, sorghum, cassava, sweet potato and bean **(figure 658)**. Returns from this field were calculated by weighing the produce as it was harvested during one cultural season. Here are the results for an area of one hectare **(figure 659)**.

The returns for the five associated crops total 4613 kg of varied foodstuffs. But all the products harvested are not of uniform quality. Some are dry, others contain a lot of water. Some are nutritious, others are filling but not nourishing. The qualitative aspects of crop returns will be discussed in Lesson 49. A distinction is made in farming between **quantitative return** and **qualitative return**. Quantitative return gives the amount of produce obtained from the land or by work effort. Qualitative return indicates the quality of the produce. The difference between the two is easy to understand. What is the good of a large rice crop if the husks prove empty ? What is the point of harvesting large quantities of cassava tubers if they are diseased and empty ?

Quality varies just like quantity. Groundnut is more nourishing than cassava, sorghum seeds are more nutritious than sweet potato tubers and plantains.

659	crops harvested in the field	yield in kilograms per hectare
	maize/dried grain	86 kg per ha
	sorghum/dried grain	565 kg per ha
	cassava/fresh roots	1157 kg per ha
	sweet potato/fresh roots	2428 kg per ha
	bean/pulses	377 kg per ha
	Total weight/foodstuffs harvested	4613 kg per ha

658

The list in **table 659** is not complete. The yields from five crops are given but some aspects of returns were left out. They are :

- vegetables such as cassava leaves ;
- haulms (stalks and stems) and straw that were used to feed livestock and poultry ;
- straw used to make mats ;
- fibres and climbers that grow wild in the field ;
- trash that is restored to the soil.

It is hard to weigh these products and value them in terms of money, but this is no reason for leaving them out when returns are considered.

Receipts and expenditure

When speaking of cash income on a farm, the question must be examined carefully from all angles before deciding that farm management is efficient. One farmer sells a lot of produce and gets poorer ; another sells little produce and gets richer. How can these seemingly contradictory situations be explained ?

Estimating increased wealth or **profitability** means calculating the difference between the money earned by the sale of produce, i.e. the **receipts** and the money spent to buy production factors, i.e. **expenditure**. The difference between the two is called the **balance** and is the sum available for the family budget **(table 660)**.

Already in Lesson 5, **figure 40**, the expenses incurred to cover family consumption, social obligations and production factors were set out separately. This distinction must be maintained in farm bookkeeping because the budget for social obligations and family consumption is

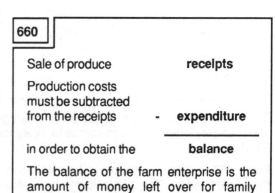

660

Sale of produce	**receipts**
Production costs must be subtracted from the receipts	- **expenditure**
in order to obtain the	**balance**

The balance of the farm enterprise is the amount of money left over for family consumption and social spending.

covered by the money contained in the balance of accounts.

Table 661 explains how to draw up a **balance sheet** summarizing the income and expenditure on a farm.

The sum of 127 500 francs is the real amount left over for family use. This balance shows how efficient the farm is from the monetary point of view. The money spent (42 500 francs) was multiplied by four to give receipts totalling 170 000 francs (42 500 francs x 4). This farm enterprise is **profitable**, meaning that farm operations brought in more money than they cost.

Table 662, however, is an example of an unprofitable farm enterprise. When the farmer sold his produce, he still had a debt of 10 000 francs. The balance is negative.

662		
total **production costs**	180 000 francs	
receipts	- 170 000 francs	
Still to be paid (balance due)	10 000	

On a farm producing food for family consumption and for sale, it is not so easy to draw up a balance sheet because produce eaten on the farm is not sold. However, the cash value of this produce is fixed at the sum it would have earned if it had been marketed. The quantity of produce is multiplied by the unit price as shown in **table 663**.

Table 663 gives the total value of produce for family consumption and for sale.

Farm expenditure, set out in **table 664**, must now be deducted from the total value of farm produce.

The balance is obtained by subtracting expenditure (production costs) from the total value of farm production. **Table 665** gives the total balance in hand for this farm. The accounts are very satisfactory because the balance in hand (740 000 francs) is more than seven times expenditure (100 000 francs).

The balance of the enterprise totals 740 000 francs but this sum is not available in cash since large quantities of produce are for family consumption. Sales totalled 224 000 francs, so cash in hand to cover consumer and social expenses totals 124 000 francs as calculated in **table 666**. But these 124 000 francs are only part of the total balance of the enterprise amounting to 740 000 francs **(table 665)**.

661

Sales	Receipts
10 bags millet at 11 000 francs	110 000 francs
6 bags beans at 9000 francs	54 000 francs
12 baskets fruit at 500 francs	- 6 000 francs
total receipts	170 000 francs

In order to obtain the produce sold, **expenditure on production factors came to :**

purchase of seeds	9 000 francs
hired labour	22 000 francs
hire of plough	7 000 francs
purchase of bags	3 000 francs
farm tools	1 500 francs
total production expenses	42 500 francs

Balance

receipts total	170 000 francs
expenditure	- 42 500 francs
balance in hand	127 500 francs

663

Farm production is as follows :

produce for family consumption

20 bags millet at 11 000 francs per bag	220 000 francs
9 bags beans at 9 000 francs per bag	81 000 francs
10 bags sorghum at 10 000 francs per bag	100 000 francs
8 baskets fruit at 5 000 francs per basket	40 000 francs
5 sheep at 35 000 francs each	175 000 francs
value of family consumption	616 000 francs

produce sold

10 bags millet at 11 000 francs per bag	110 000 francs
6 bags bean at 9 000 francs per bag	54 000 francs
12 baskets fruit at 5 000 francs per basket	60 000 francs
value of produce sold	224 000 francs

TOTAL

value of produce for family consumption	616 000 francs
value of produce sold	224 000 francs
total value of farm produce	840 000 francs

664

purchase of seeds	27 000 francs
hired labour	22 000 francs
hire of plough	19 000 francs
purchase of bags	5 000 francs
making granaries	5 000 francs
transport	4 000 francs
tools	3 000 francs
fertilizers	15 000 francs
farm expenditure	100 000 francs

665

total value of farm production	840 000 francs
expenditure (production costs)	- 100 000 francs
balance in hand	740 000

666

value of sales		224 000 francs
spent in cash	-	100 000 francs
cash in hand		124 000 francs

Economics is a complicated science that often uses different words to express the same idea.

Here are words meaning receipt : **gross production, gross income.**

These words mean production expenditure : **production costs, outlay, operation costs, enterprise expenditure.**

The balance is the difference between receipts and expenditure. It is expressed in various ways including **net margin, profit and loss account, work results, farm profits, farm accounts.**

Notes

This lesson on returns might be summed up by a go-ahead farmer in these words :

- High production levels are not necessarily a sign of efficient farming.

- I cannot say I am efficient simply because my cereal crops are good or because I market much produce.

However, the same farmer is certainly efficient if he can honestly claim that he obtains a good return from every production factor and that produce yields are high. He could then argue on these lines :

- Farm production is high and return from the land satisfactory. My family and workmen are well fed. I get a good price for my produce...therefore I am an efficient farmer.

- My cereal crops, in fact all my crops, are healthy and productive. The rate of return from the land is good, overall production is satisfactory...therefore I am an efficient farmer.

- I have plenty of farm produce and market a lot of it. Expenditure is low compared to income, return from the land is good. I maintain and improve soil fertility...therefore I am an efficient farmer.

Efficiency is not appraised by looking at an isolated aspect of farming and one set of results. It must be examined from all angles.

Lesson 47

🔲🔲🔲🔲🔲🔲🔲🔲🔲🔲🔲🔲🔲🔲🔲🔲🔲🔲🔲🔲🔲🔲🔲🔲🔲🔲🔲🔲🔲

The value of the food we eat

We know from experience that proper nourishment means eating a varied diet. In the same way, many ingredients go into the making of a good meal.

Tao is a nourishing meal served in Burkina Faso **(figure 667)** and consists of :

■ a ball of millet ;

■ a relish composed of chicken, oil and tomato paste ;

■ sombara made from locust bean seeds and okra ;

■ salt.

N'dolé is a typical dish in Cameroon. It is a relish made from herbs, onions, toasted groundnut, shrimps, fish and salt. It is served with tubers (cocoyam, cassava, sweet potato) or with plantains.

667

Why is one food different from another ?

669	**Quantity of proteins contained in some everyday foods**

cassava tubers	0.9%
carrot, lettuce, tomato	1.0%
juicy fruits (mango, pawpaw, orange)	1.0%
sugar cane	1.0%
sweet potato	1.1%
banana, okra fruit	1.5%
tubers - taro, yam, potato	2.0%
avocado	2.0%
cocoyam - tubers and leaves	2.5%
eggplant	4.0%
leaves of black nightshade and bitter leaf	5.5%
native pear fruit	7.0%
cassava leaves	7.5%
rice	7.5%
shea butter fruit	8.5%
maize	9.0%
millet and sorghum	10.0%
fonio, finger millet	11.0%
wheat	12.0%
earthpea pulses	18.0%
sesame	21.0%
pea and cowpea pulses	22.0%
bean pulses	24.0%
dried gourd and cucumber seeds	24.0%
dried groundnut seeds	32.0%
dried soya beans	40.0%

1000 kg of millet contain 100 kg of protein.
1000 kg of bananas only contain 15 kg of protein.

Just as plants are formed of many parts (Lesson 23), so the foods we eat are composed of many substances, each of which has a role to play in human nutrition. The substances called **proteins** build muscles and tissues. Sugars and starches called **carbohydrates** give the body energy (strength). So do oils and fats called **lipids**. The main function of **mineral salts** and **vitamins** is to build tissues and keep the body healthy. Lastly, all foods contain water to a greater or lesser extent.

People must eat all these substances in a wide range of foods in order to stay healthy. These elements combined contribute towards strength and good health. This is why relishes, vegetables and meat are very valuable when served with staples like couscous and tubers.

Foods and their characteristics are described in **table 668**, page 280.

Food tables

Food composition tables have been drawn up by nutritionists, scientists specializing in the composition of foodstuffs. The tables indicate the percentage of water, proteins, carbohydrates, lipids, mineral salts, vitamins and other components contained in each product.

These detailed tables can be consulted in scientific publications, but for our purpose we shall concentrate on the quantity of proteins in foods when they are harvested. The foods are listed in the order of protein content starting at the bottom of the scale **(table 669)**.

668 The value of the food we eat

Cereals

Millet, sorghum, maize, fonio (acha), finger millet, rice and wheat are some of the cereals in this food category.

These foods contain little water but much starch and some proteins. Minerals and sometimes vitamins are present in the thin outer layers of the grain. This is why flour milled from whole grain (grain only partly hulled) is always much more nourishing.

Cereal foods impart energy. They are body builders thanks to the proteins and mineral salts they contain.

Seeds

Seeds are foods with little water. They are often rich in proteins. They provide all the elements needed to build and strengthen the human body. Seeds make a full meal when served with fruit and vegetables that bring extra mineral salts and vitamins.

There are two kinds of seeds, those rich in oil called **oilseeds**, and those with low-oil content.

Oilseed plants include cotton, shea butter, castor, safflower and sesame. Some oilseeds like soya bean and groundnut are also an important source of protein. The oil is usually contained in the seeds themselves as their name implies, but oil can also be extracted from certain fruits like the fruit of the oil palm, the avocado and the native pear.

Legume seeds like bean, pea, cowpea and earthpea contain little oil but have a high protein content. As such, they are valuable foods.

Other seeds like cucumber also contain sizeable quantities of protein.

Seeds rich in protein are body builders and an essential addition to starchy foods that provide energy.

Miscellaneous foods derived from plants

Foods of vegetable origin like mushrooms, condiments and algae fall into this category.

Relishes and spices are mostly used to make food palatable. Chillies, ginger, basil, pepper belong to the relish family.

Other plants are used to make medicines, scents (for example, *Citronella* grass) and drinks like tea and coffee.

Starchy foods

Starchy foods are full of water. When the flesh is dried, it can be ground into flour or made into little round pellets containing mostly starch.

Many starchy foods form part of our everyday diet, for instance :

- cassava, cocoyam, taro, yam, sweet potato, potato and other plants with edible tubers ;
- sweet or plantain bananas ;
- fruits of the breadfruit and the baobab.

Starchy foods supply energy but do not build muscle or tissue.

Starchy foods should always be served with others rich in protein, vitamins and minerals.

Fruit and vegetables

Fruit and vegetables are usually full of water. Yet they are an important element in human diet because they provide health-giving mineral salts and vitamins. Added to starchy foods or cereals, or indeed to any kind of meal, they improve the quality of the food especially when eaten raw.

Well-known fruits are mango, lemon, orange, grapefruit, pawpaw, soursop, pineapple, passion fruit, giant granadilla, tomato, eggplant and gourd. Some are eaten raw, others are cooked and added to relishes.

Common vegetables are cassava, cocoyam, bitter leaf, lettuce, black nightshade, onion, cabbage, palm cabbage, okra, carrot and baobab.

A vegetable-rich diet usually supplies enough **fibre** to ensure that food passes through the digestive system smoothly and quickly. Sometimes, in towns, people adopt a European diet and may suffer from a shortage of fibre.

Fruit and vegetables enhance the value of food. **By supplying vitamins and mineral salts, they provide protection against disease.**

Foods of animal origin

Products of animal origin are always rich in proteins and fats and therefore supplement staples of plant origin.

These foods include meat, fish, eggs, milk, cheese, insects, worms, snails, frogs, snakes, shellfish, etc.

These foods of animal origin are essential body builders, especially for young children. Small quantities keep the body in good health.

Improving diet by growing a variety of plants

At the beginning of this book, we visited typical cropping fields. We noticed how farmers grew plants in mixture and ensured a varied supply of nourishing foods for their families. Common plant associations are photographed in **figures 670, 671 and 672**.

starchy food (banana)

cereal (maize)

starchy food (cocoyam)

legume (bean)

672

A cereal (maize) was interplanted with a tuber (cocoyam) and a vegetable (amaranth), **(figure 670)**.

On the mound in **figure 671**, a tuber (yam) is growing alongside a fruit used as a vegetable (eggplant).

This field **(figure 672)** will supply ingredients for rich, varied meals. There are starchy foods (bananas, cocoyam, potatoes, sweet potato), a cereal (maize), seeds full of protein (beans), many leafy vegetables and small fruits.

The **market gardener** has planted a variety of vegetables - tomatoes, okra, amaranth and eggplant **(figure 673)**. (Although, scientifically speaking, tomatoes and eggplants are fruits, people speak of them and use them as vegetables.)

Figure 674 shows a field of millet (cereal) interplanted with cowpea (seed rich in protein). The African locust bean in the foreground has edible seeds and the shea butter at the back produces fruit and nuts rich in oil.

cereal (maize)

leafy vegetables

starchy food (cocoyam)

670

673

fruit used as vegetable (eggplant)

starchy food (yam)

671

Thanks to all these crop combinations, the farmer is able to produce the whole range of foods that ensure a healthy diet.

Diversified foods are a source of health. Their production should be the aim of all farmers from the very moment when seedbeds are established.

Protein yields in intercropped fields

In the field described in Lesson 46, **figure 658**, **cereals** (maize and sorghum) were intercropped with tubers (cassava and sweet potato) and **legumes** (bean). This field is designed mainly to produce a variety of foods for family consumption.

The total yield from this one-hectare field came to 4613 kg of cereals, tubers and legumes combined. The products do not take the same form. Some are dry (maize, sorghum and

674

bean), others are moist (cassava and sweet potato) but all contain proteins. **A good way of judging farm efficiency is to calculate the amount of protein contained in plants for family consumption, since this is often the scarcest ingredient in the farmer's diet.**

Agricultural specialists went to the trouble of weighing accurately the harvested crops from six one-hectare fields and calculated the amount of protein contained in each crop. (Use **table 669** as a reference.) The results obtained were as follows .

- In the first field, crops were grown in mixture. The return came to 4613 kg per hectare. These 4613 kg contained 192 kg of protein.

- In the second field, bean was grown in pure stand. The return came to about 600 kg per hectare containing 144 kg of protein.

- Maize was monocropped in the third field. The return came to about 1500 kg per hectare containing 135 kg of protein.

- Sorghum, in pure stand in the fourth field, yielded 1200 kg per hectare containing 120 kg of protein.

- In the fifth field, cassava was also in pure stand and remained at least two seasons in the field. Yield for one season totalled 12 000 kg per hectare but, because cassava is low in protein, these 12 000 kg only contained 108 kg of protein.

- In the sixth field, monocropped sweet potato yielded 8500 kg per hectare for one season, with a protein content of 94 kg.

675

Good food is a condition for good health. Being well nourished entails three basic rules :

- eating enough, not too much ;

- eating a varied diet; i.e. eating a combination of foods that contain different health-giving qualities ;

- drinking clean water.

Children are in particular need of the three basic elements of good nutrition. Their development depends on the quality of their food and the proper intake of fruit and vegetables. They are also more easily harmed by dirty water. Their health and their future depend just as much on the water they drink as on a good, varied diet.

These figures prove that by growing crops in mixture, the farmer increased the quality of the foods he harvested since, by this method, he produced more protein, the element which is most likely to be lacking in the human diet.

A closer look at these results also leads to the conclusion that the farmer in question managed his land to the best possible advantage. He produced 4613 kg per hectare, whereas, if he had grown each crop in a separate plot, he would have needed an extra half hectare to produce the same quantity of varied foodstuffs. Needless to say, any farmer who can clear less land and still obtain the same amount of produce reaps the reward for such good management.

Growing crops in mixture improves the quality and diversity of food crops, economizes on land and reduces the toilsome work of land clearing.

Lesson 48

Measuring return from fertilizers

The method used to estimate returns from an are or a hectare of land can also be applied to test the effectiveness or return from fertilizers, that is to say, how much they produce in terms of farm produce and money.

The only way to test the value of fertilizer applications is to compare results obtained from several plots of the same size, say one are. Fertilizer is not spread on one plot while the other plots are spread with fertilizer in different quantities. For reliable results, the plots must be similar in many respects. If one plot is dry and the other wet, the comparison is not valid. Neither is it valid if one plot is planted with trees and the other not. Seeding and fertilizer application must take place simultaneously on the test plots. In short, the cultural situation must be identical throughout, the only difference being the amount of fertilizer spread. Two examples explain the operation step by step.

Return from fertilizer in a field of maize : the economic effectiveness

Five plots are compared.

Plot 1

No fertilizer was applied. The return was 9 kg of maize, sold for 75 francs per kg **(figure 676)**.

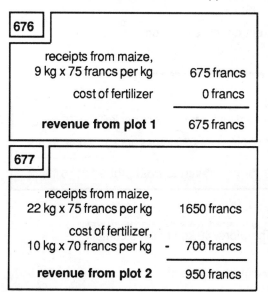

676	
receipts from maize, 9 kg x 75 francs per kg	675 francs
cost of fertilizer	0 francs
revenue from plot 1	675 francs

677	
receipts from maize, 22 kg x 75 francs per kg	1650 francs
cost of fertilizer, 10 kg x 70 francs per kg	- 700 francs
revenue from plot 2	950 francs

Plot 2

10 kg of fertilizer at 70 francs per kg were spread. The return was 22 kg. See calculations in **figure 677**.

How can the two plots be compared ?

■ Plot 2 yielded 13 kg more than plot 1. Fertilizer more than doubled the grain harvest.

■ After deducting the cost of the fertilizer, plot 2 produced a revenue of 950 francs, 275 francs more than plot 1 which only brought in 675 francs.

Fertilizer doubled the grain harvest but did not double cash income. Increased revenue came to 42%. In other words, fertilizer application increased revenue from plot 2 by less than half the income from Plot 1. However, all things considered, it was worthwhile spreading 10 kg of fertilizer on Plot 2.

However, fertilizer is like salt and sugar. It is good in itself but too much is a bad thing. What happens when fertilizer applications are increased ?

Plot 3

15 kg of fertilizer were spread here. The return was 29 kg of maize. See the calculations in **figure 678**. The revenue from this plot has climbed compared to the revenue from plots 1 and 2. What about plot 4 ?

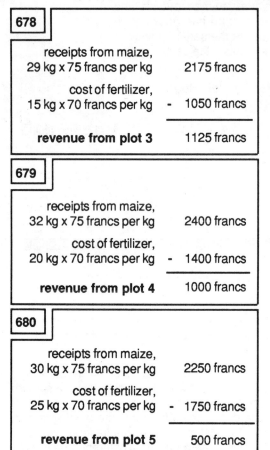

678

receipts from maize, 29 kg x 75 francs per kg	2175 francs
cost of fertilizer, 15 kg x 70 francs per kg	- 1050 francs
revenue from plot 3	1125 francs

679

receipts from maize, 32 kg x 75 francs per kg	2400 francs
cost of fertilizer, 20 kg x 70 francs per kg	- 1400 francs
revenue from plot 4	1000 francs

680

receipts from maize, 30 kg x 75 francs per kg	2250 francs
cost of fertilizer, 25 kg x 70 francs per kg	- 1750 francs
revenue from plot 5	500 francs

Plot 4

20 kg of fertilizer were spread with a return of 32 kg of maize. An extra 5 kg of fertilizer produced 3 kg more of maize, see **figure 679**.

What has happened ? Production has increased but revenue has dropped. The cost of fertilizer is growing faster than receipts from the maize crop. If the farmer wants to get return from his money, there is no point in adding the extra 5 kg of fertilizer. But perhaps it is worthwhile for the sake of the additonal food yield, as **table 684** shows.

Plot 5

25 kg of fertilizer were applied on this plot with a return of 30 kg of maize **(figure 5)** compared with 32 kg in plot 4, as is shown in **figure 680**.

Something has gone wrong. Extra fertilizer has lowered production with a loss of income. Plot 1 with no fertilizer produced only a third of what was produced on plot 5 with 25 kg of fertilizer. Yet, the total revenue from plot 5 is less than the revenue from plot 1.

Our measurements and calculations are instructive because they show how **fertilizers are only efficient (profitable) up to a certain point.** In this instance, it is useless applying more than 15 kg of fertilizer. Beyond that point, revenue from the plot declines.

- The first fertilizer applications are often very effective.
- Additional applications become less and less effective.
- Time comes when plants may be poisoned by too much fertilizer.
- When overdoses of mineral salts are applied near the roots, plants are unable to take them up and minerals are leached off by infiltrating water.

A word of warning : mineral fertilizers are beneficial when the amount is properly assessed, otherwise they drain away money. The effectiveness of using fertilizer must be reckoned accurately.

Measuring the return from fertilizer in terms of increased food production

Farming is not merely a question of money. Its first aim is to feed the family. The effects of fertilizer applications should therefore be reckoned in terms of increased food production in the same way as the yield in proteins is calculated for maize and cowpea growing in mixture (Lesson 49).

The method used to compare results from plots receiving fertilizer in different quantities can be applied to measure the food value of intercropped maize and tubers (cocoyam and taro).

Plot 6

No fertilizer was spread on plot 6 measuring one are. The crop return during the season came to

- 9 kg of maize grain,
- 48 kg of fresh tubers.

681

9 kg of maize x 75 francs per kg	675 francs
48 kg of tubers x 30 francs per kg	1440 francs
revenue from plot 6	2115 francs

Using the food composition table in Lesson 47, the protein content of these crops, i.e. the most nourishing element in these foods, can be calculated.

Maize grain contains 9% protein. 9 kg of grain contain 0.81 kg of protein (9 kg of maize x 9 kg of protein per 100 kg of maize or 9 kg of maize x 9%). Fresh tubers contain 2% protein. Therefore, 48 kg of tubers contain 0.96 kg of protein (48 kg of tubers x 2 kg of protein per 100 kg of tubers).

Protein production from plot 6 totalled 1.77 kg (0.81 kg of maize protein + 0.96 kg of tuber protein).

If the entire production is sold, the revenue from this plot comes to 2115 francs, as set out in **figure 681**.

Plot 7

20 kg of fertilizer were spread here at a cost of 70 francs per kg = 1400 francs. Harvested produce totalled :

- 12 kg of maize with a protein content of 1.08 kg (12 x 9%),
- 83 kg of tubers with a protein content of 1.66 kg (83 x 2%).

Total protein content = 2.74 kg.

By applying fertilizer, this plot produced 0.97 kg of protein more than plot 6 (2.74 kg/plot 7 - 1.77 kg/plot 6 = 0.97 kg). Revenue from plot 7 = 1990 francs **(figure 682)**.

Plot 8

Plot 8 received 40 kg of fertilizer, or 20 kg more than plot 7, at a cost of 2800 francs. Protein production was up as follows :

- 13 kg of maize containing 1.17 kg of protein (13 x 9%),
- 98 kg of tubers containing 1.96 kg protein (98 x 2%).

Total protein content = 3.13 kg. Revenue from plot 8 = 1115 francs **(figure 683)**.

Table 684 compares the difference between food production, protein production and cash income for plots 6, 7 and 8.

682

12 kg of maize x 75 francs per kg	900 francs
83 kg of tubers x 30 francs per kg	2490 francs
receipts from plot 7	3390 francs
cost of fertilizer, 20 kg x 70 francs per kg	- 1400 francs
revenue from plot 7	1990 francs

683

13 kg of maize x 75 francs per kg	975 francs
98 kg of tubers x 30 francs per kg	2940 francs
receipts from plot 8	3915 francs
cost of fertilizer, 40 kg x 70 francs per kg	- 2800 francs
revenue from plot 8	1115 francs

684

	Plot 6 no fertilizer cost : 0 francs	Plot 7 20 kg fertilizer cost : 1400 francs	Plot 8 40 kg fertilizer cost : 2800 francs
food production : maize	9.0 kg	12.0 kg	13.0 kg
tubers	48.0 kg	83.0 kg	98.0 kg
total production of proteins contained in these foods	1.77 kg	2.74 kg	3.13 kg
net cash income	2115 francs	1990 francs	1115 francs

What can be inferred from this table ?

- Fertilizer increased food production considerably, from 57 kg to 111 kg.
- There was also a sharp increase in protein production, from 1.77 kg on the plot without fertilizer to 3.13 kg on plot 8 that received 40 kg of fertilizer.
- Cash income declined from the top figure of 2115 francs on plot 5 to 1115 francs on plot 8.

The farmer is therefore faced with a choice : does he want fertilizer to produce more food or earn more money ? If food for the family comes first, then it is worthwhile spreading more fertilizer. If money takes priority, more fertilizer does not earn more money.

Let us take another look at the figures in **table 684.** Plot 6 was unfertilized and earned 2115 francs. Then revenue dropped to 1115 francs from plot 8 where 40 kg of fertilizer were spread. Loss of income = 1,000 francs, but food production increased, i.e. 13 kg - 9 kg = 4 kg of maize and 98 kg - 48 kg = 50 kg of tubers. This increased production is valued at 1800 francs (4 kg x 75 francs per kg = 300 francs for maize + 50 kg x 30 francs per kg = 1500 francs for tubers).

What has the farmer gained and what has he lost by applying 40 kg of fertilizer to plot 8 ?

- He has harvested more food valued at 1800 francs.

- He has earned 1000 francs less in cash.

- Net additonal income is 800 francs. But remember that this extra income takes the form of food, so that the 800 francs cannot be pocketed. However, if fertilizer had not been applied, food for the family would have cost 800 francs more.

Measuring return from fertilizer in terms of increased production of organic matter

Return from fertilizer application should take into account the increase in the amounts of organic matter restored to the soil.

Biomass, the total mass of living organisms in a particular area, is greater when fertilizers are spread, whether crops are in pure stand or in mixture. Fertilizer increases the production of leaves, stems and fruit returned to the soil in the form of organic matter which is essential for good soil structure.

Summing up

- Return from fertilizer must be measured in terms of farming goals.
  - □ Is food for the family the top priority,
  - □ or money,
  - □ or land conservation ?

- Fertilizer return must be calculated from all angles.
  - □ Has fertilizer application resulted in greater food production ?
  - □ Has fertilizer application increased cash income ?
  - □ Has fertilizer application increased the mass of organic matter restored to the soil ?
  - □ Has it improved soil structure ?
  - □ Has fertilizer application led to more animal feed or increased the amount of straw available for matting to make enclosures, roofing and so on ?

- Return from fertilizer depends to a large extent on the amount applied. First applications are always the most gainful. Production may decline if too much fertilizer is spread.

- Where intercropping is practised, fertilizer return must be examined in terms of all the associated plants, not just in terms of the dominant crop.

Lesson 49

Returns from intercropping

Figures in this lesson were compiled from observation of real fields but they cannot simply be applied to other fields. Every field must itself be measured. What is important is knowing how to make the relevant measurements and how to draw conclusions applicable to one's own farm.

Returns from intercropped maize and cowpea

Cereals and legumes are often intercropped, maize and cowpea being an example of this cereal-legume association.

The life cycle of maize is roughly of 120 days duration (four months). Cowpea, a bean, has a life cycle of 90 to 110 days. It is a spreading or climbing plant whereas maize is an erect plant.

In the fields for which returns are given, maize and cowpea seeds were sown on the same day at the beginning of the month of May, shortly after the first rains in April. The fields were well tended but no fertilizer was spread.

Three fields of the same size were established in order to ensure accurate measurements. Maize alone was sown in the first field, cowpea alone in the second, and a mixture of the two crops in the third. The yield from each field was carefully weighed. The results are shown in **figure 685**.

What can be inferred from this data ?

- **The are of intercropped maize and cowpea produced more grain than the are of maize or the are of cowpea in pure stand.** 56 kg of grain were harvested in the intercropped field as opposed to 29 kg of maize and 10 kg of cowpea growing in pure stand.

685	
field of maize	29 kg of maize per are
field of cowpea	10 kg of cowpea per are
field of intercropped maize and cowpea	49 kg of maize and 7 kg of cowpea per are, total 49 kg + 7 kg = 56 kg of food per are

- **The maize benefitted considerably from the presence of cowpea.** The yield came to 49 kg of maize when the crop grew in association, but only 29 kg when sown alone. Cowpea caused an improvement of 70% in maize output.

- **Cowpea itself did not benefit from the companion crop.** In fact, the yield dropped by 30% in the intercropped field but this loss was more than compensated by the gain in maize production.

These figures lead to the conclusion that, by associating maize and cowpea, the cultivator is actively pursuing farming goals since he produced a much greater quantity of grain for family consumption or for sale.

The quality and value of these products

- In 100 kg of maize, there is a little less than 10 kg of protein mixed with sugar and fat, or 9.0%,

- In 100 kg of cowpea seeds, there are almost 22 kg of protein, or 22% (refer back to **table 669**, Lesson 47).

The quantity of proteins produced on the plots can be calculated and are given in **table 686**.

What conclusions can be drawn ?

Maize and cowpea in mixture yield far more protein than the other two fields. On one are of mixed cropping, the protein yield is almost as high as from fields measuring 2.5 are where maize and cowpea are in pure stand.

686	
field of maize	2.61 kg of protein per are
field of cowpea	2.2 kg of protein per are
field of maize and cowpea in mixture	5.95 kg of protein per are

687	
field of maize	2175 francs per are
field of cowpea	1100 francs per are
field of maize and cowpea in mixture	4445 francs per are

This intercropping practice therefore enabled the farmer to increase the nutritive value of the food produced, that is to say, he improved the food quality and also the return from the land. In this instance, intercropping is a means of improving the household diet.

Cash income from each of the three fields **(table 687)** is calculated by multiplying the yield in kilograms by the price of maize grain (75 francs per kg) and the price of cowpea seeds (110 francs per kg).

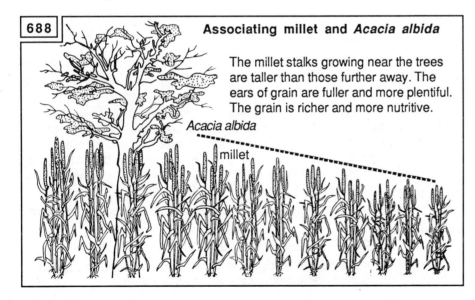

688 **Associating millet and *Acacia albida***

The millet stalks growing near the trees are taller than those further away. The ears of grain are fuller and more plentiful. The grain is richer and more nutritive.

Acacia albida

millet

Millet associated with *Acacia albida*

In earlier lessons, crop associations practised in various parts of Africa were mentioned.

Researchers in Bambey, Senegal, have carefully studied the yields obtained from a variety of millet thriving in association with *Acacia albida* and another variety not interplanted with *Acacia albida*. Their observations are illustrated in **figure 688**.

The results noted during field inspections were compared with measured yields. Millet yields per hectare were first calculated, and then their protein content.

The results in **table 689** show that

- the presence of *Acacia albida* more than doubled the production of millet (1700 kg ÷ 600 kg = 2.6),

- the quantity of protein tripled (179 kg ÷ 52 kg = 3.4). In other words, millet grain growing under *Acacia albida* is richer in protein than grain planted away from the tree.

***Acacia albida*, therefore, increases the return from the land in the form of a more abundant millet crop of greater nutritonal value.**

689	return from millet	protein quantity
hectare without *Acacia albida*	600 kg	52 kg
hectare planted with 60 *Acacia albida*	1700 kg	179 kg

Cash income obtained from the two plots is given in **table 690**, the price of millet being fixed at 80 francs per kg.

By associating *Acacia albida* and millet, the farmer earned a supplement of 83 200 francs, i.e. the difference between 136 000 francs and 52 800 francs. On top of this extra income, the trees supplied animal feed during the dry season.

Specialists also estimated the cost of the fertilizer that would have been needed on land without trees in order to produce 1700 kg of millet, the same yield as that obtained from the hectare with *Acacia albida*. Fertilizer would have cost at least 114 000 francs. Receipts and expenditure for a one-hectare field are given in **table 691**.

The balance in hand, 22 000 francs, must be compared with the 136 000 francs in hand from a hectare planted with *Acacia albida* and also with the 52 800 francs obtained from the unfertilized field without *Acacia albida*.

- It is no exaggeration to say that leaving *Acacia albida* in fields or, better still, planting the tree systematically, is an effective way of improving millet harvests and cash income from sales. It is also a cultural method within the reach of every farmer ;

- using bought fertilizers to make the soil as fertile as it is under *Acacia albida* would be an expensive,

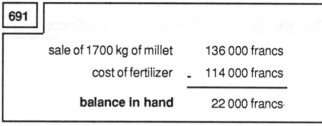

690	
hectare without *Acacia albida*	52 800 francs
hectare planted with 60 *Acacia albida*	136 000 francs

691	
sale of 1700 kg of millet	136 000 francs
cost of fertilizer	- 114 000 francs
balance in hand	22 000 francs

Reminder

Measuring the return from the land is a many-sided problem. It includes assessment of production to meet consumer demand and the extent to which land resources are protected and improved.

Where plant associations are known to benefit the land, increased yields of consumer produce may be expected. These beneficial effects influence the quantity of produce harvested and its nutritional value.

Farming systems that associate different species improve family diet and provide a good level of cash income.

Growing crops in mixture also allows cultivators in certain cases to practise mixed farming (crops and livestock) provided associated plants supply animal feed.

In order to make a complete survey of the value of plant associations, all kinds of cultural methods still remain to be observed, tested and appraised so that the farmers themselves discover what is feasible and profitable for their own farms.

But let us be wary of general statements !

What is true of millet associated with *Acacia albida* is not necessarily true of other plants. Experiments have shown that, when groundnut is sown under *Acacia albida*, it produces more leaves and fewer pods. If manures suitable for groundnut under *Acacia albida* were spread, perhaps groundnut would increase its production of seeds. Here is an opportunity for observation and experiment wide open to enterprising farmers who intercrop groundnut and *Acacia albida*.

After examining the figures above, farmers might conclude that they should not buy artificial fertilizers, but this would be a hasty decision.

We have seen that there are many ways of fertilizing the soil. Some solutions are costly, others are free. The free solutions offered by certain plants should be tried first and supplemented by more expensive solutions where necessary, but here again, farmers must decide for themselves whether they are adopting the solutions best attuned to their farming goals.

Where farmers dispose of *Acacia albida*, they could experiment by :

- planting *Acacia albida* at regular intervals throughout the field ;

- nurturing and tending the trees. A little fertilizer spread at the start of the planting season will activate growth ;

- when the trees have begun to improve the soil, adding the fertilizers best suited to whatever crops are being cultivated.

Bibliography

The books listed below are mere indicators to the vast literature on tropical agriculture.

Introduction to farming practices :

NGUGI, D.N., KARAU, P.K., NGUYO, W. (1978). W. *East African Agriculture - A Textbook for Secondary Schools*, Macmillan Publishers Ltd, London and Basingstoke, United Kingdom.

Samaka Service Centre (1962), *The Samaka Guide to Homesite Farming*, Manila, Philippines.

SOLTNER, D. *Les bases de la production agricole*, Volumes 1 and 2 - *Les grandes productions végétales - L'arbre et la haie - Planter des haies, brise-vent, bandes boisées* - Collection Sciences et techniques agricoles, Le Clos Lorelle, Sainte Gemme sur Loire, 49000 Angers, France.

On cultivated plants and their characteristics :

AUTRIQUE, A. (1981). *Principaux ennemis des cultures de la région des grands lacs d'Afrique Centrale*, I.S.A.B.U. (Institut des sciences agronomiques du Burundi, B.P. 795, Bujumbura, Burundi, and A.G.C.D., 5 Place du Champ de Mars, 1050 Brussels, Belgium).

BOHLEN, E. (1978). *Crop Pests in Tanzania and Their Control*, Paul Parey, Hamburg, West Germany.

COLLINGWOOD, E.F., BOURDOUXHE L., DEFRANCQ, M. *Les principaux ennemis des cultures maraîchères*, A.G.C.D. and Centre pour le Développement de l'Horticulture, B.P. 154, Dakar, Senegal.

GAUDY, M. (1959). *Manuel d'agriculture tropicale*, La Maison Rustique, 26 rue Jacob, Paris, France.

IRVINE, F.R. (1979). *West African Crops*, Oxford University Press, Great Britain.

LEAKEY, C.L.A. and WILLS, J.B. (1977). *Food Crops of the Lowland Tropics*, Oxford University Press, Oxford, Great Britain.

NAIR, P.K.R. (1980), *Agroforestry Species, A Crop Sheets Manual*, International Council for Research in Agroforestry (ICRAF), P.O.Box 30677, Nairobi, Kenya.

PURSEGLOVE, J.W. (1968). *Tropical Crops - Dicotyledons*, Longman Group Limited, London, United Kingdom.

PURSEGLOVE, J.W. (1972). *Tropical Crops - Monocotyledons*, Longman Group Limited, Harlow, United Kingdom.

STEINER, Kurt G. (1984). *Intercropping in Tropical Smallholder Agriculture with Special Reference to West Africa*, GTZ (Deutsche Gesellshaft für Technische Zusammenarbeit, Postfach 5180, D-62336, Eshborn, West Germany).

STOLL, Gaby (1986). *Natural Crop Protection*, AGRECOL Okozentrum, CH-4438 Langenbruck, Switzerland.

On rural economy and the environment :

CHAMBERS, Robert (1983). *Rural Development : Putting the Last First*, Longmans, London, United Kingdom.

de RAVIGNAN, F., BARBEDETTE, L. *Découvrir une agriculture vivrière*, Maisonneuve Larose, 15 rue Victor Cousin, Paris, France.

DUPRIEZ, H. (1980). *Paysans d'Afrique Noire*, TERRES ET VIE, 13 rue Laurent Delvaux, 1400 Nivelles, Belgium.

I.P.D. *Comprendre une économie rurale*, L'Harmattan, 7 rue de l'Ecole Polytechnique, 75005 Paris, France.

PICLET, G. *Notions d'économie générale et d'économlie rurale à l'usage des éducateurs et éducatrices en milieu rural d'Afrique tropicale*, F.A.O., via delle Terme di Caracalle, OO100, Rome, Italy.

RICHARDS, Paul (1985). *Indigenous Agricultural Revolution - Ecology and food production in West Africa*, Hutchinson, London, United Kingdom.

290

Data from the following books was used :

Lesson 15   PELISSIER, P. (1966). *Paysans du Sénégal*, C.N.R.S., Paris, France.

Lesson 21   ROOSE, E. (1973). *Dix-sept années de mesures expérimentales de l'érosion et du ruissellement sur un sol ferrallitique sableux de basse Côte d'Ivoire*, ORSTOM, Abidjan, Ivory Coast.

Lesson 26   DE SCHLIPPE, P. (1956). *Shifting Cultivation in Africa : The Zande System of Agriculture*, Routledge and Kegan Paul, London, United Kingdom.

Lesson 38   AJIBOLA TAYLOR, T. (1977). *Les associations culturales, moyen de lutte contre les parasites des plantes en Afrique Tropicale*, Environnement Africain, Vol II-4 and III-1, ENDA, P.O.B. 3370, Dakar, Senegal.

I.I.T.A., Annual Reports, years 1976, 1977, 1978, P.O.B. 5320, Ibadan, Nigeria.

MONYO, J.H., KER, A.D.R., CAMPBELL, H. (1976). *Intercropping in semi-arid areas*, Report of a symposium held in Morogoro, IDRC, P.O.B. 8600, Ottawa, Canada.

Lesson 42   BARLOY, P. *Le maïs*, C.D.F., Paris, France.

GILLIER, P., SILVESTRE, P. *L'arachide*, Maisonneuve et Larose, Paris, France.

Memento de l'agronome. (1980). Ministère de la Coopération française, Paris, France.

HECQ, J. (1958). *Le système des cultures Bashi et ses possibilités*, Bulletin agricole du Congo belge, XLIX, n° 4, Brussels.

Lesson 46   F.A.O. *Tables de composition des aliments*, Rome, Italy.

PELE, J., and LE BERRE, S. (1966). *Les aliments d'origine végétale au Cameroun*, ORSTOM, Yaoundé, Cameroun.

SMART, J. (1976). *Tropical Pulses*, Longman, London.

Lesson 47   CHARREAU, C., and VIDAL, P. (1965). *Influence de l'acacia albida sur le sol, la nutrition minérale et les rendements du mil pennisetum au Sénégal*, Agronomie gropicale, juin-juillet 1965.

Lesson 49   DUPRIEZ, H. (1980), Paysans d'Afrique Noire, TERRES ET VIE, Nivelles, Belgium.

REMISON, S.U. (1978). *Neighbour effects between maize and cowpea at various levels of N and P*, Expl. Agric., 14, 1978.

Botanical Names in English and in Latin 🔲🔲🔲🔲🔲🔲🔲🔲🔲🔲🔲🔲🔲🔲🔲🔲🔲🔲

The plants mentioned in this book are listed in English with their scientific names in Latin. Plant names change from country to country. Millet is called **sanio** in Senegal, **yadiri** or **youri** in North Cameroon, **dokona** in Chad. Names even change inside the same country from region to region. In Cameroon, Ewondos call yam **alok** but Fulbe people speak of **fekoa**. It would be hard to identify plants if they had to be named each time in the vernacular. Hence the use of scientific names in Latin. The Latin names are used by botanists all over the world. For example, the scientific name for millet is *Pennisetum americanum*. The first word, written with a capital letter, indicates the genus ; the second word indicates the variety.

Acacia	*Acacia albida*	**Ginger**	*Zingiber officinale*
African fan palm	*Borassus aethiopicum*	**Goa bean**	*Psophocarpus*
African locust bean	*Parkia biglobosa*		*tetragonolobus*
Aloe	*Aloe barteri*	**Gourd**	*Cucurbita sp.*
Arrowroot	*Maranta arundinacea*	**Grapefruit**	*Citrus grandis*
		Groundnut	*Arachis hypogaea*
Bamboo	*Bambusa vulgaris*	**Guava**	*Psidium guajava*
Banana	*Musa sapientum*		
Baobab	*Adansonia digitata*	**Kola**	*Cola acuminata and C. nitida*
Basil	*Ocimum basilicum*		
Bitter leaf	*Vernonia amygdalina*	**Lalang, Sword or**	*Imperata cylindrica*
Black nightshade	*Solanum nodiflorum and*	**Spear grass**	
	S.nigrum	**Lemon**	*Citrus limon*
Black wattle	*Acacia mearnsii*	**Lettuce**	*Lactuca sativa*
Bottle or	*Lagenaria siceraria*	**Leucaena**	*Leucaena glauca*
calabash gourd			
Breadfruit	*Artocarpus communis*	**Maize**	*Zea maïs*
Bush greens,	*Amaranthus spp.*	**Mango**	*Mangifera indica*
Amaranths		**Mexican sunflower**	*Tithonia*
		Millet	*Pennisetum americanum*
Cabbage	*Brassica oleracea*	**Mint**	*Mentha arvensis*
Carrot	*Daucus carota*		
Cashew	*Anacardium occidentale*	**Native pear, Bush**	*Dacryodes edulis*
Cassava, Manioc	*Manihot utilissima*	butter tree	
Castor	*Ricinus communis*	**Neem**	*Azadirachta indica*
Chillies, Bird chillies	*Capsicum frutescens*		
Citronella grass	*Cympobogon nardus*	**Oil palm**	*Elaïs guineensis*
Cocoa	*Theobroma cacao*	**Okra, Lady's finger**	*Hibiscus esculentus*
Coconut	*Cocos nucifera*	**Onion**	*Allium cepa*
Cocoyam, Tannia,	*Xanthosoma sagittifolium*	**Orange**	*Citrus sinensis*
Yautia, Tanier			
Coffee	*Coffea canephora,*	**Papaya, Pawpaw,**	*Carica papaya*
	C. robusta and C. arabica	Papaw	
Common bean	*Phaseolus vulgaris*	**Para rubber**	*Hevea brasiliensis*
Cotton	*Gossypium barbadense*	**Passion Fruit**	*Passiflora edulis*
Cowpea	*Vigna unguiculata*	**Peas**	
Cucumber	*Cucumis sp. and*	**Chick pea**	*Cicer arietinum*
	Cucumeropsis sp.	**Earthpea, Bambarra**	*Voandzeia subterranea*
		groundnut	
Date palm	*Phoenix dactylifera*	**Pea**	*Pisum sativum*
Derris	*Derris elliptica*	**Pigeon pea,**	*Cajanus cajan*
Dolichos, Lalab bean	*Dolichos sp.*	Congo pea	
		Pepper	*Piper nigrum*
Eggplant,	*Solanum incanum*		*and P. guineense*
Bitter tomato			
Elephant grass	*Pennisetum purpureum*	**Pineapple**	*Ananas comosus*
		Plantain	*Musa paradisiaca*
Finger millet, Coracan	*Eleusine coracana*	**Potato**	*Solanum tuberosum*
Flamboyant	*Delonix regia*	**Prickly pear**	*Opuntia amyclea*
Fonio, Hungry rice,	*Digitaria exilis*	**Purslane**	*Portulaca oleracea*
Fundi, Acha		**Pyrethrum**	*Chrysanthemum*
			cinerariaefolium
Giant granadilla	*Passiflora quadrangularis*		
		Radish	*Raphanus sativus*

Raffia palm	*Raphia regalis*	Spurge	*Euphorbia sp.*
Red or American mangrove	*Rhizophora racemosa*	Sugar cane	*Saccharum officinarum*
		Sunflower	*Helianthus annuus*
Rice	*Oriza sativa*	Sweet potato	*Ipomoea batatas*
Roselle, Jamaican sorrel	*Hibiscus sabdariffa*	Tamarind	*Tamarindus indica*
		Taro, Dasheen, Cocoyam	*Colocasia antiquorum*
Safflower	*Carthamus tinctorius*		
Sesame, Beniseed	*Sesamum indicum*	Tea	*Camellia sinensis*
Setaria grass	*Setaria sp.*	Teff	*Eragrostis tef*
Shea butter	*Butyrospermum parkii*	Tomato	*Lycopersicon esculentum*
Siam weed	*Eupatorium odoratum*	Tropical kudzu	*Pueraria phaseolides*
Siratro	*Macroptilium atropurpureum*	Turnip	*Brassica rapa*
Sisal	*Agave sisalana*	Umbrella tree	*Musanga cecropioides*
Sorghum	*Sorghum bicolor*	Wheat	*Triticum aestivum*
Soursop	*Annona muricata*		
Soya bean	*Glycine maximum*	Yam	*Dioscorea spp.*

General Index

Absorbing hair : 164
Adventitious root : 166
Aerial photograph : 51
Agricultural calendar : 125
Agricultural cycle : 125
Air : 15, 111
Albumen : 151
Alluvium : 88
Altitude : 15
Ammonium nitrate : 245
Ammonium phosphate : 245
Ammonium sulphate : 245
Annual rainfall : 125, 130
Anti-erosive strip : 102
Application (fertilizer) : 249
Are (area) : 271
Arid climate : 133
Artificial fertilizer : 187, 237, 245
Available water : 81

Bacteria : 11
Bactericide : 227
Balance sheet : 277
Bedding cultivation : 210
Beetle : 226
Bicalcic phosphate : 246
Biological risks : 220, 225
Biomass : 286
Boron : 237
Botany : 111
Broadcasting seed : 258
Bulb : 152
Bulbil : 152

Calcium : 237
Carbon : 237
Cellulose : 242
Centimetre : 269
Cereals : 1, 279

Chain saw : 198
Chlorophyll : 15, 111
Cleaning plant : 28, 258
Clearing land : 55, 198
Climate : 48, 125ff, 220ff
Collective reserves : 162
Complete fertilizer : 245ff
Compost : 201, 241
Compost crusher : 201
Compound fertilizer : 245
Constraint : 9
Contour line : 102
Copper : 237
Cotyledon : 148, 151
Cover plant : 243, 260
Crop rotation : 27, 35ff, 63
Cropping pattern : 27, 63
Cultivator (machine) : 215
Cultural cycle : 125

Decametre : 269
Decomposed living matter : 238
Decomposition : 238
Degree centigrade : 5
Dew : 97
Dibbling hole : 175, 259
Disc plough : 213
Dolomite : 246
Draining, drainage : 97, 204
Dry season : 125ff

Ear emergence : 114
Early variety : 143
Earthbank : 103
Earthing : 206
Earthworm : 183
Economic life : 18
Ecosystem : 76
Elaborated sap : 110ff

Embankment : 105
Embryo : 151
Erosion : 71, 93ff, 99ff,
Establishment (plant) : 114
Evapotranspiration : 98, 107

Fallow land : 28, 43, 64, 193ff
Family (home) consumption : 6, 18, 281
Fecundation : 148
Fertile : 87, 144
Fertilizer : 187, 245ff
Field space : 255ff
Flooding : 98
Flower : 149
Flowering : 114
Food chain : 11
Food composition table : 279
Formation of humus : 80, 198ff
Free water : 81
Fruit : 148
Fruit set : 114, 148
Fungicide : 227
Fungus : 11, 184ff
Furrow : 94, 98, 206ff, 249, 265

Germ : 148
Germination : 114, 151
Germination capacity or energy : 159
Gram : 270
Granary : 157ff, 161, 231
Graph : 128
Green manure : 242
Gross income : 278
Guinean climate : 130
Gully : 93, 217

Hardiness : 156
Hardy : 156

Harmattan : 15
Harrow : 215
Heat : 15
Hectare : 271
Hectometre : 269
Heliophilous plant : 120
Herbicide : 161
Hoeing : 107
Horizon : 86
Humidity : 15, 204, 217
Humus : 80, 185, 198ff
Hybridization : 146

Improved seed : 146
Infrastructure : 22
Insecticide : 227
Insurance : 231
Integument : 151
Interbeeding : 146
Intervarietal crossing 146
Irrigation : 223ff

Kayendo : 208
Kilogram : 270
Kilometre : 269
Kitty : 231
Kola pod : 149

Labour availability : 141
Land allotment : 50ff
Land authority : 60ff
Land relief : 44
Land use rights : 60ff
Late variety : 143
Laterite : 68, 84, 90
Laterization : 90
Layering : 153
Leaching : 88, 97, 238ff, 251
Legumes : 1, 37ff, 165, 198, 242, 247, 280
Life cycle : 113ff, 134ff, 248
Light : 15, 111, 116ff, 143, 257
Light energy : 15, 111
Light-demanding plant : 120
Ligneous : 242
Lignin : 242
Lime : 246
Limestone : 246
Litre : 269
Living environment : 11
Loamy soil : 84

Magnesic lime : 246
Magnesium : 237
Magnifying glass : 82
Mammal : 11, 180
Manure : 187, 237ff
Map : 10, 40ff
Market gardening : 281
Mass selection (seeds) : 158
Maturation : 115
Mature (stem) : 154
Mediterranean climate : 133

Metre : 269
Microorganism : 184ff
Microscope : 82
Millimetre : 269
Mineral fraction : 80
Mineral nutrient : 237
Mineralization : 185
Mist : 97
Molybdenum : 237
Monetary income : 272ff
Monsoon : 15
Monthly rainfall : 127
Motor pump : 225
Mould : 15
Mound : 204, 206, 249
Mounding : 206, 249
Mulch, mulching : 108
Multilocal tests : 163
Mutual savings and
credit bank : 231

Natural fertilizer : 245, 237
Natural phosphate : 245
Net margin : 278
Nitrogen : 237, 245
Nodule : 37, 165, 247
N.P.K. fertilizer : 245ff
Nursery : 162

Oilseed : 1, 280
Organic fertilizer : 241ff
Organic fraction : 80
Ovary : 148
Ovule : 148

Parent rock : 87
Perennial : 5
Pesticide : 227
Petal : 148
Phosphate : 237
Physical (abiotic) environment : 15
Pioneer plant : 28
Pistil : 148
Plant energy : 15
Plant family : 142
Plant sanitation : 226
Plant selection : 145
Plant spacing : 259
Plough : 213ff
Plumule : 148
Pluriannual cycle : 114
Pollen : 148
Potassium : 237
Potassium sulphate : 245
Power saw : 198
Primary root : 165
Production and return : 272
Production costs : 276
Production factors : 8
Profitability : 276
Prop : 122
Proteins : 279ff

Pruning : 121, 263

Qualitative return : 276
Quantitative return : 276

Radicle : 151
Rain calendar : 128ff
Rain cycle : 125ff
Rain erosion : 93ff
Rain measurement : 127
Rainfall : 127
Rainfall averages : 130
Rainfall records : 128
Receipts : 276ff
Receipts and expenditure :
276ff
Removal (minerals) : 239ff
Return from the land : 272ff
Rhizome : 168
Ridge : 207ff
Ridging : 207ff
Ringing : 112
Ripening : 115
Root cap : 165
Root collar : 165
Root extent : 164ff, 172
Root hair : 164
Root pruning : 186, 250
Root system : 164ff
Rootstock : 154
Rotary crusher : 197
Rotavator : 214
Runoff : 88, 98ff, 254
Rural settlement : 40ff

Savings : 231
Scale (map) : 41
Scion : 154
Seasonal cycle : 125ff
Secondary root : 164
Seed : 147ff
Seed bank : 231, 163
Seed field : 162
Seed rate (density) : 116ff, 259
Self-sufficiency : 6
Semiperennial : 5
Senescence : 115
Sepal : 148
Shade plant : 120
Shade tree : 4
Shade-loving plant : 120
Shape (of plants) : 256ff
Shoot : 152
Shooting : 114
Skiophilous plant : 120
Slash and burn : 28
Slope : 15, 44ff, 92ff
Slurry : 241
Social life : 18
Sodium : 237
Soil composition : 15, 80ff
Soil layer : 86ff
Soil life : 180ff, 254

294

Soil profile : 86ff
Soil structure : 83ff
Sowing and planting : 257ff
Spatial arrangement : 256
Species : 142
Square metre : 271
Stake : 122
Staking : 122
Stamen : 149
Starchy foods : 280
Stem elongation : 114
Stockage : 231
Stolon : 154
Stomata : 113
Stone bank : 104
Straight fertiliser : 245
Straw chopper : 201
Sudanian climate :132
Superphosphate : 245

Temperature : 15
Temporary homestead : 55
Thinning : 175
Tie-ridging : 103
Tonne : 270
Tontine : 231
Trace element : 237
Transect : 42
Transplanting : 173
Tree stock : 163
Tuber : 1, 168
Tuber formation : 115

Unelaborated sap : 112
Urea : 245

Valley bottom drainage : 97
Varietal characteristic : 143
Variety : 142ff
Vegetative growth : 114

Vegetative seed : 152
Virus : 11
Vitamin : 279ff

Weed : 178, 253
Weeding/hoeing : 26, 108, 178, 213, 253
Weekly rainfall : 126
Weevil : 157
Wet season : 130ff
Wet tropical climate : 132
Wilting : 113
Wind : 71ff
Wind erosion : 71ff
Windrow : 198
Winnow : 158

Yeast : 11

Zinc : 237